AMERICA IN SPACE: Pioneers or Aggressors?

AMERICA IN SPACE: Pioneers or Aggressors?

An Editorials On File Book

Edited by Oliver Trager

PROPERTY OF

Facts On File Publications
New York, New York • Oxford, England

31334000 030 110.

AMERICA IN SPACE: Pioneers or Aggressors?

Published by Facts On File, Inc.
460 Park Ave. South, New York, N.Y. 10016
© Copyright 1986 by Facts On File, Inc.

All rights reserved. No part of this book may be reproduced in any form without the permission of the publisher except for reasonably brief extracts used in reviews or scholarly works. Editorials reprinted with permission of cooperating newspapers, which retain their copyrights.

Library of Congress Cataloging-in-Publication Data
Main entry under title:

America in Space.

(An Editorials On File Book)
Includes index
1. Astronautics--United States. 2. Astronautics, Military--United States. I. Trager, Oliver. II. Series.
TL789.8.U5A646 1987 629.4'0973 86-32775
ISBN 0-8160-1596-1

629.4
TRA

International Standard Book Number: 0-8160-1596-1
PRINTED IN THE UNITED STATES OF AMERICA
9 8 7 6 5 4 3 2 1

Contents

Preface

Space exploration has held a special fascination for Americans since the days of the first unmanned rocket launches in the late 1950s. The U.S. Moon landing in 1969 was the crowning glory of the space program's first era. The scientific and technological breakthroughs that made the feat possible became a source of great national pride. But that elation was short-lived. Within three years, the Apollo Moon missions ceased. Subsequent U.S. manned missions were sent to the space station, Skylab, but they too, ended after three years in 1975. It was six years before the the next U.S. manned venture, the shuttle Columbia, was launched in 1981. Unmanned flights fell victim to the same national apathy experienced by the Apollo program. And although the photographs and information collected by the Voyager 2 spacecraft on its visits to Jupiter, Saturn and Uranus were beautiful and enlightening, no unmanned U.S. probe has been launched in the last ten years.

The reusable space shuttles took center stage in America's space program in the 1980s and carried out their missions with such flair that excellence and efficiency seemed routine. However, the tragic 1986 explosion of the space shuttle Challenger altered the country's perception of the space program and brought many questions to the surface that had been taken for granted since the U.S. space program began twenty-five years ago. Just what are our purposes and goals? Is space a frontier to learn from in Man's unquenchable thirst for the unknown? Or is it an environment to exploit for both military and business options? If scientific knowledge and understanding are the true goals, aren't there cheaper and more efficient means of achieving these ends?

As the United States and Soviet Union continue testing space weapons and as other countries begin space programs of their own, the militarization of space has become a vital concern. The use of the shuttle to carry military payloads has drawn NASA further from its original orientation toward basic research and development. The Reagan Administration's "Star Wars" designs have evoked fears of a celestial universe studded with nearly as many weapons as stars.

The U.S. space program is at a crossroads. *America in Space* examines the accomplishments and failures of the space program through words and pictures of the leading editorial writers and cartoonists.

December, 1986 — Oliver Trager

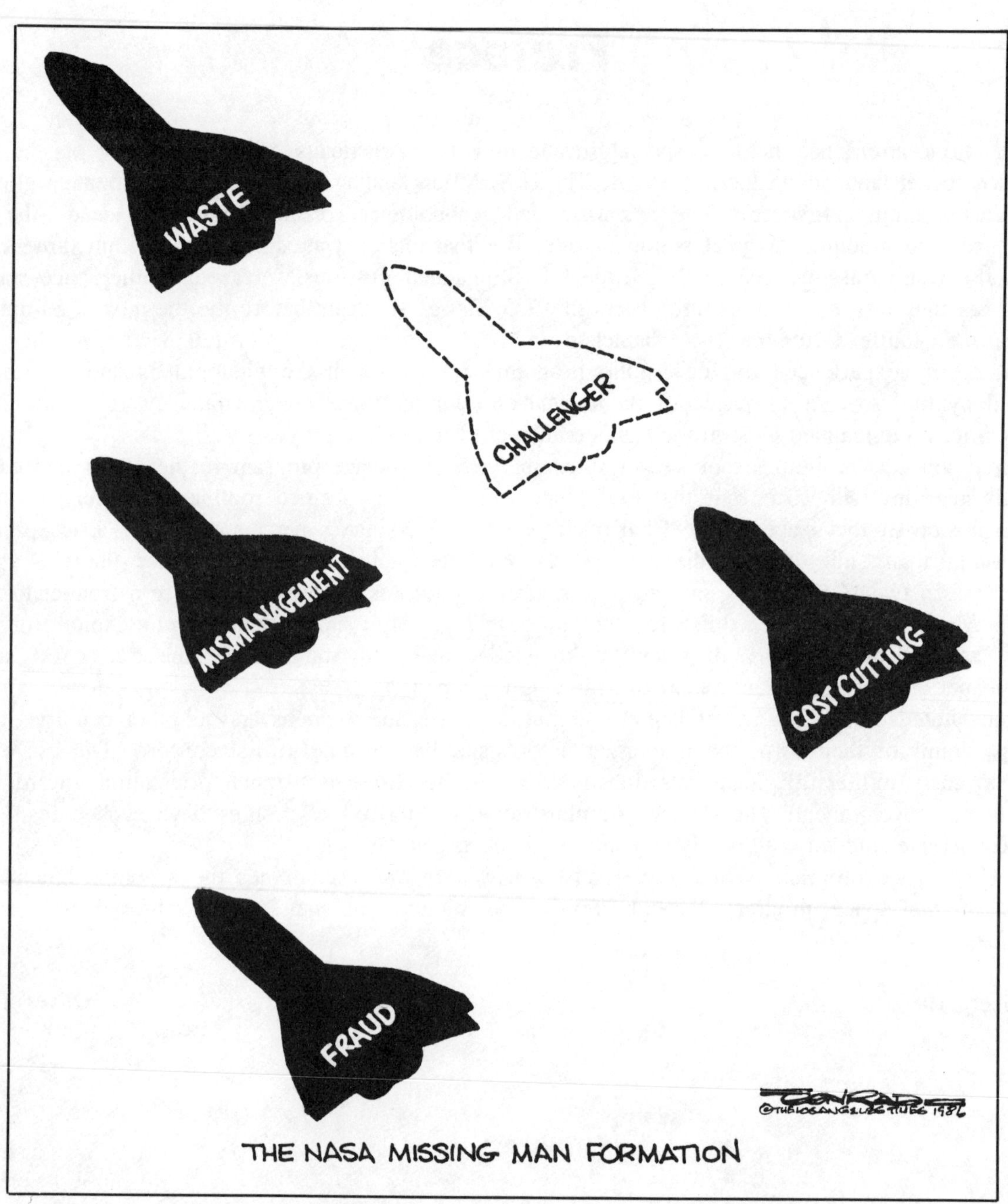

THE NASA MISSING MAN FORMATION

Part I: NASA: A Brief Overview

The National Aeronautics and Space Agency (NASA) has faced adversity from its inception in 1958. Whether it was from the branches of the U.S. military, which have competed with and pressured the civilian agency, or from the sometimes skeptical American public, the problems and controversies facing and surrounding NASA have never been short-term. Initiated by President Dwight D. Eisenhower as a means of centralizing the efforts being undertaken by the military or private research and development, NASA tried matching the early successes of the Soviet Union's space program. The first years were dismal at best. But through the inspiration and vision of President John F. Kennedy, NASA mobilized behind the quest for a moon landing by the end of the 1960s. The Mercury, Gemini and Apollo programs culminated with the triumphant landing in 1969.

After a lull of cynicism in the 1970s, the shuttle missions of the 1980s brought NASA to nearly the level of its previous stature as a symbol of American determination and ingenuity. "The Right Stuff," as popularized in Tom Wolfe's epic neo-journalistic account of NASA's early days, became a term which many Americans identified with. But when the shuttle Challenger blew-up in the blue sky over the Atlantic Ocean, it was an explosion of more than just a spacecraft. The space shuttle had become the most visible symbol of American leadership in space technology. The country's faith in NASA seemed affirmed by its success. Suddenly, after an investment of more than $30 billion and 14 years, we were left full of doubts not only about the shuttles and NASA's fabled competence, but about the very fundamentals of our national space policy. Revelations of cost overruns, faulty workmanship and mismanagement at the highest levels gave cause to reassess the entire space program. The military's continued interest and involvement with NASA has also been a great cause of concern. Since NASA was originally envisioned and commissioned as a "civilian" agency, any connection with the military can be viewed as a compromise of those original ideals and further reason to reassess NASA's purpose and intent.

NASA's Twenty-fifth Anniversary Marked

In 1983 The National Aeronautics and Space Agency (NASA) celebrated its 25th anniversary. President Dwight Eisenhower asked Congress April 2, 1958 to authorize the creation of a civilian space agency to administer the U.S.'s nonmilitary space research and exploration projects. He proposed that the Defense Department retain space projects of a military use. The bill was passed by both Houses of Congress and approved by President Eisenhower later in the year. NASA would be headed by a Presidentially-appointed $22,500-a-year administrator.

NASA officially began its existence October 1, 1958 and Dr. Keith Glennan immediately, under President Eisenhower's direction, took over the 150-man team operating the Navy's Vanguard satellite program. Glennan said the program would continue, at least temporarily under the supervision of Dr. John Hagen and would use Navy facilities "for the time being." NASA announced Oct. 4 the appointment of Dr. Abe. Silverstein as director of space flight development and of John Crowley Jr. as director of aeronautical and space research.

The Pittsburgh PRESS

Pittsburgh, PA, October 1, 1983

In early April 1958, President Dwight D. Eisenhower asked Congress to establish a "National Aeronautics and Space Agency."

Only eight weeks earlier, with the launching of the Explorer I satellite, the United States had finally entered the "space race" so unexpectedly begun by the Russians the previous October with their Sputnik I, and quickly followed by Sputniks II and III.

Congress came through with a bill creating the National Aeronautics and Space Administration, which the president signed July 29.

It wasn't until Oct. 1, 1958, however, that NASA officially went into business.

For those millions of Americans under 25, for whom the space age has always existed, it may be difficult to appreciate the doubts and controversy that surrounded the developing space program, the despair over every setback and the pride in every triumph.

Or the boldness of President John F. Kennedy when in 1961, even before the United States had equaled the Russian feat of orbiting a manned capsule, he committed the nation to sending a man to the moon and returning him safely to earth before the end of the decade.

Rather than attempting to recount all the events of the past quarter-century of space exploration, we think the most fitting way to observe NASA's 25th birthday would be to recall Kennedy's words in defending the goal he announced:

"Well, space is there," he said, "and the moon and planets are there, and new hope for knowledge and peace are there . . . The exploration of space will go ahead whether we join it or not, and it is one of the great adventures of all time.

"Whether (space exploration) will become a force for good or ill depends on man, and only if the United States occupies a position of pre-eminence can we help decide whether this new ocean will be a sea of peace or a new terrifying theater of war."

The Times-Picayune The States-Item

New Orleans, LA, October 1, 1983

The National Aeronautics and Space Administration turns 25 this weekend, and rarely if ever has a technological endeavor gone so far so fast. NASA was created on Oct. 1, 1958, a year after the Soviets orbited the first artificial satellite.

Beginning behind and having to invent as it went along, NASA in 11 years had surpassed the Soviets and put the first human on the moon. In all, NASA has sent dozens of men and one woman into space, 12 of them to the moon and back; sent unmanned probes to five planets and has one heading to two others; made significant scientific discoveries; tied space to Earth through such devices as communications, weather and survey satellites; and is now equipped, by the space shuttle fleet, to use space both as a permanent human outpost and as a springboard to distant reaches.

Birthdays are times for looking forward as well as back, and NASA administrator James M. Beggs has plenty to offer. "If we continue to pursue the program out beyond the 21st century," he says, "I believe we will be back on the moon, probably with a base. We will begin to develop the resources of the surrounding bodies of the universe, perhaps even doing things like grappling and studying and even mining asteroids." By the end of the next 25 years, he says, we may be ready for a manned expedition to Mars.

The space agency's budget has been cut back considerably from the days of the Apollo moon program, but President Reagan has shown interest in approving money for a space station project in the 1985 budget. Mr. Beggs says the United States could have a six- to eight-person station operational by 1991. This is clearly the next step to the routine use of space, and the Soviets have been working on a small station for some years.

Given the current level of our ability, Mr. Beggs says, enormous infusions of money need not be required. "If we have sustained funding at a reasonable level," he says, "I think where we are today or maybe just a tad more money will take us very, very far in the next decade."

The key is "sustained funding," for space programs have long lead times and require adequate staffing by specialized professionals. Such programs can be seriously hampered by feast-or-famine or slow-starvation budgeting.

Space operations can no longer be thought of as entertaining adventures or nationalistic grandstanding. They are now a substantial factor in the mix of basic national interests and must be pursued with appropriate conscientiousness.

The Dispatch

Columbus, OH, April 5, 1983

It may be hard to believe, but this country's space program is 25 years old this year and the scheduled launch Friday of the space shuttle *Challenger* demonstrates just how far the program — and the nation — have come during the last quarter century.

The National Aeronautics and Space Administration introduced to the world this country's first batch of astronauts on April 9, 1959. They were the elite: bright, quick, courageous and willing to take risks few others had dared to imagine. The country greeted them as heroes and their early exploits attracted wide interest, especially John Glenn's orbital flight — the first by an American.

But space exploration has become routine. Space flight has become relatively common and it no longer captures the world's imagination. The machines that NASA has constructed perform so well that much of the drama has been removed from space activities. It is not surprising that the next flight in the shuttle series is generating little interest, even though it is the first working-repair mission designed to restore crippled satellites to service — a task that is essential for a full exploitation of space potentials.

This week offers a good opportunity, however, to salute the women and men who have made the space program so successful during the last 25 years and to wish them well in the next 25 years.

AKRON BEACON JOURNAL

Akron, OH, July 29, 1983

TODAY is a landmark in the revolution in science and technology that has accompanied the post-World War II boom: The National Aeronautics and Space Administration — the highest of all high-technology enterprises — is 25 years old.

On July 29, 1958, President Eisenhower signed a law that created the space agency and paved the way for some of mankind's most notable technological achievements. The United States sent men into orbit around our own planet and put the American flag on the moon. Unmanned space flights have added tremendously to man's knowledge of his own solar system and, just this summer, distant traveler Pioneer 10 became the first manmade object to hurtle beyond the solar system's outermost known planet.

The establishment of NASA was a reaction to the launching of the Soviet spacecraft Sputnik. In this case, the Red Scare had positive results — America entered the space age and revitalized its education system to meet new demands.

Today, new challenges face NASA and the United States. The space race has tapered off and America's unquestioned lead in communications technology is being threatened. While positive reactions to such ventures as the space shuttle and unmanned space probes show high public interest in space flight, the commitment of 25 years ago appears to be missing.

A 1981 poll found that 60 percent of those polled want to see advancements in space technology, but only 24 percent wanted to spend more money to fund them.

And perhaps the major underpinning of technological advancement — strong public schools — gets little Reagan administration support.

All this leaves NASA at a crossroads on its 25th birthday. The initial investment in the space agency — paralleled by improved public education funding — was high. But the investment has more than paid off in medical advances, in technology, in new products and in American prestige and pre-eminence.

A similar commitment to excellence will be needed in the next quarter of a century, not just to NASA, but to education. With that commitment, NASA — and, symbolically, America's leadership in world technology — can mark a new beginning at the young age of 25. Without it, American dominance may face a premature senility.

DESERET NEWS

Salt Lake City, UT, October 2, 1983

When the National Aeronautics and Space Administration was established 25 years ago — Oct. 1, 1958 — its attempts at space exploration captured the imagination of the whole world.

But as NASA reaches its 25th birthday, it has become less adventurous. The heroics and grand plans have fallen victim, in part, to mundane things like reduced budgets.

Most of the space agency's eggs are now in one basket — the space shuttle. Admittedly, it is a considerable basket, indeed. However, even shuttle operations have been scaled back somewhat.

The shuttle, which makes going into earth orbit with a variety of payloads and passengers seem routine, has opened the door for what some have called the "commercialization" of space.

This business-like approach, with an eye on profits instead of exploration, promises dramatic technological advances in the coming years, including the building of orbiting space stations to serve as industrial production centers.

At least three U.S. companies already have approached NASA with deals for joint or partial ownership of manned space stations.

Until the space program, most of mankind's technological leaps forward were associated with wartime. But going into space has raised an equivalent kind of challenge to science and industry. Computers, miniaturization and many new materials owe their development in large part to the space effort.

As long as the U.S. has a viable space program, it is guaranteeing itself world leadership in many areas of technology.

But where profits are possible, competition is not far away and NASA does not have the field all to itself.

Russia, of course, has its own space program. A consortium of European nations has developed its own launch rocket; the Japanese are moving into that area, and many countries are building devices that could manufacture various products in space.

Purely scientific projects are not all dead. Sometime in the next year or so, NASA will launch a 12-ton telescope into orbit. This will give astronomers their first clear and detailed look into the universe without the picture being distorted by the earth's atmosphere.

But by and large, the U.S. space program will be looking more toward earth and business than thrusting outward into exploration. At least that is the picture until into the next century.

But tucked away in obscure corners of NASA are those who are plotting projects to bring back the excitement in perhaps another 25 years — projects like sending a manned spacecraft to land on Mars.

Now there's something to really capture the imagination and propel the U.S. light years ahead of the rest of the world in technology.

The Orlando Sentinel

Orlando, FL, July 30, 1983

It seems only yesterday that space exploration was confined to comic books and the silver screen, featuring Buck Rogers and his buddies. But, how time flies: Friday was the 25th birthday of the National Aeronautics and Space Administration.

And how it has grown. Starting as the successor to a space advisory committee, it had the task of pulling together a program in disarray. Quickly, it became the hallmark of proficiency.

Through those 25 years NASA has thrilled the world by tossing men and machines to the moon or to circle Earth. Through NASA, satellites compress the world in picture and word with a simple dial tone. Lost sailors and flyers are found, new earth resources are discovered and products ranging from an improved frying pan to microsurgical instruments are commonplace.

Best of all is the future. A new pharmaceutical process that will open doors to better drugs is near the production stage. And within a year or two, industrial processes will be at work in space, opening an entirely new industrial revolution. Truly, in space is our future.

Moonwalk Remembered Fifteen Years Later

1984 marked the fifteenth anniversary of one of the American space program's greatest achievements: the Apollo Moon landing and walk by two astronauts, Neil Armstrong and Edwin (Buzz) Aldrin. The two astronauts flew a fragile four-legged spaceship to a safe landing on the moon July 20, 1969. With most of the civilized world watching by TV from a distance of 241,500 miles, the spacesuit-clad Americans climbed out of their lunar module 6 1/2 hours later. Then for 2 1/4 hours, they walked, worked, performed experiments, collected samples of lunar rock and took photos in a barren, airless world to which no man had ever gone before them. This was the first landing of a man on an astronomical body other than the earth, and the two astronauts stayed on it for 21 hours 36 1/4 minutes before blasting off in their spaceship July 21 for a safe journey to their home planet.

The first man to put his foot on the Moon was Neil Alden Armstrong, 38, civilian commander of the Apollo 11 spaceship, which had been launched from Cape Kennedy, Fla. July 16 on the flight that brought the astronauts to the Moon. After gingerly backing out of the spaceship's hatch, Armstrong slowly climbed down a ladder and placed his clumsy left spaceboot on the dry lunar soil and uttered his famous words: "It's one small step for [a] man, one giant leap for mankind." He was followed by Air Force Col. Aldrin, 39, pilot of the lunar module, which the astronauts had code-named "Eagle." The lunar module had descended to the moon from Apollo 11's orbiting command and service module, code-named "Columbia" where its pilot, the third member of the Apollo 11 crew, Air Force Lt. Col. Michael Collins, 38, remained behind.

The manned landing on the Moon—and the astronauts' safe return to Earth four days later—climaxed an eight-year program set in motion by the late President John F. Kennedy May 25, 1961 when he called on the United States to achieve the goal, "before this decade is out, of landing a man on the Moon and returning him safely to earth." At least 400,000 people and some 20,000 business organizations and universities were employed in the effort that finally set the two astronauts on the moon. Various estimates were given as to the money cost, but the most widely accepted figure was $24 billion.

The Arizona Republic

Phoenix, AZ, July 20, 1984

FIFTEEN years ago today, residents of a moonstruck world gazed skyward as if in hope of seeing one of man's greatest achievements — the first landing on the moon by three American astronauts.

That was on July 20, 1969. For the Earthlings, it was a vicarious pursuit of a long-sought dream. For the trio of astronauts, Edwin "Buzz" Aldrin, Michael Collins and Neil Armstrong, it was, as Armstrong said in his famous first words from the moon's surface, "one small step for a man, one giant leap for mankind."

That giant leap continued for three years, during which time 10 other Americans walked on the moon. Then, the once much-celebrated occasion got to be kind of old hat, and the Nixon administration, the public and Congress turned their attention to other national needs.

Now, the moon again is a focal point. It is seen as an ideal site for a base for space operations.

Earlier this year, President Reagan pumped new life into the National Aeronautics and Space Administration by giving the agency a new mandate to develop a permanent manned space station within 10 years and to follow a grand vision into the future.

Aldrin makes a good case for scheduling the moon as one of the first stops after the space station, which now has top priority. Aldrin, who believes the nation should have committed itself to a moon base before a space station, contends that it would be the next logical step.

"The solar system's most desirable space station already has six American flags on it," he said. "That's the moon. Let's use it and not turn it over to foreign pioneering frontiersmen."

This makes sense. The moon, a quarter-million miles from Earth, is a mere hop, skip and a jump as far as space travel is concerned. The six missions to the moon found that there are considerable minerals and building material there, meaning that the base could be built largely from resources at hand.

Encouraged by Reagan's renewed interest, NASA is now making long-range plans. In the years ahead, it could well be that a moon base will be the first stop on a manned flight to Mars.

The Commercial Appeal

Memphis, TN, July 19, 1984

FIFTEEN YEARS ago this week Richard Nixon was in the White House, the United States was in Vietnam, Sen. Edward M. Kennedy (D-Mass.) was in a car accident that killed Mary Jo Kopechne, and astronauts Neil A. Armstrong, Edwin Aldrin and Michael Collins were in space for 195 hours, 18 minutes and 35 seconds.

Memphis and the nation anxiously watched as the astronauts were blasted into 30 moon orbits in Apollo 11 on July 16, 1969. We sat glued to television sets into the wee hours of July 20, 1969 to see Armstrong become the first human to touch lunar soil.

"That's one small step for man," he said, "one giant leap for mankind." We clapped and cheered and toasted the U.S. gaming triumph over space and the Russians.

And as if in a classic burnout, public interest in the outer limits was just so much spent fuel after that. No flight since has stirred the public excitement and commitment of that Apollo 11 mission to the moon.

So today Americans are the target of a national marketing campaign. At Cape Canaveral, Fla., about 100 nostalgic space agency workers launched a three-foot Saturn 5 model rocket at 8:32 a.m. Memphis time Monday, the same time that Armstrong, Aldrin and Collins lifted off on their history-making flight 15 years ago.

The model flew 500 feet up, then gently landed with the aid of parachutes, signaling the ignition and liftoff of Space Week 1984.

Space Week 1984 is a national effort to celebrate America's space program and, no doubt, to re-establish and expand the space program in the public mind and federal budget.

The problem is that the people know that the spending of billions of dollars for space exploration has risen above the level of interplanetary gamesmanship between nations. The money goes to dig deeper trenches aligning USA capitalistic and USSR communistic philosophies.

The hope of July 1975 has vanished. That's when the crews of America's Apollo 18 and the USSR's Soyuz 19 linked-up in flight, ate together, shaved together, held news conferences together and conducted scientific experiments together in a world apart from Earth. The melding of space experience fertilized the seed of hope that animosities between the two nations would subside.

Instead, the two countries have expanded their space travel and equipment and ensured that warfare between the two big Earth powers now would not be relegated to simple sea, ground and inner atmospheric maneuvers, but would encompass outer space weapons and tactics as well.

Fifteen years ago, every week was Space Week in the United States. Today, Space Week 1984 is a stark reminder that technological success in space must go hand-in-hand with moral responsibility on Earth.

WORCESTER TELEGRAM.

Worcester, MA, July 22, 1984

Man's time on this earth is said to represent only a moment in the history of the universe. If so, then the 15 years that have passed since the first day that mankind set foot on another natural object in space is just a millisecond in time.

It has been 15 years since Neil Armstrong stepped from the lunar lander "Eagle" and put man's first footstep on the moon's surface.

Since then, 10 more Americans walked on the lunar surface. More important, the space shuttle was designed, built and has been launched repeatedly to service satellites and conduct experiments in space orbit.

Now the NASA planning focuses on the shuttle and on a space station that, when built, will provide a launch station for lower-cost travel in near space.

The debates go on. Is space exploration worth the price? Cast an eye down the list of devices and products and scientific knowledge associated with space. That list is long and growing longer. Many every-day items got their start from the demand for new knowledge and new skills required by the space effort.

Computer science has kept up and in part is a beneficiary of the commitment to space.

There are questions to answer that lie beyond economics. What role should space and the moon play in defense, warfare and the competition between nations? What impact will man have on the moon's surface and how can that be held to a minimum?

The view of earth that man now has, that of a fragile blue-green sphere, has changed our thinking and continues to change our future.

The many questions will be answered. The view of NASA and the people who work toward space exploration is summed up by NASA administrator James Beggs. "We still have much to learn and much to benefit."

That's a succinct view of the future — and one worth supporting.

THE DENVER POST

Denver, CO, August 4, 1984

FIFTEEN years after Neil Armstrong took his "one small step for a man," the United States faces some crucial decisions on where to take the "giant leap for mankind."

In the weeks surrounding the 15th anniversary of humanity's most inspiring space adventure, two possible orientations have become apparent. One is seen in the squabbling between Moscow and the White House over talks in Vienna next month aimed at limiting the militarization of space.

The other has been less widely publicized — a report by the Office of Technology Assessment discussing ways in which the United States can "involve its private sector effectively in applications of space technology."

From July 20, 1969, until Dec. 19, 1972, 21 Americans orbited the moon and 12 actually walked, frolicked and even drove upon those ancient sands. Since then, the space adventures — except for two robots sent to Mars in 1976 — have been much closer to home: a few maneuvers involving large space laboratories and, more recently, the beginning of thrift in space with reusable shuttles that fly home to land like so many slightly scorched DC-9s.

But the great vastness beyond this small blue planet, even the apparent desolation of the nearby planets, remains mostly a mystery. A more frightening unknown is future space policy. Will space be rekindled as a beacon to the imagination, or will it devolve into a battleground for circling machines of war?

Military tacticians are lobbying for the philosophy that space should be considered the "high ground" in strategic planning. The Reagan administration is seeking $150 million to begin study of an orbiting space station and pushing a program for space-mounted defense systems — derisively called "Star Wars" — that would cost billions more. A space station, too, could be used as a military installation — but it needn't be. A space station is simply a location — for scientific research, public service, defense or other endeavors, including private industry.

Philosophically, cooperation, not confrontation, seems much more in keeping with the all-enfolding nature of space.

The moon landings were the greatest adventure of earthbound humanity, a stirring triumph. We tarnish that memory by emphasizing military exploitation over scientific exploration.

THE MILWAUKEE JOURNAL

Milwaukee, WI, July 22, 1984

How quickly is the extraordinary absorbed into the commonplace. In 1957, the Russians sparked worldwide publicity by lofting Sputnik into orbit; today, the launch of a space satellite gets two paragraphs on Page 88, if that. Flights of the US space shuttle are generously — but not excitedly — reported.

Space triumphs of the past are only dimly remembered. It was 15 years ago last week that Neil Armstrong became the first person to walk on the moon; eight years ago last week that the Viking I lander reached Mars; nine years ago last week that US and Soviet spacecraft docked together.

The anniversary of those tremendous achievements was little noted by the public. Apparently Americans have learned to ignore the extraordinary.

Fortunately, they still seem willing to fund it generously. In 1969, the year of the first moonwalk, the National Aeronautics and Space Administration had a budget of about $4 billion; in fiscal 1984, it is a little more than $7 billion. Inflation distorts the difference, of course; as a percentage of total federal outlays, the agency's budget is declining.

But NASA says it is doing better science today, and involving private industry in space projects. That contrasts with the space stunts of the past. As a NASA spokesman commented regarding moon exploration, "I won't relate it to sex, but let's just say that it gives you a good feeling but it doesn't last very long."

Currently, NASA is working to develop a low Earth-orbiting space station to create products with commercial application. Scientists could work in the station's shirt-sleeve environment for three to six months.

It would be a mistake to underestimate the importance of the contemporary US space program. Some scientific projects can be done best — or done only — in space. Although the cost of some projects is extreme, so are the potential rewards.

The United States is the unquestioned leader in space technology. For this country to shortchange space research would be to abandon that leadership role. And to abandon or ignore the mysteries of the space frontier would be to deny humankind's curiosity about the unknown and ability to satisfy that curiosity. It would be to accept ignorance about an infinity of fascinating unknowns.

U.S. Commercial Program Launched; Shuttle Venture Weighed

NASA began discussions with private investors seeking to build a shareholder-owned space shuttle orbiter, it was reported March 20, 1983. Congress had funded four shuttles in all, but NASA had argued that the projected volume of military and commercial use justified a fifth. The new $2 billion proposal was by the Cyprus Corp., a Pittsburgh-based investment and operating concern headed by William F. Rockwell, a founder and director of the Rockwell International Corp., the original prime contractor for the government's space shuttle program.

President Ronald Reagan February 24, 1984 signed an executive order authorizing the Transportation Department to coordinate a program that would help enable private corporations to launch satellites into space. The order allowed Transportation Secretary Elizabeth Dole to clear away what she described as a "thicket of clearances, licenses and regulations that keep industrial space vehicles tethered to their pads. It was part of a broad set of recommendations, following up on a theme set in President Reagan's January 25 State of the Union address, for facilitating the commercial exploration of space. In addition to the changes in NASA's liscensing requirements and regulations, the Journal of Commerce reported Feb. 17, the Administration was seeking changes in such areas as tax laws that were seen as inhibiting the commercial development of space.

At a White House ceremony, Reagan declared, "Private enterprise made America great, and if our efforts in space are to show the same energy, imagination and daring as those in our country, we must involve private enterprise to the fullest." At the ceremony, Dole proclaimed the U.S. commercial space industry "open for business." NASA had picked a firm to take over the marketing of satellite launches with the Delta rocket, the Wall Street Journal reported January 10, 1984. The firm, Transpace Carriers Inc. of Greenbelt, Md., would serve as a middle-man seeking customers who wanted to launch small or medium weight satellites aboard the expendable Deltas, the workhorse of the satellite program. NASA in December, 1983 had chosen General Dynamics Corp. to play a similar role with the heavier-lifting Atlas/Centaur rocket combination. Both actions were in fulfillment of President Reagan's policy to expand commercial use of space.

Lincoln Journal

Lincoln, NE, July 16, 1984

A retired Navy engineer is trying to raise $1 million or so to put a non-government astronaut into orbit, or sub-orbit. He has in mind creating his own space program, competing with NASA for the business of launching satellites.

To Robert Truax, all one can do is tip one's hat and acknowledge the glories of private enterprise. To Fell Peters, the prospective astronaut, wishes of good luck are in order. He may need it.

Nobody says the space shuttle has to be a monopoly. This side of outer space, in the friendly skies above our earth, privately owned planes, both commercial and personal, co-exist with military and other government aircraft.

But those private planes do have to meet certain regulatory standards and the operators must be licensed. If the government hasn't considered the matter of similarly controlling commercial exploitation of space, it should be doing so.

There is nothing inherently wrong with engineers and investors trying to make a buck by offering space-oriented services of various kinds. But they should be subject to no less rigorous government criteria than airlines or shipping or rail and truck transportation.

If Truax is serious about shooting Peters into sub-orbit, he should be prepared to demonstrate first that his scheme is sound and offers reasonable safety for his lone astronaut. If something goes wrong, he should have ready a plan by which Peters can be returned to earth.

Otherwise, if the latter is stranded up there somewhere, you know who will be called upon to rescue him. Government, of course — the National Aeronautics and Space Administration. NASA doesn't have the time or the resources to run around fetching private enterprise's marooned astronauts and clearing its debris out of the space paths.

Los Angeles Times

Los Angeles, CA, July 31, 1984

A new direction in America's space program is emerging after nearly a decade of uncertainty. Its goal is the practical utilization of space—a shift that brings with it a host of questions and doubts.

The most important questions involve the role of private industry in space. If human beings are to live and work in space, business must be involved in some manner—at least in the United States. The questions are how much and how fast.

A new report by the Office of Technology Assessment and recent comments by President Reagan bring the issue into timely focus. Reagan invited business to move in as quickly as possible when he offered incentives for investment in space activities. The office's report was more cautious, warning that the United States faces a key decision in the immediate future on the degree of private industry's involvement in space.

The report compared the developing role of business in this country with the relationship of European and Japanese industries to their nations' space programs. It concluded that in many ways the United States has failed to develop "ways to involve its private sector effectively in applications of space technology," citing "high cost and high technological and economic risk" as barriers to investment.

The office's outline for correcting this situation is well-reasoned. In each area of space technology, the report concludes, the transition to private investment will involve precise government policies. To begin this process, the office says, will require two developments: a national consensus on the direction of America's space programs and a broad coordination on the development of policy between a range of government agencies. The sooner that U.S. policy-makers unify behind the colonization of space, the better. Too long a wait could bring U.S. space activities to a quick and fatal halt. The space program is on the edge of a new era; it requires only a good push to send it off.

However, two notes of caution are well advised at this critical moment: First, the drive for practical utilization of space must never eclipse the desire to explore the universe simply to satisfy a thirst for knowledge. Research-oriented projects such as Voyager and Viking must not be left behind and forgotten. Second, U.S. policy-makers should not forget, in their haste to unburden the government, that government's role remains vital. An overeagerness in committing to private industry in the wrong places before the technology and necessary economic structures are fully prepared could lead to trouble. Government must continue to construct the foundation of the final frontier.

Private enterprise can build on that foundation. As Americans unify around a new vision of space exploration, forethought, vision and cooperation should be their watchwords.

The Evening Gazette

Worcester, MA, August 27, 1984

It seems like science fiction but researchers are talking about using materials found on the moon to support further space exploration. One possibility is mining oxygen for rocket fuel. After all, the moon is half oxygen and the possibility of "filling up" makes more sense to some than carting spare fuel from Earth and storing it up there.

A panel of National Aeronautics and Space Administration deep thinkers just spent 10 weeks debating whether it makes sense to set up a base on the moon and use indigenous resources. Some still think it's more reliable to carry all supplies up to a space station and warehouse.

The fantasy is that the argument is concerned with "after 1992," the date when the first space platform is expected to go into orbit. Those faraway dates have a habit of coming up sooner than expected.

In addition to using the moon as a gigantic filling station in the sky, some scientists propose mining metals and bulk soil to build shelters.

Geologist David McKay says that since the pioneers used local wood and stone to build shelters, roads and wagons, the same principle should apply in space. He is supported by former astronaut Buzz Aldrin who says that even asteroids can be used since these would yield billions of tons of water.

But other scientists disagree. Dr. Rocco Fazzolare, an energy expert at the University of Arizona, calls a lunar base a waste of time. "We can make oxygen on earth for pennies. We are a water planet. I think we should concentrate on building a better transportation system to deliver what we want in space." He's suggesting space-bound buses and trailer trucks.

Fazzolare says we should study the moon and scout around for security reasons but it's a waste of time to build a huge base there. He also points out that while the moon has plenty of oxygen, it has little hydrogen, the other vital component of rocket fuel, so that would have to be shuttled up and mixed there.

There are also some legal minds at work. The United States has said no one owns the moon nor should. But the Soviet Union appears to be developing some big booster rockets supposedly to explore the moon. Whoever gets there first may not "own" the moon but will have a say in how it is used.

For us mortals here on Earth, it may seem more fantasy than reality to consider the moon as a filling station. But in our own time, fantasy has had a way of becoming reality. And that is why NASA's daydreaming deserves a second look.

THE PLAIN DEALER

Cleveland, OH, April 20, 1984

Extraterrestrial mining is a long way into the future, but already, earth-bound politicians and businesses are seeking to mine the space program for something that is almost as precious as rare minerals—lucrative contracts and thousands of jobs.

Cleveland and the Lewis Research Center, operated by the National Aeronautics and Space Administration (NASA) near Cleveland Hopkins International Airport, are in the midst of the battle for research contracts, jobs and prestige. The very survival of Lewis may be at stake, along with 3,600 public and private jobs and an annual government payroll of $109 million.

Lewis officials are convinced that the center should be the primary research and development unit for part of the $8 billion manned space station proposed by President Reagan in his State of the Union message last January. Lewis wants to work with private industry in designing, building and testing the largest orbiting solar power collector ever—a mammoth array of solar cells large enough to cover a football field.

Lewis, even though it has done extensive work on electric power systems, has competition from the Marshall Space Flight Center in Huntsville, Ala., which already has been designated the primary center for three of the six other components of the space station. Alabama's political and business leaders point out that Marshall designed the largest solar panels built to date, for the Skylab space station.

Cleveland officials, a coalition of Northeast-Midwest U.S. representatives and a group of Great Lakes governors are attempting to head off the Alabamians. They quite properly want to ensure that Lewis, the only northern NASA field center, retains a leadership role necessary to guarantee its survival through the turn of the century.

If the Reagan administration is sensitive to the needs of the older industrial cities, it will make sure that a major and vital portion of NASA's next major space venture is developed at Lewis. Most states receive more from the federal government than does Ohio. Ohioans want more of their tax dollars coming back to the state, not for welfare benefits and unemployment payments, but for jobs, such as those involved in the continued exploration of space.

□

In terms of federal programs, Ohio may lag behind other states partly because of its disjointed and relatively ineffective congressional delegation. The state's 21 House members and two senators rarely meet to discuss strategies for competing with other states and regions. The last time a delegation meeting was held, the senior member—Rep. Delbert R. Latta, R-5, of Bowling Green—didn't even attend. The Lewis situation was recently discussed at a meeting in Sen. John Glenn's office, but only a handful of congressmen attended.

That's hardly the way to put together an effective team. Voters, as they visit with their congressmen during this election season, ought to be asking why their representatives can't work closer together.

The Dispatch

Columbus, OH, December 10, 1984

The National Aeronautics and Space Administration is opening the door to the galaxies and inviting private industry to share in the cost as well as the benefits of doing business in space.

NASA plans to make it easier for private companies to get involved in space-related technologies by granting seed money for new enterprises and reduced flight prices for certain research and development ventures.

With NASA controlling the shuttle program, the government can steer clear of the frivolous and impractical while defraying some of the mounting costs of space exploration and utilization that up to now have been borne almost exclusively by the taxpayers.

NASA expects to broaden its relationship with the aerospace industry and the science community to encourage high-tech enterprises of interest and value to other industries and academic institutions.

Guidelines for future ventures require that NASA's private partner have "significant capital at risk" and a venture with substantive "potential benefits for the nation." Technological advances possible through space programs could, in NASA's words, "help conquer diseases, produce smarter and faster computers, develop lighter metals, increase communications and information availability and enhance understanding of our environment and its resources."

All the problems of the world cannot be solved "via satellite," but significant contributions are being made and can be expanded upon. Satellite communications continue to bring the nations of the world closer together. The recent *Discovery* shuttle mission demonstrated that multimillion-dollar communications satellite ventures that go awry can be salvaged.

NASA's space shuttle program has more than proved itself. Expanding the role and contribution of private enterprise in future ventures is a logical step and one the government should vigorously pursue.

TULSA WORLD

Tulsa, OK, August 6, 1985

FORGET Star Wars and elaborate space shuttles to explore the planets and stars.

A growing number of scientists, businessmen and space officials see the profit motive as a more important factor in future space operations — to the expected tune of $25 billion annually by the turn of the century.

The market forces have already been set in motion, but it will be some time before the bottom line is determined.

In an industrial park away from earth's gravity, it is possible to achieve greater manufacturing precision, particularly useful in electronics, communications and medicine.

The first product to be on sale commercially is a vial of 30 million perfectly round polystyrene beads that sells for $384. The beads are precise enough to be a means of calibrating instruments such as microscopes.

In 1988 Johnson & Johnson is expected to market a still secret made-in-space product that is believed to be a treatment for diabetes. A McDonnell Douglas study has identified 12 pharmaceutical products that might profitably be produced in space.

In response to expressions of interest by several hundred companies, the National Aeronautics and Space Administration has established a new division to promote private-enterprise space ventures, offering cut-rate shuttle prices for commercial experiments.

The risks and uncertainties are still great, discouraging most private investors or corporations from risking great sums of money. The satellite industry has been an economic success, but it filled a waiting demand in the international communications market.

The next industrial revolution — space — may sharpen America's competitive edge.

The Dallas Morning News

Dallas, TX, June 14, 1984

Federal policy, planners say, has always been directed toward turning the race into space over to private enterprise. But just where the lines will be drawn is the debate that's under way now.

The National Aeronautics and Space Administration and Fairchild Industries Space Co. are involved in such a debate now. Fairchild plans to orbit a private space platform — Leasecraft — on which other companies can lease space for production and research. However, periodically Leasecraft and a NASA shuttle would have to exchange payloads. Where would they meet? NASA wants the private platform to descend to the shuttle's 160-mile-high orbit, while Fairchild officials want to meet the shuttle halfway between that orbit and Leasecraft's 260-mile orbit.

NASA officials, responding to complaints about slow negotiations, say that they must be very careful because each decision marks a precedent in the transition of space effort. That's frustrating for private firms that move at a faster pace.

Nevertheless, it's encouraging that space progress now centers on who'll meet whom where. Not if the meeting will take place at all.

THE LINCOLN STAR

Lincoln, NE, March 28, 1984

Sometimes the federal government can be such a contradictory irritant that it makes you mad just to get up in the morning and go to work. In a short little news story Monday, the National Aeronautics and Space Administration said the federal government intends to encourage private investment in space by offering tax incentives and reducing the technical and financial risks involved.

That really makes the blood boil, just as we, as a nation, are trying to claw our way out of annual deficits in the $200 billion range and a national debt into the 13-digit range. That range, in case it's a little rough for you to fathom, is the trillion-dollar range, where interest rates on your debt are into the billions per year.

THAT'S ONE of the nice little things about our numerical system — at whatever digit range you are indebted, your annual interest will simply be somewhere in a digit lower category. Thus, if you owe trillions, your interest is in the billions. If you owe billions, your interest is in the millions. If you owe millions, your interests is in the thousands and if you owe thousands your interest is in the hundreds.

It's all a kind of play on the number 10, but enough of that.

James M. Beggs of NASA told a business gathering that the White House "has some tax incentive plans to encourage business in space. . . . For our part, NASA will establish a high-level office to serve as a focal point for our activities in stimulating commerce."

He said the agency will support research aimed at commercial applications, permit access to NASA experiment facilities and schedule flight opportunities as space shuttle missions.

HOW ABOUT Republican candidates in the current election taking the stump to blast that federal largess out of the water? No, they would rather talk about cutting federal spending for people's subsistence than to talk about federal giving of your and my assets to big corporations running up huge profits.

No wonder the prime rate is creeping up and the going consumer credit interest rate is so high it ought to be an embarrassment to those with the nerve to assess it.

I have come to have contempt of our characteristic saturation of use of the phrase, but you want to know where the beef is? It's with the fat cats who know that the slickest deal around is not in what you can get off Uncle Sam but in what you can avoid paying him in the first place.

All those personal pronouns used above by Beggs are you and me, pal, because we are the taxpayers whose payroll deductions keep NASA in business so it can fox up those tax writeoffs for the corporate giants.

Chicago Tribune

Chicago, IL, November 14, 1985

In a recent report to the National Space Commission, Peter E. Glaser of Arthur D. Little, Inc., and Albert Kelly, former project manager of the Apollo program, urged the federal government to turn its attention and tax dollars to economic development in outer space, providing federal incentives to make the moon, the planets and the larger asteroids accessible for industrial parks to be used by both humans and robots.

This would certainly be what real estate people like to call a "big score," especially if the government were to pick up a lot of the transportation costs. In the case of President Reagan's Strategic Defense Initiative alone, the truckage charges for carting all that weaponry up into space are expected to total more than $600 billion.

Land on the moon or Mars would be, well, dirt cheap, and there'd be no real estate taxes or zoning laws to worry about. No one would be around to complain about hazardous waste dumping except nonvoting Martians. And, though school teachers, senators and journalists are being signed up for space voyages, no one has said a word about OSHA inspectors.

Not to be a stick-in-the-mud, if these dreamers would lower their gaze a little, they'd find plenty of vacant space for government-sponsored industrial development right down here in America's big cities. Unlike the moon and Mars, Chicago has ample roads, railways, water, electricity and willing workers already in place. And supply trucks would not have to run a gauntlet of death rays to get to their destination.

The Evening Gazette

Worcester, MA, July 24, 1985

The first products manufactured in space for commercial use have gone on sale. Millions of microscopic latex beads were created aboard the space shuttle Challenger using a chemical process developed by the National Aeronautics and Space Administration and Lehigh University in Pennsylvania.

Before anyone gets the idea of buying a necklace of space beads for his lady love, however, be advised that even a vial-full is almost too tiny to be seen. Thirty million of the beads, each 12,500th of a inch in diameter, fit in a 5-milliliter (about one-fifth of an inch) vial that will sell for $384.

The National Bureau of Standards will sell the beads to companies that can use them to calibrate special instruments that make or measure fine particles. The beads will be used to improve microscopic measurements in the electronics industry, high technology and medicine.

The bureau will split the proceeds of the 600 vials available with NASA. No butterfingers will be permitted to handle the vials, in which the beads are suspended in water.

In the months and years ahead, NASA hopes to provide other made-in-space products including perfect crystals for electronic components, new drugs and new alloys. Someday we may have industrial parks among the stars.

Houston Chronicle

Houston, TX, August 22, 1985

For decades, the U.S. space program has produced a bounty of industrial "spinoffs" here on Earth. The era is fast approaching when industry will have its own private presence in space.

NASA has agreed to launch a privately built, unmanned space station envisioned as the first step in an orbiting industrial park. Room in the station will be leased to companies for research and manufacturing, although no customers have been signed up yet.

The unmanned station will be serviced by space shuttle crews and, later, astronauts aboard a permanent U.S. space station. Those astronauts may be using a rocket tug currently being designed by NASA.

The unmanned station, to be launched in 1989, is particularly significant for Houston, home of the Johnson Space Center and Space Industries Inc., the private company that will design and build the space platform.

Not long ago, factories in space and rocket tugs were concepts that belonged more to science fiction than real life. Reality, however, has shown a startling capacity to catch up with the wildest speculations. Houston's space center and the private companies budding around it are more than a little responsible.

THE DAILY HERALD

Biloxi, MS, August 26, 1985

American businessmen deplore the loss of some export markets and feel the hot breath of increasingly fierce foreign competiton in the high-tech industry.

But U.S. entrepreneurs abound and the success of NASA's space shuttle program is giving them an edge in developing the possibilities of that big market in the sky.

As a matter of priority, astronauts and scientists have formed an exclusive club of space adventurers in the first decades of the burgeoning Space Age.

But they're destined to be joined by the likes of the first teacher in space, manufacturing specialists, well-heeled tourists and, of course, hucksters for every conceivable product. Coke and Pepsi blazed that sky trail with well-publicized products-testing on a recent shuttle flight.

The pace of the rapidly-developing commercialization of space is emphasized by other developments in the news last week, one of particular import for the Mississippi Gulf Coast.

In case you missed them:

► The Institute for Technology Development, a little-noted private, non-profit organization created to bring high-technology jobs to Mississippi, will establish a National Center for the Commercial Development of Space at the National Space Technology Laboratory in Hancock County.

► The Houston-based Space Industries Inc. is well on its way to becoming the first industrial landlord in space. It has won NASA's approval to build a space platform for lease to manufacturers of products which require a gravity-free environment to obtain an unworldly perfection.

► And, for jaded travelers who've seen it all and have the bank account to accommodate their whims, a Seattle-based travel agency is planning the ultimate, an out-of-this-world luxury excursion aboard a module designed to fit in the shuttle's bay.

It adds up to increased income for the taxpayer-funded National Aeronautics and Space Administration and an exciting opportunity for private enterprise to strut its stuff.

NASA already has agreements with 20 companies and is negotiating with 24 others for commercial activities in space.

Mississippi's potential to share in research and new products development is greatly enhanced by the established NSTL facility and the foresight of the state's Legislature and Economic Development Administration.

State funding enabled the Institute for Technology Development to begin operations last February and, subsequently, to win NASA approval to establish a space research center at NSTL.

Identification of commercial applications for remote sensing of earth and oceans from space is expected to be a primary program of the center.

The institute, headed by Dr. David L. Murphree of Jackson, is one of only five national centers to win NASA incentive grants for research to stimulate commercial applications for space technology. Contributions from private companies seeking technological benefits will be sought to augment research funding.

It's a significant step toward identifying Mississippi with research and enhances opportunities for the state to attract high-tech industries and the skilled, well-paid jobs they provide.

General Dynamics Accused of Fraud; NASA Chief, 3 Others Indicted

A federal grand jury in Los Angeles December 2, 1985 indicted General Dynamics Corp. and four current or former executives of the company on charges of conspiring to defraud the Army on a weapons contract. The most prominent of the four was James M. Beggs, the current administrator of NASA. Beggs, 59, had been a General Dynamics executive vice president and a director before he left the company in 1981. The weapon in question was the DIVAD (division air defenses) Sergeant York mobile antiaircraft gun. Defense Secretary Caspar Weinberger had canceled the DIVAD production program in August, 1985 citing cost overruns and poor test results. The alleged wrongdoing had occurred between July 1, 1978 and August 31, 1981, when General Dynamics was developing a prototype of the DIVAD in competition with Ford Aerospace & Communications Co. (a unit of Ford Motor Co.). At the time all four indicted individuals had been executives at General Dynamics' Pomona (Calif.) Division, which developed the prototype.

If convicted, General Dynamics faced a maximum fine of $70,000. The individuals faced maximum penalties of five years in prison on each count and a $10,000 fine. General Dynamics issued a statement Dec. 2, saying that "the company and the individuals intend to contest these charges vigorously. General Dynamics knows that the individuals were honest in their judgement and acted in complete good faith." The individuals Dec. 2 all denied wrongdoing.

The Pittsburgh PRESS

Pittsburgh, PA, December 4, 1985

The National Aeronautics and Space Administration, an agency that has basked in popular approval after racking up success after success in its space shuttle program, has come to a crisis not of its own making.

But the crisis, the indictment of NASA administrator James M. Beggs on charges that he and three others tried to defraud the government, reflects poorly on the agency. It places NASA in the position of losing credibility, an absolutely necessary ingredient for continued success.

Mr. Beggs was indicted for allegedly trying to defraud the government by illegal billing on the trouble-plagued Sgt. York anti-aircraft gun, the first major weapons system to be canceled in production in 20 years. He was an executive vice president of General Dynamics Corp. during the three years that the costs allegedly were hidden.

Officially known as the Division Air Defense gun, the Sgt. York system consisted of two 40mm anti-aircraft guns mounted on an M-48 tank chassis. It had one major flaw. It didn't work.

During one test that embarrassed the Army, the system's radar locked onto the exhaust fan of a latrine. In another test, the gun fired at target drones that flew lower and slower than usual and the drones were blown up by remote control, falsely making it appear that the weapon had destroyed them.

The misadventure cost the government $1.8 billion for 65 guns that were dismantled for their parts.

The indictment, though, dealt only with the $40 million Army contract for production of the Sgt. York prototype. It charges that $7.5 million was knowingly mischarged so that losses would be reduced.

Even though Mr. Beggs is only accused and far from convicted, the charges are serious enough that NASA could be directly harmed.

Still, we can accept the adminstrator's position that he will not resign. He has distinguished himself in the four years he has headed NASA and charges still are only charges at this point.

But because his agency and its program are too important to be placed under even a shadow of a doubt, his decision to take an immediate leave of absence is a prudent one.

The Hutchinson News

Hutchinson, KS, December 6, 1985

Can it really be true that the Justice Department is going after crooked defense contractors?

Let us hope so.

Defense giant General Dynamics is the latest in a dreary string of allegations about greed and looting in the national defense industry.

This time, however, some high brass in the industry will be brought before a judge on criminal, not civil, charges. That's a big difference. It's also a much-needed change.

The charges against four present or former General Dynamics executives are that they conspired to loot the U.S. treasury. The federal grand jury charged that the four executives rigged the billing to get back some of the money the company was going to lose on the ill-fated Sgt. York anti-aircraft gun.

The men and the corporation will get their days in court, and, if the men are found guilty, in jail as well. If they are found innocent, they will be accorded the treatment that all other falsely accused Americans receive.

Either way, the precedent should be set for accountability of executives within corporations. Criminal actions within corporations should be treated just as criminal actions outside corporations are treated.

That means crimes are treated as crimes, not merely as indiscretions by inanimate corporations.

• If criminal conduct is found, civil suits also should be brought by the taxpayers against General Dynamics. A good jury to pick for a civil trial would be the Texas jury that socked it to Texaco for its corporate dishonesty in the deal with Pennzoil and Getty.

The Hartford Courant

Hartford, CT, December 4, 1985

James M. Beggs, who has been indicted on criminal charges of defrauding the Army on a weapons contract, should take the broad hint dropped by the White House and step aside for the time being as NASA administrator.

Mr. Beggs is known as a forceful executive whose absence could hurt the agency in 1986, when NASA will attempt six more space shuttle missions than in 1985, and will seek to pin down congressional support for its permanent space station project.

NASA's need for strong leadership is a good reason to regret the indictment against Mr. Beggs. But he can't continue to perform his duties under a shadow. White House spokesman Larry Speake's suggestion Tuesday that the NASA chief should do "the right and proper thing" is widely interpreted as meaning Mr. Beggs should take a leave of absence. That's good advice.

A federal grand jury in Los Angeles accused Mr. Beggs of illegally billing the Army for millions of dollars in cost overruns on the Sgt. York anti-aircraft gun when he was senior executive vice president of the General Dynamics Corp. Three other company officials have also been indicted. They and Mr. Beggs were also charged with several counts of trying to cover up the allegedly illegal costs.

Mr. Beggs denies any wrongdoing. The company promises to "vigorously" contest the charges, claiming that the "issue is a highly sophisticated regulatory and accounting matter which should be resolved in a civil forum, not in a criminal case."

Nonetheless, it's a criminal matter. Although an indictment is only a finding of evidence, Mr. Beggs could face a prison term or a large fine if he's found guilty.

Indictments against high-ranking government officials are not easily brought; grand juries don't act casually in such cases. We remember the indictments that are handed down because there are so few. Because of the notoriety, a stigma attaches to the accused until the case is disposed of. That may be unfair, but it comes with the territory.

In addition to the public's loss of trust, the criminal charges raise practical considerations. Should NASA contract for any work with General Dynamics while the agency's administrator and company officials are under indictment? Can Mr. Beggs adequately perform his job while preparing his defense?

It would be better for the agency and the nation's space program if Mr. Beggs temporarily stepped aside, with pay, and let his senior assistants run NASA until there's a decision in his case.

Richmond Times-Dispatch

Richmond, VA, December 9, 1985

DIVAD, the anti-aircraft gun infamous for missing its targets and ultimately abandoned by the Pentagon, seems finally to have hit an unintended one: The nation's third-largest defense contractor is accused of having defrauded the Defense Department in the course of developing two DIVAD prototypes. General Dynamics has been indicted, and suspended, temporarily at least, from new federal contracts. Three current executives, and a former vice president who recently became administrator of NASA, have also been indicted.

All deny any wrongdoing. Said James Beggs, in taking a leave of absence from NASA: "There is nothing that I did in the case involved that I would not do again . . . We acted in an entirely ethical, legal and moral manner."

Maybe. Forty million in tax money was budgeted for those two prototypes. After $40 million, the DIVAD account was essentially closed. But, according to the indictment, General Dynamics spent $7.5 million more, which it included in Pentagon billings for R&D on new weapons and preparations to sell them. A federal grand jury saw possible fraud in that. General Dynamics sees "a highly sophisticated regulatory and accounting matter which should be resolved in a civil forum, not in a criminal case."

We see two matters to be resolved, in whatever suitable forum. First, the governing regulations. Are they so "sophisticated" that a giant corporation's accountants can't tell whether it's OK to shift billings at will from one account to another? Are the rules so "sophisticated" as to allow shifting the cost of a contractor's every miscalculation to the taxpayer? Are they so "sophisticated" as to provide a handy niche in which to stash any unanticipated expense?

The argument that defense contractors must be free of the worst vicissitudes of the marketplace has merit. That hardly means they must be freed of all customary costs and risks of doing business. It hardly means that the usual and useful connection between decisions taken and profits made or lost must be severed. And it doesn't mean that the bounds of ethical conduct must be elasticized. To the contrary: Defense contractors are *more* bound by them precisely because they are largely protected and subsidized — and because they are dealing with not only the nation's money but the security it must buy.

If contracting regulations are so "sophisticated" that defense contractors can get confused as to right and wrong, and the Pentagon can't enlighten them, then the regulations are in dire need of de-sophistication.

The second matter to be resolved is the guilt of actual persons. The four indicted in this case say they did nothing criminal, and they may well have not. General Dynamics says nothing criminal was done, and that may be. But if something criminal was done, somebody did it. Prosecuting him, or them, may help pierce the notion that the "defense industry" is an amorphous entity, in which everyone — therefore no one — is responsible when that other amorphous entity, the "taxpaying public," is ripped off.

A matter of "cost-accounting judgments," says General Dynamics of this case. Indeed, it is one of many cases suggesting a pattern in defense contracting: too much left to the judgment of too many people without the guidance or ethics to exercise it well.

Los Angeles, CA, December 4, 1985

The latest chapter in the ongoing tale of defrauding the Defense Department was written this week when four executives — including James M. Beggs, who now runs NASA — were indicted for bilking the U.S. out of millions through phony billings. The indictments, along with reports of Pentagon rip-offs, point up a fundamental problem with our defense program: the waste of taxpayers' dollars on weapons that too often don't work.

This week's indictments were leveled against former and current officers of General Dynamics, which had already been accused of overcharging the governmment some $244 million for defense work. Now, the firm has been accused of trying to recoup another $7.5 million in nonrefundable overcharges on the Sergeant York anti-aircraft gun prototype — *plus* the $41 million it spent and charged to the government in attempting, unsuccessfully, to win the contract for building the gun. (The contract went instead to the Ford Aerospace Company.) By the time the project was abandoned earlier this year, the Sergeant York had cost the U.S. $1.8 billion, and the weapon still didn't do what it was supposed to do.

Other outrageous white elephants have been uncovered in the defense establishment by the Reagan administration over the last two years, including nickle screws costing $100 each and 60-cent lamps costing $511. These, along with loose Pentagon accounting procedures, untimely tax cuts and profligate social programs, have contributed to the massive federal deficit. But, in defense, as in all other expenditures, the American taxpayers deserve their money's worth — especially with the drive for still more defense-spending increases.

A strong defense is essential to America. Waste and fraud must, however, be eliminated — and not always after the fact. Reassuring as the administration's steps in this direction are, more needs to be done. As the president's Grace Commission noted two years ago, government could buy not only more guns but more butter, too, if its spending were continuously and scrupulously monitored.

The Honolulu Advertiser

Honolulu, HI, December 4, 1985

General Dynamics Corporation and its executives, including present NASA Administrator James Beggs, are innocent until proved guilty of charges of conspiring to defraud the government with overcharges in constructing a prototype of the ill-starred Sergeant York anti-aircraft gun.

Still, Beggs' decision under congressional and administration pressure to take an indefinite leave of absence while he fights the charges is the least he could do under the circumstances.

In fact, considering the National Aeronautics and Space Administration, a resignation might have been more in order

The year ahead for NASA will be the most ambitious since the Apollo moon-landing program over a decade ago. Fourteen shuttle flights, a robot spacecraft flight past Uranus, advanced probes of Jupiter and the sun and orbiting of a space telescope are all planned in '86.

With that much going on, NASA would be better served by a properly nominated and confirmed administrator than by an acting caretaker appointee.

The Pentagon's decision to suspend General Dynamics from new government contracts during the legal proceedings is also appropriate and the least that should be done under the circumstances.

The huge company, with almost $7 billion in defense contracts last year, has been the object of the biggest investigations in Defense Department history and was suspended briefly in another case involving "unallowable expenses" earlier this year.

These federal investigations are largely an outcome of outside reporting and revelations about defense-contract abuses, and of demands for greater scrutiny of the fortune this country spends in the name of national defense.

THE PLAIN DEALER

Cleveland, OH, December 4, 1985

Defense Secretary Caspar Weinberger drove a stake through the heart of the Sgt. York gun last August, but the weapon continues to haunt the Pentagon and defense contractors. Just last Monday, the York project arose from its crypt in the form of indictments against four present or former officials of General Dynamics, including current NASA Administrator James M. Beggs. You could make a movie out of this: The Weapons System That Wouldn't Die; The Contract's Curse.

The indictments charge that the four officials wrongly billed the government for cost overruns incurred while General Dynamics was competing for the contract to build the York gun. The contract went to Ford Aerospace, which built the weapon with cost overruns of its own. After all that, the gun didn't work, which prompted Weinberger's summary enspikement. He was too late: Like a cranky mummy, the project carries a curse —or at least some very bad vibes.

In itself, the indictment of General Dynamics officials should do little more than raise your eyebrow a notch. The huge contractor ($6 billion in 1984) has been in trouble before: It was suspended last May from signing new government contracts for three months. In an earlier case, two different officers were accused of accepting kickbacks from a subcontractor. General Dynamics has also been accused of abusing government expense accounts and violating security regulations.

If the charges against General Dynamics have a familiar ring, the indictment of Beggs is problematic. His stewardship at NASA has been very impressive—marked by strong leadership and success. Yet his reign at NASA is at least temporarily tainted. That won't mean much to the Reagan administration, which has tolerated the appearance of impropriety before, and which probably will sustain Beggs' decision not to step down now. Still, he should be encouraged to take a leave of absence until the case is resolved.

No such hiatus should stall the reform movement now gathering steam in the Pentagon. If anything, the most recent indictments should provide further impetus for the drive to revamp the Defense Department's entire systems-acquisition process. Pentagon contracting blunders are not restricted to an occasional $700 hammer. When it comes to procurement, the Defense Department is a poorly managed bureaucracy that wastes millions of tax dollars every year through faulty oversight and inadequate project controls.

The Sgt. York gun, which had to be scrapped after the government spent $1.8 billion on it, seems to be immortal, or at least hard to kill. But maybe its unnatural life will serve as a useful symbol—even better than toilet lids—of the need to reorganize and reform military procurement procedures. Without further Defense Department vigilance, other horror shows could remain uncovered, untreated and undead.

ST. LOUIS POST-DISPATCH

St. Louis, MO, December 4, 1985

To the federal grand jury in Los Angeles, the case involved a conspiracy to defraud the government of millions of dollars. To General Dynamics and its four present and former officials who were charged in the seven-count indictment, it was a simple case of "cost-accounting judgments" involving no wrongdoing whatsoever.

The indicted parties, of course, are entitled to a presumption of innocence, but the General Dynamics statement would be more persuasive if this had been the first allegation made against the giant military contractor. But in fact, these indictments follow a long series of charges and counter-charges, congressional and Justice Department investigations and press reports of waste, fraud and abuse by General Dynamics and no doubt contributed to the Pentagon's decision yesterday to suspend the firm's right to bid on new contracts.

Ironically, this particular case involves one of the biggest military procurement fiascos in recent memory, the Sgt. York Division Air Defense gun. General Dynamics built an unsuccessful prototype. (The other bidder, Ford Aerospace, "won" the contract, which since has been canceled.) The federal indictment charges that company officials as early as 1978 realized that there would be a "substantial overrun" — roughly $7 million — on the prototype's cost and that the four General Dynamics officials then conspired from 1978 to 1981 to shift the cost overruns to other government accounts, thus limiting the company's losses while defrauding the government of "several million dollars."

In its defense, General Dynamics hedges by arguing that "the issue is a highly sophisticated regulatory and accounting matter which should be resolved in a civil forum, not in a criminal case." In a sense, the company is correct: This is a very sophisticated accounting issue, and that is no doubt part of the problem. The sheer complexity and confusing nature of Pentagon procurement accounting not only make for loopholes on a grand scale but leave the door wide open to abuses of all kinds. Reforming this aspect of the problem will be difficult and time-consuming.

But in another sense, General Dynamics is wrong when it argues that this case is one for "a civil forum." If there is to be any real hope of cleaning up waste, fraud and abuse in military contracting, there must be personal accountability by company officials for acts that can be proven to have broken the law — and that accountability must include criminal as well as civil penalties. To do otherwise would institutionalize the double standard for so-called white-collar crime. Moreover, there must be serious penalties for the companies as well. In this case, General Dynamics, which had $7.8 billion in sales in 1984, faces a maximum corporate fine of only $70,000. That is no deterrent.

The case has another noteworthy aspect. Among those indicted is James M. Beggs, a former General Dynamics official who is now the head of NASA. While stoutly maintaining his innocence, Mr. Beggs has done the proper thing for the space agency by taking a leave of absence from his post until the case is settled.

The Birmingham News

Birmingham, AL, December 4, 1985

The indictment of NASA administrator James M. Beggs on a charge of conspiring to defraud the government while he was an executive of General Dynamics Corp. is merely an accusation. He may yet prove to be innocent.

In the meantime, though, the man who now presides over nearly $8 billion in federal spending should consider stepping aside until the criminal matter is resolved. A source in the Reagan administration has told the Associated Press that Beggs intends to do just that.

Beggs said Tuesday he would not resign, but he apparently will follow the example of former secretary of labor Raymond Donovan, who took a leave of absence during the early stages of a criminal prosecution against him. Donovan is charged with defrauding the New York City Transit Authority on a subway construction contract awarded before he joined the administration.

Donovan, who has pleaded innocent, remained on that leave of absence from the time of his indictment until it became clear that he actually would have to stand trial in the case. When a judge refused to dismiss the charges, Donovan resigned his Cabinet post on March 15 of this year.

Beggs was faced with at least three options — holding onto his job while battling the fraud charge that is unrelated to his performance at the space agency, distancing himself from the agency at least temporarily while the charge is pending, or resigning completely in the face of what may prove to be an unsubstantiated accusation.

The middle course, at least for now, looks the best. Beggs is accused of conspiring with other General Dynamics executives to hide cost overruns in the ill-fated Sgt. York anti-aircraft gun. That development program was much like the high-technology development efforts that NASA funds for the space program, so allegations of Beggs' misconduct on a defense contract could possibly raise questions about his administration of space contracts.

It may be, as General Dynamics has asserted, that all those indicted "were honest in their judgments and acted in complete good faith." If so, it would be unfair for this accusation to cost Beggs his job.

At the same time, for the good of NASA, he will do well to put some distance between himself and the administration of the agency until the matter is resolved.

'PEACE THROUGH STAR WARS, GOODWILL TOWARD DEFENSE CONTRACTORS.'

The Indianapolis Star

Indianapolis, IN, December 9, 1985

The case of James M. Beggs, head of the National Aeronautics and Space Administration, points up the need for established guidelines to follow when top-level federal officials are indicted for criminal activity.

Beggs, a former vice president of General Dynamics, is charged with trying to defraud the government by hiding cost overruns on the now-canceled Sgt. York anti-aircraft gun.

His indictment precipitated a flurry of speculation and rumor about whether he will resign, get the ax, stand firm or take a leave of absence. Beggs, as it turns out, is taking a leave but the multiple-choice guessing game that preceded the decision is ridiculous.

An important federal agency is involved here. Its efficient operation should not be jeopardized by the lack of a set formula or process regarding what is proper under such circumstances. Personal preference, even that of the president, should not be the crucial factor in determining what is proper.

There is a debt of loyalty between a president and the people he appoints to top-level government posts. A resignation under fire, whether voluntary or forced, always carries implications of guilt. Many people assume the White House has some inside dope and is trying to cut its losses. The assumption is unfortunate but real.

On the other hand, if the appointee stands pat, the White House is accused of political favoritism and partisan criticism is galvanized. It's a no-win situation for all concerned and it ought to be replaced by an impersonal, impartial mechanism.

Once a top-level appointee is formally charged, specific procedures should be automatically instituted. They might allow for an paid leave of absence and the appointment of a deputy with full power to carry out official obligations and duties. If an appointee is involved in a criminal investigation, a congressional committee could decide whether an enforced leave is appropriate or necessary.

The trick will be to frame the legislative guidelines so artfully and to impose them with such equity that they won't be construed as punishment or repudiation or imply guilt. The presumption of innocence must be preserved.

Having a mechanism in place would in itself give a sense of neutrality and dispassion. The accused or the investigated would, in part at least, be removed from the whirlwind of publicity. The president would be spared decisive involvement and sniping for political advantage might be minimized.

Controversy surrounding the charges wouldn't disappear, but they would be separated from the day-to-day workings of government. That alone ought to be sufficient incentive for the Congress to consider a legislative mechanism for future Beggs cases.

Audits Show NASA Wasted Billions, Saved on Safety

The waste by NASA and its contractors amounted over the years to at least $3.5 billion, according to the audits reported by the New York Times April 23-24, 1986. And that was "the tip of the iceberg," according to one spokesman, Francis La Rocca, attorney for the NASA inspector general's office, one of the auditing agencies. In many instances, the waste was not assessed in dollar terms. The Times material, much of it obtained through the Freedom of Information Act, included more than 500 audits of the space agency by its own inspector general, Congress's General Accounting Office and the Pentagon's Defense Contract Audit Agency.

Some of the more glaring examples of waste were that:

- NASA paid $3.6 million in rent for a building recently owned by the U.S. government that had been sold to private investors for $300,000.
- NASA was charged $159,000 for an electronic cooling fan that cost the supplier $5,215. Another contractor charged $12,000 each for parts available on the open market for $2,000. The agency paid $120 each for bolt assemblies available elsewhere for $3.28, $315 each for three-cent metal loops, $80 each for $1 metal washers, $86 each for $2 clamps and $1,621 for a $70 bolt.
- As early as 1970, more than $320 million worth of equipment had not been accounted for at the Kennedy Space Center in Florida for as long as four years.

The Times article April 24 concerning safety spending said the agency had reduced or delayed half a billion dollars in spending on safety testing, design and development of the shuttle program, from the beginning to explosion of the Challenger in January, 1986. The cutbacks and delays in spending covered work on the main engines, the orbiter, the external tank for liquid fuel, the solid booster rockets and landing equipment.

Los Angeles Times
Los Angeles, CA, April 28, 1986

Nothing succeeds like success, it has been said, which is an accurate description of the National Aeronautics and Space Administration before the Challenger accident three months ago. While it was running up an unparalleled string of breathtaking space spectaculars, few noticed that it was also running up a bill that was padded with waste, fraud and abuse.

As detailed in two articles last week in the New York Times, federal auditors throughout the 1970s and '80s told the agency repeatedly that millions of dollars were being wasted, that management controls were inadequate and that NASA contractors made defense contractors look like pikers. Millions of dollars of equipment disappeared. Work was done by contractors before a price was agreed on, and the government then paid what the contractors asked for. Spare parts were bought from subcontractors who bought them from other vendors and then jacked up the price. The Rocketdyne Division of Rockwell International in Canoga Park, which makes the shuttle engines, was paid $120 each for bolt assemblies worth $3.28, $315 each for metal loops that cost 3 cents, $80 for $1 washers, $1,621 for a $78 bolt, and on and on.

The auditors documented waste of at least $3.5 billion, which the attorney for the space agency's inspector general calls "the tip of the iceberg," adding, "There is probably much more out there."

But the agency sloughed it all off. After all, every time it lit up one of the shuttles, the thing took off, didn't it? So what's a little vigorish here and there to assure good contractor relations?

One effect of the waste of money was that the space agency tried to economize by cutting back its testing program. It saved $68 million, the auditors found, by not testing the shuttle's solid-fuel booster rockets, the failure of which caused the Challenger disaster.

The absence of financial controls at NASA greatly increased the cost of the shuttle program. Congress originally approved it on the strength of the argument that a reusable vehicle would be cheap and that these remarkable spaceplanes would pay for themselves and make money for the government to boot. But the shuttle wound up costing more and flying less than the rosy predictions made to Congress. Everyone, including NASA, has known for years that the shuttle will never make a profit.

The man who ran NASA from 1971 to 1977, while the shuttle was being developed and the taxpayers were being robbed, was James C. Fletcher. He is the James C. Fletcher who is now being brought back with great fanfare to pick up the space agency's tattered reins.

There must be somebody better for the rudderless agency to turn to. Fletcher's nomination is now before the Senate. His previous stewardship should fill all of the senators with misgivings about his ability to straighten things out.

AKRON BEACON JOURNAL
Akron, OH, April 24, 1986

IT'S ALL depressingly familiar: A government agency being billed $315 for 3-cent wire fasteners, billed for fancy oak furniture and private beach condominiums, and paying to fly a business executive's pet dogs coast to coast.

But in this case the agency is not the Defense Department; it's NASA.

The waste was revealed this week by reporters for Knight-Ridder Newspapers, who studied government audits obtained through the Freedom of Information Act. Though the waste has no direct tie to the loss of the space shuttle Challenger, investigations after the tragedy suggest that NASA might have cut corners on safety because of lack of money.

George Spanton, a government watchdog formerly with the Defense Contract Audit Agency and supervisor of contract billings at Kennedy Space Center, estimates one-third of NASA's budget is fat. NASA officials say that is a "ludicrous statement." But problem areas point to a too-familiar government pattern.

The records show NASA spent $159,000 for a cooling fan that was built for $5,000; it is paying $6 million to lease private warehouses next to similar government buildings that are standing empty; it allowed contractors to recover business expenses that include "self-serving promotions" and charity contributions.

The documents also show NASA wasted money by paying large markups on some parts from private contractors, even though many of the items — especially nuts, bolts, washers — could have been bought cheaply through the government's bulk supply system.

It can add up to large sums when you consider that nearly 90 percent of the NASA budget now goes to private contractors. And NASA officials ignored suggestions that the million-dollar bonuses of at least one contractor — Rockwell — be withheld until "inefficiencies" are corrected.

It goes without saying that these abuses should stop, through better oversight among all government agencies. But the pattern will continue until someone in government gets serious about the problem. Ronald Reagan campaigned on a pledge to eliminate waste, fraud and other abuses, yet whistle-blowers are still not given the proper hearing or esteem; they are often lucky to even get their findings released to the public.

All of this may be unrelated to the shuttle disaster and may be no reflection on the overall job done by NASA and other agencies. But if these abuses can be identified, they can be stopped by concerted government action. Taxpayers have a right to expect the best use of their money. Yet too much evidence suggests that the government is failing them.

WORCESTER TELEGRAM.

Worcester, MA, April 27, 1986

It has been suggested that the Pentagon was the champion of wasting money and obfuscating figures in keeping track of spending, supplies, research and tests. Now it turns out that a civilian agency, the National Aeronautics and Space Administration, runs circles around the military in the waste and coverup derby.

If it had not been for the lives lost in the Challenger disaster, NASA probably would have continued with mismanagement and the slight-of-hand manipulation of reams of money. But the investigation of what went wrong with the shuttle, especially the testimonies of those in middle management, has caused close scrutiny of NASA's operating procedures. Recent reports, based on more than 500 federal audits, disclosed massive mismanagement, waste and fraud. Painfully, they also revealed that NASA took chances with safety, time and again.

Suddenly, the Challenger disaster no longer seems like an isolated, unavoidable accident. And NASA's once impeccable reputation has taken a nosedive, along with its carefully maintained image of a trouble-free, glorious super-agency.

With each new revelation a horror story emerges. The details are astounding. NASA didn't waste pennies; it wasted millions of dollars. When the waste resulted in a financial pinch, the agency began cutting corners and jeopardizing safety. The audits document $500 million worth of tests eliminated or delayed to save time and money, including $68 million for testing the shuttle's booster rockets. There was fraud and coverup; for example, there is evidence that a subcontractor falsified records to cover up faulty welds on Challenger.

While the agency was besieged by problems, NASA officials continued to mislead Congress and oversell their programs. For example, former administrator James Fletcher told Congress that the cost of lifting cargo into orbit would be $100 a pound; the actual cost was $5,264. Congressional overseers whose job was to keep tabs on NASA — Sen. Jake Garn and Rep. Bill Nelson, among others — became cheerleaders for the agency.

NASA acknowledges the validity of the audits which conclude that more than $3.5 billion was wasted but denies that all that mismanagement and malfeasance caused major safety problems. In view of the disclosures, that claim rings hollow. Obviously, many contractors have made fortunes off shoddy and defective work or over-charge. In fact, the Pentagon's $400 coffee pots and $70 toilet seats look like bargains compared to NASA paying $12,000 each for parts the contractor had agreed to sell for $5,000 while they were available on the open market for $2,000.

Some say it's not nice to kick NASA while the agency is down. Indeed, NASA had been a superb organization for years, and its contribution to the U.S. space program should not be forgotten. But, after the immense success of the moon flights and the departure of agency head James Webb, things began to deteriorate. The Challenger disaster merely lifted the lid off a box of horrors.

If NASA is to continue to serve as a viable agency, Congress should take a close look at its practices, policies and programs for the future. It should also question the wisdom of bringing back Fletcher, who presided over a good part of the agency's decline. NASA must do the utmost to change the image of an agency where waste, incompetence, fraud and the concealment of facts became a matter of routine.

The Chattanooga Times

Chattanooga, TN, April 29, 1986

The image of the National Aeronautics and Space Administration as a well-run, safety-conscious agency has been slowly crumbling since the explosion of the space shuttle Challenger three months ago.

First we learned that NASA had ignored repeated warnings about potential failure of the very shuttle component believed to have caused the tragic loss of the Challenger. Then we learned that higher-ups in NASA were never even told about engineers' 11th-hour recommendations to delay the ill-fated launch.

Now we learn, through *The New York Times*, that the agency has been grossly mismanaged over the last 15 years, that documented waste exceeds $3.5 billion, that NASA managers ignored auditors' recommendations to correct abuses and that budget constraints flowing from waste and mismanagement resulted in curtailment of safety-related programs.

The *New York Times* reports were based on hundreds of audits of the space agency over the years — by NASA's own inspector general, Congress's General Accounting Office (GAO) and the Pentagon's Defense Contract and Audit Agency. They reveal that many of the management procedures and abuses originated during the time (1971-77) that Dr. James C. Fletcher headed the program. Mr. Fletcher is President Reagan's nominee to become the new administrator of NASA.

It was he who sold Congress on the shuttle program, promising it would pay for itself through commercial business, such as deploying and repairing private satellites. But costs have exceeded Mr. Fletcher's estimates in every case — in some by a factor of 10. And his promise of economic viability was based on gross overestimation of the frequency with which the shuttle would fly. "We're all disappointed it didn't come out the way we calculated," Mr. Fletcher said. What he didn't say was that the miscalculations add up to failure of the shuttle program, measured in terms of the basis on which it was "sold."

Mr. Fletcher's response to the auditors' devastating documentation of mismanagement was that the need for correction of management practices was a matter of judgment — his against the auditors'. Many of the abuses cited by auditors, however, are beyond any question of judgment.

Many violate federal law and NASA regulations. Consider, for instance, that from 1975 to 1981, NASA allowed one of its contractors, Rockwell International, to buy parts on the open market when millions of dollars of identical parts were available at much less cost from federal sources. Example: NASA was found to have paid $315 each for 3-cent metal loops.

Auditors found that the Marshall Space Center in Alabama authorized Rockwell to buy $1.6 million in spare parts that weren't needed. Officials agreed to rent a building for another contractor, Morton Thiokol Inc., at $3.6 million over 13 years although NASA had recently sold that building for only $300,000.

At the Ames Research Center in California, nearly $1 million in property was lost between 1977 and 1982 because required inventories weren't done. The next year auditors found another $2.4 million worth of computer equipment had disappeared. At the Langley Research Center in Virginia failure to change a reclamation filter meant that over three years, $214,000 worth of silver literally went down a laboratory drain.

The problems are clearly spread throughout NASA.

At the Johnson Space Center in Texas, the number of employees monitoring contract compliance was reduced from 28 during the Apollo moon program to two for the shuttle program. There was, for instance, one person to monitor the 44,700 purchase orders Rockwell issued in 1981.

NASA's contractors were given such free rein that, at times, more than $750 million in work was being done without price agreements, meaning contractors were free to spend as much as they wanted and bill the agency later.

The astronomical cost overruns and tremendous waste combined to put the agency in budgetary straits. NASA's response was to cut back on testing of shuttle components. This decision not only resulted in higher costs due to the greater expense of making changes after components were in actual production or use, but also clearly de-emphasized safety considerations.

Congress must accept its share of the blame for allowing NASA's fiscal folly to go unchecked for so long. Claims of congressional supporters that they had to rely on the judgment of NASA officials are disingenuous. Congress had access to the many audit reports on mismanagement over the years. It ignored them.

Shortly the Senate will rule on Mr. Fletcher's nomination. Recognizing that mismanagement underlies NASA's failure to maintain a safety-first attitude, and thus contributed to the Challenger tragedy, this is not the time to reinstate the man who oversaw the deterioration of the space agency during the '70s. Congress should reject his nomination.

RAPID CITY JOURNAL

Rapid City, SD, April 29, 1986

It now appears that not only did the National Aeronautics and Space Administration run up a string of successful space spectaculars, it also ran up a bill padded with waste, fraud and abuse.

Through the 1970s and 1980s, federal auditors repeatedly told the agency that millions of dollars were being wasted, that management controls were inadequate and that NASA contractors made defense contractors look like pikers. Million of dollars worth of equipment disappeared, work was done by contractors before a price was agreed on, and the government then paid what the contractors asked for. Spare parts were bought from contractors who bought them from other vendors and then jacked up the price.

Auditors documented waste of at least $3.5 billion which the attorney for the space agency's inspector general calls "the tip of the iceberg."

One effect of the waste of money was that the space agency tried to economize by cutting back its testing program. It saved $68 million, the auditors found, by not testing the shuttle's solid-fuel booster rockets, the failure of which caused the Challenger disaster.

The absence of financial controls at NASA greatly increased the cost of the shuttle program. Congress originally approved it on the strength of the argument that a reusable vehicle would be cheap and that these remarkable spaceplanes would pay for themselves and make money for the government to boot. But the shuttle wound up costing more and flying less than the predictions made to Congress promised.

In light of these revelations, it's strange that James C. Fletcher, who was in charge at NASA in the early and mid-1970s when the shuttle was being developed, has been chosen to resume his former position.

It would seem a better choice could be made.

The Birmingham News

Birmingham, AL, April 28, 1986

Charges that the National Aeronautics and Space Administration has wasted billions of dollars by mismanagement are not astonishing. Name a federal or state bureaucracy which does not waste upwards of 10 percent of its appropriation each year.

One cannot excuse NASA any more than one can excuse the Department of Health and Human Services or the Defense Department for the fraud, waste and mismanagement which have dogged them throughout their history.

Organizational dynamics and human nature guarantee that waste will occur in any large bureaucracy, whether it's NASA or General Motors. The people at the top may wage a continual war against waste, but their efforts usually prove futile because of a lack of reliable information from below.

Writing in *The Washington Monthly*, editor Charles Peters says, "In any large government or business organization, there exists an elaborate system of information cutoffs, comparable to that by which city water systems shut off large water-main breaks, closing down first the small feeder pipes, then larger and larger valves. The object is to prevent information, particularly of an unpleasant character, from rising to the top of the agency, where it may produce results unpleasant to the lower ranks. Thus, the executive at or near the top lives in constant danger of not knowing, until he reads it on Page One some morning, that his department is hip-deep in disaster."

Even so, NASA's brass does not deserve all the blame. With the proliferation of federal programs and agencies, congressional committees charged with oversight of these agencies have become notorious for their lack of interest in the day-to-day business of the agencies, once they are functioning.

The executive department is also remiss in this regard. No administration wants to court scandals by uncovering waste and fraud. President Reagan has made some progress by appointing inspectors general for many departments, but this is only a symbolic move at getting at the very real problem.

The answer is to pare government agencies to a number which can be given proper oversight by congressional committees and the General Accounting Office. Special protections should be institutionalized for whistle-blowers. The executive department should also study methods to increase upward movement of vital information.

None of these measures will end waste and corruption, because the federal government has reached unmanageable proportions. They may, however, be able to reduce waste and abuse, and that is not to be despised

The Pittsburgh PRESS

Pittsburgh, PA, April 27, 1986

Over the past 15 years, most Americans have taken immense pride in the National Aeronautics and Space Administration shuttle program.

NASA's image was one of an agency and a program that was squeaky-clean, devoted to overall excellence and absolutely unyielding in its demand for total safety.

In reality, federal audits show, NASA was wasting billions of dollars and compromising safety with cold-hearted regularity.

The deterioration of the agency with the gleaming image was revealed in shattering detail in two stories last week in The New York Times, which reviewed more than 500 federal audits of NASA. From the audits emerged a pride-puncturing picture of gross mismanagement and inattention to pleas for better reliability and safety testing.

The articles were published during two days of congressional confirmation hearings for James C. Fletcher, the University of Pittsburgh engineering professor nominated by President Reagan to once again head the agency he guided from 1971 to 1977. He headed NASA during much of the time covered by the startling audits and helped develop the shuttle program. A vote on his confirmation is scheduled Tuesday.

We have been wary of Dr. Fletcher ever since he characterized the investigation of the Shuttle explosion as "a witch hunt." And, he said, "I don't like witch hunts."

It is possible that he doesn't like witch hunts because boiling cauldrons sometimes are found, for instance:

• At least $3.5 billion was wasted during NASA's shuttle years.

• About a half-billion dollars worth of tests, a good many of them critical safety tests on the booster rocket which apparently caused the shuttle explosion, were eliminated or delayed.

• NASA often allowed contractors to start — and even finish — projects before costs were calculated, leaving the contractor virtually free to determine them. In just such a situation, the Rockwell International Corp., main contractor on the shuttle, spent $20 million on a shuttle propulsion system that NASA had expected to cost a little more than $3 million.

• Dr. Fletcher told Congress in 1972 that shuttle launches would cost $10.45 million each. He didn't say that figure represented just operating costs and didn't include construction costs. Even so, today's launches cost $151 million each just for operations. With construction included, each launch costs $279 million. Allowing for inflation, the operations cost alone is 500 percent more than he predicted, although NASA had said a rise of just 75 percent would make the shuttle uneconomical.

• Dr. Fletcher also predicted that the cost of launching each pound of shuttle payload would be $100. It now costs $2,849 per pound just for operations, $5,264 per pound including construction costs. Discounting for inflation, the actual cost is 9 to 19 times his projections.

These and many other instances of mismanagement, misinformation and misapplication during the Fletcher years at NASA are excruciatingly distressing.

Even though Congress is understandably in a mood to quickly restore confidence in NASA, it made a major mistake last week when it ended its confirmation hearings without asking some of the tough questions posed by the audits and without pursuing further some of the pablum-quality answers Dr. Fletcher provided to other, feather-soft queries.

The confirmation hearings should be reopened so that Dr. Fletcher's past NASA performance can be examined more closely.

He should be made accountable to the people who paid — seven of them with their lives — for the shuttle program.

We need better questions, and better answers, about NASA. Without them, either he or President Reagan should withdraw his nomination.

FORT WORTH STAR-TELEGRAM

Fort Worth, TX, March 20, 1986

It appears that the National Aeronautics and Space Administration is trying to compete for headlines with the Defense Department's $600 hammers and $1,800 coffee pots. Most readers will recall the adverse publicity that accompanied revelations of the outrageous DOD purchases.

Now, *The Miami Herald* has turned up copies of NASA audits showing the space agency paid $30 for pins that should cost 3 cents and $159,000 for a $5,000 cooling fan. NASA also foot the bill at the tune of $256 to fly a contractor's dog coast-to-coast.

The exorbitant misuse of taxpayer funds is bad enough, but coming on the heels of revelations that NASA had cut back on its safety staff because of lack of funds puts the space agency in line for some heavy criticism.

The Miami newspaper said it had obtained dozens of NASA and defense contract audit reports to substantiate wasteful spending claims. One auditor's report said Rockwell International employees at Johnson Space Center in Houston were found playing cards on the job and wasting one-third of their work days. Another showed Rockwell's Rocketdyne Division paying its subcontractors up to 78 times the price for parts offered by the government's own supply system. Rocketdyne then sold the same parts to NASA for up to 3,000 times the government's price.

One report said Rockwell sold NASA a fan to cool electric parts on the space shuttle for $159,000. The price tag included $153,785 in overhead costs by the manufacturer, Sundstrand Corp., and middleman fees from Rockwell. A NASA audit later determined the actual cost of materials and labor was $5,215.

This smacks of the same questionable practices that turned up some months back in contracts involving some of the nation's biggest defense contractors. Such practices amount to stealing from the taxpayers, and it is the duty of government investigators to ferret out such questionable practices and prosecute those found to be in violation of the law.

The problem with NASA is that the space agency tends to get too cozy with its contractors, who do about 90 percent of the agency's work. That kind of activity cannot be tolerated ever, but it is especially bad during these times of heavy government deficits. A full-scale investigation of NASA's purchasing practices appears to be in order.

The Miami Herald

Miami, FL, April 19, 1986

WOULD YOU pay $315 for a 3-cent wire-end connector? How about $159,000 for a $5,000 fan? Or $6.3 million to rent a building near a vacant building that you already own? You wouldn't let yourself be fleeced this way — but you're not the National Aeronautics and Space Administration. NASA wasted millions of dollars on these and other outrageous expenditures at the very time that it was cutting back its flight-safety programs because of lack of funds.

To compound its profligacy, NASA time and again rejected its auditors' advice to refuse payment of questionable billings from its suppliers. At best the agency withheld only part of the amounts that auditors advised. Equally foolishly, NASA routinely special-ordered noncritical hardware — such as the wire fasteners, bolts, and the like — that it could have obtained from Government stocks at a tiny fraction of the cost.

These examples of NASA waste appear in documents that *The Miami Herald* obtained from the agency under the Freedom of Information Act. The documents show that NASA, like the Pentagon, allows its contractors to order parts and equipment from subcontractors and then add their own middleman's charges for "related costs" to the price charged the agency. Thus the $5,000 cooling fan had $154,000 of these costs added on by the time it reached NASA.

Absurd? That's an understatement. Like the Pentagon, NASA tolerates unconscionable gouging in part because its officials develop cozy relationships with space contractors with whom they often work closely for years.

NASA unquestionably will undergo changes in the aftermath of the *Challenger* tragedy. The Rogers Commission's hearings already have shown that safety procedures need tightening, and some reorganization of NASA may be justified as well. Part of that post-*Challenger* reform must address NASA's purchasing procedures and the agency's tendency to reject its auditors' well-founded recommendations to deny exorbitant billings. The sky should be the limit only of NASA's aspirations, not of its tolerance for profiteering by its contractors.

I DON'T KNOW--
IT WAS HERE A
MINUTE AGO!

Part II: The Space Shuttle

The triumph of the Apollo and Skylab missions was a true demonstration that the United States had mastered the technical challenges of space flight. But justifying the huge costs of even simple missions, however, proved almost as difficult as developing the complex technology itself. Until some way of reducing costs was implemented, future space progress was stymied. Because modern space boosters are all "one-shot" vehicles, their individual rocket stages burn out as they are discarded, falling back into the ocean or breaking up high in the Earth's atmosphere. The precision-made rocket engines, their huge casings and the complex electronic systems are lost forever. The development of reusable launchers was the obvious answer. The U.S. Congress voted in May, 1972 funds to NASA for the development of the space shuttle, a reusable orbital spacecraft capable of ferrying people and equipment into space cheaply and safely.

The space shuttle consists of three main elements: an orbiter spacecraft, a large external propellant tank and twin solid fuel boosters. The orbiter is a hybrid with features common to both a spacecraft and an aircraft. It is designed to operate for both the orbital flight and maneuvers inside the Earth's atmosphere. It was, therefore, the first of NASA's space vehicles to have a distinctly aerodynamic shape like many orthodox high-performance aircraft. The orbital flight sequence for the shuttle begins when its main engine and twin boosters ignite on the launch pad. The vehicle lifts vertically off the pad and begins accelerating toward orbital velocity.

The actual insertion into orbit is accomplished by the orbital maneuvering system which takes over from the shuttle's main engine. The external tank is jettisoned immediately after this vital step, plunging to destruction in the Earth's atmosphere. Once in orbit, the commander and pilot are responsible for precision maneuvering and operating the payload bay doors. Most importantly, the jobs of the flight commander and pilot center on the launching and landing of the shuttle. Mission specialists are responsible for carrying out scientific experiments, manipulating the shuttle's robot arm and performing other tasks while the craft is in orbit. The bay itself is a large utility area occupying much of the fuselage of the orbiter. Inside are manipulating arms which can be operated remotely from the flight deck. TV monitors in the bay and manipulators enable the pilot to see exactly what he is doing as he deploys whatever the shuttle has carried into orbit. Perhaps the most important application of the shuttle has been placing satellites into orbit. Other missions have included rendezvous with existing satellites to carry out on-the-spot-repairs. Additionally, the shuttle has been used by the scientific community to carry out various laboratory experiments that may open the way for new avenues of research and development. NASA has not been the sole benefactor of the shuttle's relative success and part of the agency's original plan for the shuttle involved the launching of satellites for private enterprise. It was hoped that this service would help offset the costs of the space shuttle program. This, however, has failed to transpire. The military's increasing involvement with the shuttle program combined with the "Star Wars" program pushed by the Reagan Administration have thrown the goals and use of the space shuttle into a more critical light.

The January, 1986 explosion of the space shuttle Challenger, which killed all seven astronauts aboard, grounded the fleet of the three remaining shuttles and signaled a new evaluation of the program's objectives. The tragedy, in the long run, however, may lead to a sharper definition of American space policy.

Space Shuttle Columbia Orbits Earth Successfully

The U.S. space shuttle Columbia, the world's first reusable spacecraft, lifted off from its launching pad at Cape Canaveral, Fla., on April 12, 1981. After 36 orbits and 54 hours in space, the craft glided to a perfect landing on a dry lake bed at Edwards Air Force Base in the Mojave Desert in California April 14. Manning the craft were John W. Young, 50, a veteran of four space flights including two missions to the Moon, and Robert L. Crippen, 43, making his first trip into space after 15 years of waiting and training as an astronaut. The mission, which came nearly three years later than originally scheduled, was the first manned U.S. space flight in almost six years. The blast-off occurred on the 20th anniversary of the first flight into space by Soviet Cosmonaut Yuri Gagarin. The point of the flight was to determine how well the shuttle—which represented a radical departure in design from earlier spacecraft—performed. The key elements of the mission were that the shuttle demonstrate that it could take off and land safely, and open the large doors over the cargo bay while it was in orbit. Donald Slayton, the orbital test manager for the National Aeronautics and Space Administration (NASA), commented after the Columbia landed, "We consider it a 100% successful flight."

While in orbit, Crippen and Young opened and closed the cargo bay doors, finding that they performed as planned. As it orbited, the craft passed rapidly from day to night and back again. These changes were accompanied by rapid shifts in temperature of up to 500 degrees Fahrenheit , creating concern that the doors might bend or expand as a result, and consequently not close tightly. Since the shuttle was designed to carry satellites and other such payloads into orbit, it was vital that the doors function correctly. Also, the doors had to be opened to allow the shuttle to radiate away excess heat while in orbit. Otherwise, the heat would build up and curtail the length of any mission.

A NASA official said April 14 that, going by an "optimistic" estimate, the Columbia would be ready for a second orbital flight in "less than six months." This schedule was predicted on the assumption that a close inspection of the craft revealed no major problems. The mission was also intended to test the shuttle itself, rather than perform independent tasks in space.

THE SUN

Baltimore, MD, April 9, 1981

The shuttle is a technologically complex hybrid of rocketship and airplane that will blast-off straight up like the rocket it is, enter orbit and then return to earth to land like the airplane it also is. It is by far the most sophisticated machine man will have put in space, and it may help Americans recapture a sense of their technological superiority and national purpose once again. But the launch, scheduled for tomorrow, is riskier than most space shots; the shuttle, unlike earlier vehicles, has not been tried in space unmanned.

The shuttle is no less controversial on the eve of its launch than it has been throughout its existence. It has been perennially starved for funds, and thus its construction often was delayed; some critics say that it also was plagued by military meddling. But finally, a year ago, only two major engineering problems remained to be solved: how to design and apply thermal tiles to prevent burnup during re-entry and how to assure that the sophisticated new rocket engines would work properly. These are now solved, NASA says.

But even if the shuttle should be a resounding technical success, that will not allay all of the criticism. Some scientists contend that the costs of the shuttle have bled NASA's budget so dry that projects promising more yield for the dollar in scientific knowledge—a 1986 close-up study of Halley's Comet, for instance—had to be aborted. "The single most important purpose of NASA as defined by its charter," says one critic, "is exploration of the solar system. But now the institutions that control funds clearly regard science as not useful, and believe that applications, especially military applications, are sacred."

Yet the shuttle will make undeniable scientific contributions. One of the most important of these will be putting the space telescope into orbit. This telescope, to be operated from the Johns Hopkins University, will expand the known universe a thousandfold. Another possibility is using the shuttle to help construct permanent space stations from which both manned missions to the farthest reaches of the solar system and solar-energy collecting units will be deployed—to mention one "scientific" and one "applied" goal.

The shuttle, in short, may turn out to be nearly indispensable for *all* of man's space activities, of whatever kind. Optimists are even saying that it heralds a new beginning for the space age, a time when men truly become "spacefaring" creatures. This may indeed be the case as we resume America's ventures into space. Our best wishes go with the first shuttle astronauts, John W. Young and Robert L. Crippen.

Rocky Mountain News

Denver, CO, April 16, 1981

IT was a long time coming, but it was worth the wait.

Three years behind schedule, the problem-plagued space shuttle Columbia did everything they said it would do, and did it so flawlessly there can be little doubt that a significant new chapter has opened on human exploration and utilization of space.

It was an unplanned computer glitch that postponed Columbia's launch so that it coincided with the 20th anniversary of Yuri Gagarin's historic one-orbit flight. But how fitting that was.

Gagarin, the first man in space, proved that space travel was possible. In Columbia, the first true spaceship as opposed to a capsule, John Young and Robert Crippen have shown the feasibility of transporting crews and equipment into Earth orbit and back again on a routine, scheduled basis with reusable shuttle ships that are a hybrid between spaceship and airplane.

The only thing that could prevent this from happening would be if Americans, having achieved a dramatic national goal, turned their backs on space again as they did after the Apollo moon landings.

In this regard, it is too bad that so much has been made of the military potentialities of the space shuttle, although that seems to be the magic formula for getting funding from Congress these days.

Certainly, the shuttle has military applications. But so did the biplane the Wright brothers built for the Signal Corps in 1909, and that is not why they are remembered.

The hundreds of missions envisioned for four shuttle ships over the next two decades foreshadow the day when it becomes commonplace for engineers, meterologists, astronomers, perhaps even poets and journalists, to leave Earth for sojourns on permanently inhabited, orbiting space stations.

One series of shuttle missions alone — the orbiting of a telescope above Earth's interfering atmosphere — promises to expand our knowledge and understanding of the universe beyond present imagining.

In the meantime, Columbia's flight had just enough suspense and potential danger to remind Americans of those exciting days when the nation first embarked on what John Kennedy called "the new ocean." And it had such stunning success as to make all Americans proud again.

BUFFALO EVENING NEWS

Buffalo, NY, April 8, 1981

If all goes well this Friday, the spacecraft Columbia will rocket into space for a two-day experimental orbital flight and a soft, gliding landing in the Mojave Desert.

The project promises to be a prestigious gain for the American space program and the beginning of a new era of space travel through use of the space-shuttle concept.

The shuttle concept moves space travel a step closer to the kind of routine flights seen in the airline industry today. The Columbia spacecraft, now more like a small airliner than a space capsule, can be used over and over, rather than being burned up in the atmosphere on re-entry. The two rocket boosters that propel it into orbit also will be recovered.

The obvious advantage of this concept is the reduction in the cost of putting payloads into space. Much more ambitious projects will become feasible, such as space stations, more elaborate surveillance and communication satellites, space telescopes, solar-energy stations and other forms of scientific experimentation.

In the military field, outposts in space will make possible improved surveillance of Soviet intercontinental missiles, and the hunting of Soviet killer satellites, now under development, would be assisted. The Soviet Union is ahead of the United States in the technology of such killer satellites, which could destroy U.S. satellites and impair U.S. surveillance capability. The United States, on the other hand, is believed to be perhaps a decade ahead of Russia in developing the space shuttle.

The Columbia is two years behind schedule, and its cost, through inflation and technical problems, has risen from $5 billion to $10 billion. Some scientific critics of the project feel that other scientific ventures, such as the study of Halley's Comet in 1986, have been shortchanged in favor of the space shuttle. In any case, the Columbia has now reached the moment of truth, and, if all goes as scheduled, the United States can expect to begin reaping some of the scientific, military and economic benefits of this mammoth project.

THE ATLANTA CONSTITUTION

Atlanta, GA, April 10, 1981

There is a theory that human beings on Earth have ancestors out there in space. No one can say for sure that's the case. But if it is, it's one explanation for what appears to be endemic in human nature here, a desire to explore the unknown beyond our atmosphere. Perhaps it is some sort of homing instinct.

Whatever the reason, humans always have peered skyward in amazement and wonder. It is only in recent times that science and its practical application sidekick — technology — have advanced to the stage where exploring can be done in that direction. Now, human beings in general and this nation in particular embark on a whole new phase, a venture into that cold darkness with a reusuable spaceship, the Columbia.

This is exciting and a bit risky — more so than previous efforts because it is the first time that a spacecraft has not been tested in space without astronauts in it. An unmanned flight in the Columbia would have been impossible given the nature of the reusable craft.

The orbiter is scheduled for 36 trips around the world, but officials of the National Aeronautics and Space Administration may cut them short if problems developed. There have been several serious problems to date. One has to do with insulation tiles on the outside of the craft. When it starts homeward these tiles are supposed to absorb the heat. If one of them falls off or otherwise fails (as they have done on Earth) the heat caused by re-entry could disintegrate the entire spaceship.

The Columbia has a cargo bay about the size of a railroad boxcar and the idea is that it can be used for experiments or to carry things to be dropped into space, such as satellites, thus saving the cost of expensive multiple rocket launchings. But no one can be sure the cargo doors will open out there. There are also new heating systems, navigation systems and engines.

It is a natural extension of the space program begun in the early 1960s when President John F. Kennedy set the goal of a man on the moon before the end of the decade. How impossible that sounded. He didn't live to see it, but Americans were proud when the United States made it — with five months to spare.

Just as the Wright brothers' first flight at Kitty Hawk, N.C., was shorter than the wing span of a Boeing 747, the Columbia's hardware and potential is so much greater than what has gone into space before.

The ancient Egyptians, the Minoans, the Greeks, the Vikings, and all the rest set out for the unknown to find out who they were and what they were. So then must modern mankind go a step further to add to the body of knowledge in the hope it will somehow benefit those who wait and watch back home.

As with those earlier explorers there is the element of risk, of danger that cannot be completely erased, not even with all the safety devices and procedures available to us. That is the nature of exploring. And it is what makes the explorer — be he ship captain, trail blazer or astronaut — a hero.

Sentinel Star

Orlando, FL, April 9, 1981

THAT CONTRAPTION — as it is often referred to — sitting there on Launch Pad 39A at Kennedy Space Center is something of a vision become reality.

In this time of practicality, it is a 184-foot monument to the realization that NASA's space program is an investment that can more than pay its own way. So, when the space shuttle Columbia lifts off, it launches more than just two astronauts tucked away in a plump DC-9 with delta wings — it's a full-fledged leap from space exploration into an important era of space exploitation.

Not since the last Apollo moon landing more than a decade ago has there been such excitement within the scientific community. The hundreds of thousands of people expected to jam highways for miles around Cape Canaveral are testimony to a sense of national pride and the rediscovered thrill of a launch.

COUNTDOWN ONE

No doubt about it, this has been a troubled effort that has tested our commitment to science and to space. Congress provided $8 billion for the 12-year project, but NASA still has found itself continually strapped for dollars. What was produced is a smaller and less revolutionary shuttle than originally envisioned, not the airplane of tomorrow that would make endless low-level flights into space.

There have been miscues with the engines, and — within the past month — two deaths. But nothing has raised as many questions as have the 31,000 heat-resistant glass tiles glued to felt pads that are, in turn, glued to the shuttle's aluminum skin. There is an uneasiness with the knowledge that the safety of astronaunts John Young and Robert Crippen depends on such a troublesome, seemingly jury-rigged, piece of equipment.

But it is a bold adventure. It has the potential of becoming one of our most successful applications of science. A Government Accounting Office analysis says the U.S. space program has already produced a 27 percent return on investment; a study by the highly regarded Chase Econometrics says the return may be as much as 43 percent. Now, the shuttle could launch us into another era of frequent, lower-cost flights in which space is made to work for us as never before.

While its potential as a cargo hauler — of things like satellites — is good for business, the shuttle also stands to be good for science. It can open the doors to an array of scientific experiments and advanced applications of technology. Then there are the military applications.

Once the Apollo program was behind us, the country really wasn't in the mood for another big space effort. But a good show this time, along with growing evidence that a space industry is there to be developed, may get us back in the mood rather quickly.

When those engines finally fire and the roar thunders through the Florida palms and pines, we can only hope that our commitment has been enough. Godspeed, Columbia.

Space Shuttle Columbia Complete 2nd Flight

The space shuttle Columbia landed safely at Edwards Air Force Base in California November 14, 1981 after a mission lasting 54 hours and 36 orbits of the Earth. One of the craft's electricity-producing fuel cells had started to malfunction shortly after the Nov. 12 launch from Cape Canaveral, Fla., and the mission controllers decided to cut the flight short rather than run the risk of a breakdown that might pose a major threat. Originally, the second flight of the Columbia—the first vehicle to make multiple flights into space—had been planned to run for 124 hours and 83 orbits. Piloting the Columbia were Col. Joe Engle of the Air Force and Capt. Richard Truly of the Navy. Despite the shortened flight, NASA officials insisted the mission had accomplished its major objectives. They also said Nov. 15, after they had had an opportunity to make a preliminary inspection of the Columbia, that the craft had weathered the flight in much better condition than it had after the first flight. "We accomplished so many of our flight objectives that I envision a minimum tuning on the next flight to catch up," said Donald Slayton, manager of the orbital test program. "Damage to the bird is minimal compared to what we'd seen after the first test flight," Slayton added. On the first flight over a dozen of the Columbia's heat-shielding tiles had been so badly damaged that replacement was necessary. The craft's surface was covered with some 31,000 tiles. By contrast the, the shuttle emerged from the second flight with no tiles missing and only a dozen requiring replacement, because of damage. Overall, its condition was described as "superb."

While in space, Engle and Truly carried out a successful test of the mechanical "arm" that would be used in later flights to move satellites and instrument packages from the shuttle's cargo bay out into space, and also to retrieve these objects from space if necessary. Successful manipulation of the 50-foot robot arm ranked as one of the highest priorities for the mission. Although shortened, the mission produced valuable data, not obtained on the first flight, about both the Earth and the shuttle's flight performance. The Columbia carried remote scanning instruments that studied the Earth for indications of mineral, petroleum and other resources. The need to gain flight performance data, particularly data concerning the shuttle's turbulent re-entry into the atmosphere, was one of the reasons for NASA's decision to take a cautious course after one of the three fuel cells malfunctioned.

The Pittsburgh Press

Pittsburgh, PA, November 17, 1981

The space shuttle Columbia had a perfect liftoff from Florida and, thanks to the flying skills of Astronauts Joe Engle and Richard Truly, a perfect landing in California.

But before takeoff and in flight it suffered a worrisome number of troubles.

Officials of the National Aeronautics and Space Administration have been quick, perhaps too quick, to declare the mission a success. But their satisfaction, real or feigned, should not stop them from investigating and correcting the many things that went wrong.

★ ★ ★

It is true that the astronauts and the technicians on earth were able to cram into two days of flight most of the important experiments that had been planned for five days in space.

Included were the test of Columbia's 50-foot mechanical arm, designed to place payloads in orbit and to retrieve them, and an instrument survey of earth for mineral and petroleum resources and fisheries.

It is also true, however, that this first flight of a "used" space vehicle was plagued by mechanical failures, some of which could have endangered the crew.

For example, the launch was delayed by such mishaps as a spill of nitrogen tetroxide, which damaged heat-protecting tiles; a fouled oil filter, and a faulty data-processing unit.

The worst problem cropped up shortly after liftoff when an electricity-producing fuel cell failed, posing the risk of an explosion and forcing a shutdown of the unit.

Fuel cells have been used on manned space flights since 1965 and are considered highly reliable. Yet this one fizzled right off the bat.

★ ★ ★

NASA officials must re-examine their own procedures and also determine if they are receiving dependable equipment from the contractors.

A reusable space shuttle in a $10-billion program ought to be more mission-worthy than, say, a used car.

The Boston Globe

Boston, MA, November 5, 1981

The frustration over cancellation of the second launching of space shuttle Columbia, just 31 seconds before scheduled liftoff, should be nothing more than a fleeting sensation in comparison with the realization that safety still enjoys priority over spectacle in the nation's space program.

The shuttle spacecraft is still something of an unknown quantity. Its first flight last spring was an unquestioned triumph, but there are still fears over the tiles that must protect Columbia as it burns its way back into the atmosphere. A troubled launch could shake some of them loose, unnoticed before the craft reached orbit, and endanger pilots Joe Engle and Richard Truly.

The specific problem, a clogged filter, is a vivid illustration of the need for absolute attention to detail at even the most trivial level. The delay will cost money and is a reminder of the cost of manned space flight, but it is a cost the country is certainly willing to bear.

In the whole scheme of space exploration, though, it is an even more pointed reminder of the need to make maximum use of unmanned space programs that can return vast, detailed information of great scientific value. The two spectacular flights past Saturn are perhaps the most spectacular. The orbiting telescope is not so well known to the public but is of enormous importance for gaining better understanding of deep space.

Other such programs, including a flight to intercept Halley's Comet in 1986, have been jeopardized or cancelled by the Reagan Administration's intensifying budget problems. The cutbacks do more than merely scrub this project. They dismantle teams of skilled personnel who are lost to those projects still in place. Those programs, like Columbia, deserve all the careful and expensive attention they can get.

The Kansas City Times

Kansas City, MO, November 13, 1981

Some astronauts have traveled in space more than once but never before had one of their spacecraft — not until Columbia rose from the launch pad at Cape Canaveral Thursday on a tail of fire for a second trip into orbit, just as it did last April. And if the vehicle continues to perform, it could make as many as 98 more such flights.

Always before, in the days of the Mercury and Apollo missions, when the spacecraft plunked into the ocean after its return it was retrieved (one sank after the astronaut got out) only as a relic, a museum piece. But Columbia is merely the first of a projected fleet of scores of reusable shuttle craft that can be rocketed into orbit and later — mission completed — glide back to earth for an airplane-type landing.

This is the new threshold in the nation's space program whose crossing was signaled by Columbia's second liftoff. The first manned space flights, reaching ever higher, the moon landings — these were voyages of exploration, experiments to test the limits of our technological capabilities. But the shuttle is a calculated program using known abilities to usher in a new era of space *utilization*. Columbia and its sister ships are intended to be space trucks and buses, with roomy cargo bays and the 50-foot crane or "arm" being tested on this flight, to place in orbit, repair or retrieve a variety of communications, military and industrial satellites. They can build stations or factories in space and haul people and materials to them.

The shuttle demonstrates how the skills and experience that the National Aeronautical and Space Administration teams have accumulated make a launch more and more routine. On this latest mission, when hydraulic trouble developed, the oil and filters were changed; when a fuel tank leaked, it was sealed; when a data relay system malfunctioned, a replacement part was borrowed from the second shuttle craft in California — an old Air Force trick — and flown by jet to Florida. Each problem is no longer a challenging new frontier; NASA knows how to fix these birds.

The unmanned space probes to the outer planets, sending back their spectacular color photographs, still are quests for pure knowledge. But the shuttle seeks to create a payoff in more than just industrial applications of the technology developed by the space program. The industries and governments which will buy load space on future flights, after the concept has been fully developed by this series of test flights, will be getting tangible returns for their investment. It is meaningful that the second Columbia flight isn't named for some ancient, legendary Greek god; it is mundanely dubbed STS-2, for space transportation system.

The Morning News

Wilmington, DE, November 13, 1981

One person around this shop suggested the other day that the current voyage of the Columbia required little or no comment because, after all, yesterday's liftoff was for the space shuttle's second flight, not its first. The suspicion here is that this person was trying to put us on, having a little fun at the expense of our enthusiasm. It is the very fact that this is the *second* shuttle flight that is significant.

The first flight, April's spectacular success when we thrilled to see the great white bird landing on a runway, to outward intents and purposes like an airplane — a glider — was merely to see if the ship could fly.

This five-day mission which began, like the first, at Cape Canaveral, Fla., and is to end at Edwards Air Force Base, Calif., is to demonstrate that the shuttle can *fly again.*

That's the "shuttle" part of the program, you see, developing a craft that can go back and forth between Earth and what a pre-launch story by New York Times science writer John Noble Wilford felicitously referred to as "the orbital frontier."

What a stirring phrase! It conjures visions of sprawling, brawling towns in space, with pioneers of science and technology arriving daily, opening up new worlds.

As in the settlement of the American West, those journeys may never become routine; each may be an adventure. Science magazine writer R. Jeffrey Smith last month speculated that the Columbia's fantastic complexity — demonstrated by minor problems which made between-flight preparations agonizing at the Cape — "will resist attempts to operate with aircraft-like efficiency." The first attempt to get Columbia off the pad for this flight proceeded to within 31 seconds of liftoff before such problems developed as to cause another week's delay. Even then there were those 150 knuckle-whitening minutes between scheduled 7:30 launch and actual liftoff at 10:00.

But that may well be what will make each and every launch of the Columbia significant.

And this flight helps permit another evaluation by Science: that space science is one of the most vital and productive fields in the United States, with fresh data flowing in to researchers, bulging the journals of many fields of study, and providing high technology already "being used in a rich variety of utilitarian applications."

That enthusiastic appraisal by James A. Van Allen of the University of Iowa is tempered, however, with notations on the space calendar for 1980 and 1981. This author bemoans these scanty schedules. He calls the current outlook "bleak" and says it requires desperately a "critical and dispassionate reappraisal" of our national policy in space.

The Columbia is carrying instruments that will monitor pollution, map the planetary surface and pinpoint mineral and fishery resources. And that's just a start of its "rich variety of utilitarian applications."

No, the mission being flown by astronauts Joe Engle and Richard Truly is no less worthy of comment — and no less historic — than that first voyage last spring.

Fort Worth Star-Telegram

Fort Worth, TX, November 18, 1981

The shortened flight of the space shuttle Columbia in no way can be considered a mission failure. To the contrary, it was a great success. It proved that the space craft is reusable.

And it will be used again and again. No sooner had the Columbia left the launching pad at Cape Canaveral than work crews were getting the pad reconditioned for another launching of the Columbia in March. And there will be another flight in June.

The Columbia could have stayed in space for its full 124 hours and 83 orbits, but it returned to earth after two days, six hours and 13 minutes. The craft has three fuel cells and one of them failed shortly after takeoff. Columbia can operate effectively on two cells but flight officials chose to be conservative and safe. There is so much yet unknown about the Columbia just as there is much unknown about this planet and its surroundings.

The next tests will go a bit further, getting more information about the craft's performance and more information about space and the earth.

Instruments on Columbia are able to examine ocean life, mineral deposits, air pollution and other earthly conditions. And it is hoped that one day scientists will be able to forecast dramatic earthly changes through electronic readings taken from space. Scientists already are able to measure ocean water temperature through an infrared radiation monitor in space.

The shuttle is giving us the ability to place informational instruments in orbit, service them and return to earth. Space is becoming a station for humankind's future. That Columbia was shot into space a second time and returned with even less wear than was experienced on the first shot is another star on the United State's space program's record.

The success of Columbia perhaps will awaken this country's drowsy interest in space and cause the federal budget makers to reconsider this country's space investment.

What can be accomplished to improve living on Earth may well be determined by what information we obtain from space.

Des Moines Tribune
Des Moines, IA, November 17, 1981

Astronauts Joe Engle and Richard Truly earned a modest place in history last weekend when they became the first to fly a used spacecraft into space and back again.

The second flight of the space shuttle Columbia was not perfect. One of its three fuel cells failed and had to be shut down, causing the flight to be cut from a planned 124 hours to 54 hours.

Despite the shortened flight, Engle and Truly carried out 90 percent of their intended missions. One of these was to test the mechanical arm that will be used in the future to lift satellites out of the cargo bay and into orbit and to retrieve satellites from space for servicing.

Another achievement was that the shuttle's return to Earth was virtually perfect.

The cost of the space-shuttle project — $10 billion and rising — is a matter for concern, but the payoff could be the birth of a technology that holds the promise of revolutionizing man's relationship to space.

Sentinel Star
Orlando, FL, November 3, 1981

WHEN it's a second time around for anything, people tend to become pretty blase. How many of us remember the names of the astronauts who went on the second moon landing trip? And so it is with the second launch of the space shuttle Columbia. It has not exactly captured the nation's undivided attention.

But Wednesday's planned launch of Columbia does deserve attention because this is the voyage that counts. This is the one that will demonstrate the spaceship's uniqueness: its ability to be reused.

This one will also last longer than the first voyage; it will go on for five days instead of 2½. And in this one, Columbia will be able to experiment with new methods of measuring air pollution and with a Canadian-built 50-foot mechanical arm that some day will be a critical part of the shuttle's role as something of a cargo train into space.The astronauts will even be able to take photographs of electrical storms, an experiment that may help develop a warning system for severe weather.

The second flight of Columbia should help determine the scope of its future missions and tell us how valuable a tool the space shuttle really is. The business, military and scientific applications of the U.S. space program are finally becoming much more than just a dream. Columbia's second mission is just as historic as the first.

The fact that launches of any sort have become so ho-hum for the American people, particularly for those of us in Central Florida, is something of a testimony to just how successful this nation's ventures into space have been.

By the way, Richard Gordon, Alan Bean and Charles Conrad were the second group of astronauts to go to the moon. That was a dozen years ago. We can only imagine where a shuttle-supported space program will be a dozen years from now when the world looks back to the historic flight of Richard Truly and Joe Engle.

THE TENNESSEAN
Nashville, TN, November 18, 1981

ALTHOUGH shortened by a malfunctioning fuel cell, the second journey into orbit for the space shuttle Columbia was still a highly successful mission and one which opens a new era for space flight.

The Columbia journey marked the end of the extremely expensive practice of using a space vehicle once and discarding it. Already the National Aeronautics and Space Administration is looking toward a third flight of the Columbia — probably sometime next spring.

There were a number of positives about the second flight that pleased NASA officials. One was the protective tiles which shield the craft on re-entry from extreme heat. Although six tiles were blistered on the Columbia II flight, none fell off as they did the first time. That would indicate the tile problem is getting to be under control.

And despite the shortened trip, Astronauts Joe Engle and Richard Truly managed to perform some of the planned experiments. One was an imaging radar for mapping earth's geological formations and the other was an infra-red sensor. The results were pleasing and "good data" were obtained.

There was also the use of the "robot arm" device which could be used in the future to grasp satellites from the space shuttle's hold and put them into space, or to recover from space those satellites that may have gone dead for some reason.

Some of the secondary tests were cancelled because of the time reduction, but these will be conducted by the next shuttle flight. They include exploring reception and transmission patterns of the Columbia's radio antennas, which could be done almost any time.

The engineers will also gain some knowledge of what happened to the faulty fuel cell. Normally the cells are extremely reliable and the malfunction of one caught everybody by surprise.

But what was proved was that two cells were sufficient in this case and there may not be a need for more redundancy than having three.

In any case it is evident that the U.S. space program has the capability for a transportation system that can provide invaluable service through the rest of the century.

Other than the transportation part, however, the space science program is being endangered by budget cuts. There is the possibility that the Voyager program might be turned off and that the Deep Space tracking network might be scrapped. It would be unfortunate if, after having gone to great expense in investment and setting up the capabilities, that the U.S. would default on vital portions of its space program at the time when the Soviet Union is stepping up its own efforts.

The U.S. ought to maintain the parts of the program that are practical and financially prudent. It is hoped the Reagan administration will see it that way in the future.

The Dispatch
Columbus, OH, November 19, 1981

THE SPACE SHUTTLE *Columbia* returned safely to Earth Saturday marking the first time that a craft had made a second round trip to space and back. The failure of a power unit on board the shuttle caused the flight to be shortened by three days, but the problem — and others that have troubled the craft — should be put into proper perspective.

Hitches and glitches are what test flights are all about. Experiments with a vehicle are designed to identify problems and to solve them before the vehicle is put into full operation. The only way to test the shuttle is to fly it.

The *Columbia* is the most sophisticated transportation vehicle ever built. It has 300 major electronic boxes on board. It contains 300 miles of wiring and 2.6 miles of tubing. It has a thousand valves and more than 2,000 dials and switches.

Scores of tanks hold more than half a million gallons of fuel. A problem with any one of these elements can cause a mission to be delayed, scrubbed or shortened.

Test flights are intended to find problems, find ways to overcome them, and to develop an understanding of the seriousness of problems when they arise.

The shuttle's track record is a brief one and NASA officials are rightly reacting cautiously to problems that develop. As more in-flight knowledge is gained, officials will be able to better evaluate the hitches and glitches and to develop a confidence to know when a problem is flight-threatening and when it can be ignored.

Until that confidence is developed, it's wise to play it safe. The shuttle is an extraordinary vehicle designed to accomplish an extraordinary task: the development of space for human benefit.

There's no reason to hurry the mission — space will be there when the shuttle is ready.

CHARLESTON EVENING POST

Charleston, SC, November 17, 1981

Sandwiched between the perfect launch and the perfect landing of the shuttle Columbia was an unprecedented return to space that had to be abbreviated because of a faulty fuel cell. The problem was typical of those that have afflicted the project for a couple of years, causing delays in the overall schedule. Sight should not be lost of the fact, however, that Columbia's second flight, like its first, was a test. In test stages, some things can be expected to go wrong with a craft as complicated as Columbia.

There is a brighter side to the Columbia performance. Even though their mission time was halved, astronauts Engle and Truly accomplished, in truncated form, about 90 percent of what they were supposed to accomplish by way of scientific tasks. The shuttle itself was pronouncd in "solid" condition upon return — meaning preparation for its third flight in March should be easier. And meaning, also, that the "reuseable" concept is sound.

NASA officials say they are confident that within five years the shuttle system will be offering commercial service and paying its own way. That is when judgments fairly can be made on its success. Meanwhile, Americans have reason to applaud the courage of the men who took Columbia back into space, the prudence of the men who ordered the mission shortened, and the technological skills that have enabled those in NASA to sort out and overcome the problems that have caused the scrubs and the delays.

The Salt Lake Tribune

Salt Lake City, UT, November 17, 1981

As adroitly as its participants, managers and partisans excused space shuttle Columbia's interrupted second flight, the project has nonetheless suffered a setback. Fortunately, time and opportunities remain for redemption.

It's perfectly normal and proper for Columbia crewmen Joe Engle and Richard Truly, National Space and Aeronautics Administration spokesmen and the White House to shower compliments on Columbia's performance under stress. But for all concerned, it would have been exceedingly more satisfactory had last week's premier U.S. space spectacular worked as advertised.

In fact, none of the post-landing rave reviews were inaccurate. Again, Columbia proved its stunning capacity for gaining earth orbit, exercising its machinery there and, although among the more clumsy appearing of glider craft, settling serenely, safely onto an awaiting Earth runway. In stressing such achievement, in praising the shuttle's reproven potential, no liberties with either truth or relevancy were taken. However, there may have been a conscious effort to soft-pedal a Columbia vulnerability, also made conspicuous by the shortened second test run.

The nation's space program has always been costly. Naturally, it has attracted criticism. A reusable space shuttle, as practical sounding as it is, hasn't been immune from spoken doubts. A $10 billion undertaking, Columbia's testing can be defended as the advent for a genuine space age, of working, pioneering and advancing in reaches beyond Earth's atmosphere But critics are afforded a wider, easier hearing with every stumble.

When failed fuel cells led NASA officials to order Columbia's return three days ahead of schedule, reasons for wondering about the project were strengthened. The intricacies in space vehicle launch and astronaut return have long been acknowledged. They have a particular significance as far as Columbia's ultimate purpose is concerned, however.

The reusable space shuttle is supposed, eventually, to largely "pay its own way." That is, cargoes of the future will be carried for government and industry for a fee. Convincing businesses of the prospects has been about as slow-going as could be imagined. If the Columbia's image becomes one of costly unreliability because of minor "snafus," enticing industry "on board" could become even more difficult.

But Columbia is not through proving itself. Confident NASA authorities say the March take-off is a certainty. It wasn't failure of a new or especially sophisticated piece of equipment which ended last week's flight disappointingly early. The fuel cell system, although improved, is essentially the one used successfully during most of NASA's past, triumphant space missions. Moreover, even crippled, Columbia was able to carry out between 90 and 95 percent of crew and mechanical assignments. Good enough under the circumstances.

The space shuttle still is very much a part of this country's space exploration and exploitation future. Necessary preparation for that must continue without hesitation.

Rockford Register Star

Rockford, IL, November 15, 1981

Americans, who sometimes see themselves beset with "the slings and arrows of outrageous fortune," have something to cheer about in Columbia II. The space shuttle proved it was every bit as space-worthy this second time around as it was the first.

True, there were hitches — and a truncated trip. With one of three fuel cells on the blink, NASA opted for caution rather than bravado — and brought the ship back three days early. But its scientific mission was complete. The critical test involving a 50-foot crane, known as a mechical arm that will place future satellites in space, went off smoothly.

From lift-off to dry-lake landing, Columbia II gave us a superb show.

Few will forget that magnificent take off from Cape Canaveral.

A beautiful morning, the launch almost three hours delayed but perfect, the jet fires looking like pistons from Dante's Inferno, propelling two men and their space vehicle and two propellant tanks and one large fuel tank into the sky, a feat seemingly without weight or flaw, merely greased lightning conceived here on earth.

And in the orbiting space vehicle American astronauts Joe Engle and Richard Truly were ecstatic as their mission began. They tended to say words like, "You wouldn't believe this; this is fun." (That, from Truly.) Or to say things like, "Very smooth." (That, from Engle.)

Surely, they did not exaggerate. Nor were their earthling fans, especially those assembled in Florida for an on-site view of take-off, disappointed by the spectacular breadth of the achievement.

But "fun" and "smooth" do not adequately put the mission in focus.

Americans paid for this flight from their hard-earned incomes, an investment that few begrudge, considering the skill and efficiency and scientific advantages that attend such an undertaking.

Indeed, only a national government as big as our own, having access to the vast sums needed for such an endeavor, could have carried it off. Columbia II should be a prod to those in Washington who want us to retrench from space.

At the same time, however complicated and confounding the launch became, let us not forget those who made it possible, some of whom gave their lives. When Cape Canaveral was known for the late president as Cape Kennedy, three men lay aboard an Apollo capsule anchored high above the launch pad, going through a practice space take-off when the oxygen they were breathing exploded.

We should recall their names and their sacrifice here, because all three died that Friday night, Jan. 27, 1967: Virgil (Gus) Grissom and Edward Higgins White, both veterans, joined by Robert B. Chaffee.

Finally, aside from our obvious pride in Engle and Truly, what about those tireless technicians before the computer screens, checking and double-checking 'round the clock to make sure that these pilots were ready for their next big test, bringing their rocketship safely back to earth?

And, perhaps one more thing. What about American pride as NASA teams performed before the acuity of long-distance television lens, every aspect of this monumental and endlessly tricky venture exposed to a monitoring world, goofs and all, everything hanging out. Is there another nation gutsy enough, expert enough, yes, generous enough to say, "Look, come join us for a ride that'll give you an emotional high."

Think it over.

Then let 'er rip, "Hail, Columbia!"

Space Shuttle Columbia Completes First Operational Flight

The U.S. space shuttle Columbia landed safely at Edwards Air Force Base in California November 16, 1982 after a five-day mission that was its first operational flight after four test missions. The Columbia lifted off from Cape Canaveral, Fla., Nov. 11 after a virtually flawless countdown. In the earlier missions, the craft had been manned by just two astronauts, but there were four astronauts for the first operational flight. They were flight commander Vance Brand; Col. Robert Overmyer, the pilot; and mission specialists William Lenoir and Joseph Allen. A key objective of the mission, and the task that earned the flight its designation as "operational," was the delivery of two satellites into orbit for commercial customers. Both satellites apparently achieved their desired orbits. The ferrying of satellites into orbits would account for the bulk of the operations of the shuttle fleet in the coming years.

The mission was generally hailed as a success, but there was one major disappointment: a planned "spacewalk" outside the shuttle by Allen and Lenoir had to be canceled because of malfunctions in the newly designed spacesuits. The astronauts did not wear the special suits while they remained inside the shuttle. Nevertheless, James Beggs, the administrator of NASA, said after the Columbia landed, "We don't do them any better than this." Following the flight, the Columbia was to undergo a refitting that would take it out of service for some time. The next three missions would be carried out by the shuttle Challenger. Mission controllers had originally planned to have the Columbia land on automatic pilot, but difficulties with the on-board computers caused this plan to be abandoned.

Two satellites were carried into orbit from Columbia's cargo bay by springs. One was a $30 million communications satellite operated by Satellite Business Systems (SBS) of Virginia. The other, also a communications satellite, was owned by Telesat Canada, a state-controlled company. The shuttle distanced itself from the satellites, which then used strapped on rockets to move up from the shuttle's 185-mile-high orbit to an orbit 22,000 miles above the Earth. The SBS satellite moved to its higher orbit in rocket firings Nov. 11 and 12, while the Canadian satellite was lifted into place Nov. 12 and 16. The objective, in the case of both satellites, was to attain a geostationary orbit—that is, an orbit that would keep the satellites permanently over one spot on the Earth. The astronauts and NASA officials were elated at the success of the strapped-on rockets, since such operations were to be a mainstay of future satellite-launching missions. The shuttle, at least in its present form, was not capable itself of reaching the high-altitude orbits for which many satellites were designed.

The News American

Baltimore, MD, November 17, 1982

The Columbia astronauts need new tailors. They were unable to accomplish planned spacewalks because their $1.5 million space suits weren't ventilating properly. The pity is that the suits — new, sleek improvements over the old, clumsy wide-lapel models — were developed over a period of 10 years at a mere cost of $250 million.

Now Buck Rogers never wore space threads designed by Ralph Lauren or Pierre Cardin, and he made the trip safely every time. So did the 36 Apollo astronauts who wore the old bulky numbers that resembled the wardrobe of a deep sea diver. The difference, however, involves more than durability. There is a certain amount of snob appeal attached to the garb of a space-walker.

The suits worn by the Apollo astronauts were custom-made for each wearer at a cost of $12 million each and were not reusable.

But the cheaper models designed for the Columbia space travelers were no-class off-the-rack suits that were intended to be hand-me-downs.

There is a certain amount of sympathy for cutting the defense budget in non-essential areas, but, heavens, not at the risk of providing our astronauts with garments that are made to be worn by just anybody. It's like buying a space suit in a department store.

Another thought that comes to mind is this: Wasn't it just the other night that President Reagan defended the quality of our military equipment against charges that some of it was inferior and didn't work?

But that's another issue, a secondary one compared to making certain that American astronauts are properly haberdashed. After all, we wouldn't want the Russians to overtake us in the area of proper dress for space walks, too.

Gentlemen's Quarterly would never approve of wearing second-hand space suits.

ARGUS-LEADER

Sioux Falls, SD, November 18, 1982

Americans may take pride in the technical excellence of the space shuttle Columbia, demonstrated by five successful flights totaling more than 10 million miles over the last 19 months.

Columbia flight five was the first operational use of the world's first reusable spacecraft. Columbia launched two commercial communications satellites in space, a better way than firing them from earth.

There was disappointment that Columbia's space walk had to be canceled because of a technical fault in new $2 million space suits. But that can be accomplished on another flight after the difficulty is corrected.

Columbia's fifth flight was the first commercial cargo use of the space shuttle, leading to a NASA (National Aeronautics and Space Administration) motto, "We Deliver," and the crew's sign, "Fast and Courteous Service."

Columbia's perfect landing in California Tuesday was uneventful — something Americans have come to expect from their astronauts and Mission Control in Houston.

During the next 10 months, Challenger, the second shuttle in the U.S. spaceship fleet, will fly NASA's missions while Columbia undergoes overhaul.

Challenger is scheduled to deploy a Tracking and Data Relay Satellite in space on Jan. 24. The satellite will serve as a relay station between the ground and as many as 100 orbiting satellites.

Challenger flights in April and July are scheduled to carry communications equipment for Telsat Canada and the Indonesian and Indian governments.

Challenger's first three flights are expected to send the first American woman into space, complete a genuine roundtrip with a landing at Cape Canaveral, the first nighttime landing and the first space trip by a black astronaut.

Columbia and Challenger, America's new spaceship fleet, are extending both U.S. technical skill and the availability of communication satellites in today's information age. Both are positive contributions to a better world.

Well done, Columbia! Good luck, Challenger!

THE TENNESSEAN

Nashville, TN, November 13, 1982

THE National Aeronautics and Space Administration has chalked up another exciting technical achievement by launching two commercial satellites into space from the orbiting space shuttle, Columbia.

The agency is being paid $17 million to launch satellites for Satellite Business Systems of McLean, Va., and Telesat Canada. This is just a minor share of the $250 million cost of a shuttle flight. But the space crew is performing other experiments and NASA hopes to improve on the commercial returns in the years ahead.

The space crew is carrying out a pair of experiments selected by NASA from hundreds submitted by students around the nation. One is the study of the growth of sponges in a gravity-free environment. The other is the formation of crystals in this environment.

As for the commercial prospects, NASA says it doesn't expect to break even on such launchings until 1986 or 1987 when launch costs will be lower and rates for launchings will be higher.

Even that deadline may seem a little optimistic to some. No one doubts NASA's technical abilities. But making a business success of a shuttle flight is something else. If NASA manages this achievement, the taxpayers will be both gratified and surprised.

Herald News

Fall River, MA, November 18, 1982

The first operational mission of the space shuttle Columbia ended successfully Tuesday morning. There was some disappointment because the scheduled space walk had to be cancelled when the spacesuits two of the astronauts were to wear malfunctioned.

The purpose of the space walks would have been to test the capacity of astronauts to repair imperfections in satellites or an emergency on the space shuttle itself.

It is now reported that the space walk will be rescheduled either for the next mission of the shuttle in two months or the one after that.

Apart from the cancellation of the space walk, the Columbia's first commercial flight into space was a complete success.

For the next year the space agency has an accelerated schedule with five separate missions planned. Three of them will be performed by the Columbia's sister ship, the Challenger.

In the course of these missions, commercial satellites and tracking and data relay satellites will be left in space. The first liftoff at night will take place, and one of the crews will include a woman.

All these planned activities and developments point in the direction of establishing space travel, not as an isolated event, but as something regularly instituted and carried on in the normal course of world affairs.

Within a relatively short span of time, regularly manned stations will be established in space, and the shuttles will go back and forth between them and earth.

All this has been brought measurably nearer by the Columbia's successful mission last weekend.

There is a vast difference between space exploration as such and missions such as the one just completed.

Space exploration in itself means no more than visiting areas of space where no one been before. The Columbia is setting up the advance guard for the settlement of space.

The difference is similar to that between the explorers who first visited the Atlantic coast line of this country and the settlers who followed them.

What we are witnessing and, to some degree, participating in is man's first tentative efforts to break through the confines of this planet and find a foothold for himself in space.

It is for this reason that the space shuttle's flights back and forth are so important.

They are certainly spectacular, but their value does not lie in their appeal to the eye.

Their real value is that they are proving that the same space vehicle can be safely used for regular journeys, that, within limits, it is possible to establish a schedule that will make given points in space regularly and readily accessible.

All this is the initial activity preceding the gradual establishment of a station in space.

That station will be the first human settlement in space, and will be the true inaugural of the Space Age.

What we are seeing now is a prelude leading to that momentous event.

The shuttling back and forth to and from space is by no means simply an exercise of skill for its own sake.

It points toward the future, when man will have outposts in space, then fully constituted settlements.

The superbly coordinated flights of the Columbia are excursions into the future for us all.

The atmosphere here on earth is so cluttered with hostilities of all kinds that it often seems suffocating.

By contrast, uncluttered space offers man another chance to redeem his past errors and make some more of his dreams come true.

THE DENVER POST

Denver, CO, November 13, 1982

FLAWLESSLY, the space shuttle Columbia embarked on its fifth voyage Thursday morning. It was a routine launching. That anything as complex as a mission into space could be performed routinely is a tribute to American technology.

But the mission itself is anything but routine. For the first time ever in space exploration, the shuttle carried a crew of four. And for the first time, Columbia was doing what it was designed to do, haul a payload.

The four previous flights were tests. This time, stowed in Columbia's ample cargo hold were two communications satellites, each about the dimensions and weight of a full-size automobile, which were to be tossed into orbit. Moreover, these were really payloads; an American and a Canadian company together would pay $17 million to have the satellites delivered. Columbia on this voyage is carrying some 16 tons of cargo. Future shuttle flights will haul up to 32.5 tons.

This nation's manned flights into space began with Alan B. Shepard's brief non-orbital 300-mile ride in a rocket in 1961. John Glenn flew into orbit in 1962. Neil Armstrong and Edwin Aldrin set foot on the moon in 1969. These were tentative exploratory milestones on the road to building a true spacecraft that could be brought back in a controlled landing.

The shuttle program has taken a decade and cost $11 billion. It was plagued with problems, some as seemingly picayunish as heat-shielding tiles that refused to stick to Columbia's skin, but each of them was solved in time. Columbia's commander, Vance D. Brand, a Colorado civilian, is scheduled to bring the ship down Tuesday morning to Edwards Air Force Base in the California desert. Until then, good sailing, Columbia, and happy landing.

The Providence Journal

Providence, RI, November 10, 1982

For the bargain price of $9 million apiece, two commercial satellites are hitching a ride as the space shuttle Columbia is launched tomorrow. This is the first strictly-business cargo to be lofted in America's "space truck."

The liftoff-for-pay from Cape Canaveral marks the start of an era. One of the communications satellites, owned by a Canadian firm, is called "Anik," meaning Little Brother in Eskimo. Its companion, property of an American consortium known as Satellite Business Systems, is marked SBS-3 (and, fortunately, has *not* been nicknamed Big Brother).

This autumn event in the United States space program is anything but humdrum, although shuttle flights — this is the fifth — are becoming routine. The satellites are to pop out of Columbia's cargo bay like jack-in-the-boxes, then ignite their own rockets to climb into orbits more than 20,000 miles high.

Columbia's fiery takeoff and the customary media hoopla before, during and after the flight will provide the usual good show. However, this should not obscure the sobering military implications of the U.S. and Soviet space programs.

The Reagan administration continually declares it has no plans to put actual weapons in space. Next fall, however, a U.S. shuttle vehicle is scheduled to carry into orbit the first entirely military (and highly secret) cargo. Thereafter, 113 of the 311 shuttle flights planned through 1994 will have military payloads. Most of these cargoes are expected to be sophisticated electronics packages, the latest in a series of U.S. reconnaissance satellites that goes back to the 1960s. Gen. Robert T. Marsh, head of the Air Force Systems Command, said recently, "The space shuttle will change the way we do business. We will depend on it for launching virtually all of our national security payloads."

That's plain enough. So is the implication of "Cosmograd," the city in the sky being planned by the Soviet Union. An enormous rocket, bigger even than the Saturn 5 that launched the U.S. Apollo astronauts to the moon, is expected to be ready by 1985 to lift into orbit a 110-ton space station, which could be enlarged by adding modules. Enough connected stations and modules could serve as building blocks for a sky village, if not quite a city.

While the purpose of this planned Soviet space station is unclear, the United States cannot ignore its possible military use. Some U.S. analysts even are suggesting that "Cosmograd" might be developed into a platform for Soviet laser weapons aimed at U.S. satellites.

The Soviet Union has more satellites in orbit than does the United States. Yet in much space technology, the United States is believed to enjoy an edge in sophistication over the Soviet Union. Soviet electronics are sometimes rudimentary and their satellites occasionally poop out after a half-dozen months in orbit, while most American satellites keep flying.

Regrettable as it may be, this space competition appears likely to remain vigorous for many years to come. The United States would be derelict not to pursue its own space projects, learning more about the physics, chemistry and mathematics of this utterly alien environment. The more American scientists can learn from these extensive shuttle flights, the more capably the United States can prevent the Soviet Union from gaining military supremacy in space. If it should settle into a military competition, the space race will be one that the United States simply cannot afford to lose.

The Wichita Eagle-Beacon

Wichita, KS, November 10, 1982

Thursday will be yet another historic day in space, if no last-minute hitch delays the launch of space shuttle Columbia's first commercial mission. For the first time in history, a National Aeronautics and Space Administration crew will use its rocket-it-out, fly-it-back space truck to hang two communications satellites in orbit for paying customers. Columbia's mission this time is to make launches from space of one satellite that will serve corporate business systems in the United States and television watchers in Canada.

The first, to be rocketed to an imaginary shelf several thousand miles above the Pacific Ocean, in an orbit scheduled to keep pace with Earth's rotation, is owned by Satellite Business Systems. This is a partnership of Aetna Life and Casualty Co., Communications Satellite Corp., and International Business Machines, relaying what amount to long-distance telephone calls and, we assume, data transmissions, from one Earth station to another.

The second, scheduled to be positioned on Friday, is owned by the Canadian government, and will bounce television programs back to roof-top antennas in Canada.

In its four test missions earlier, Columbia's crews have demonstrated various kinds of specialized expertise at performing space chores. But this will be something even more impressive — launching other space orbiters from an orbiting "pad" 185 miles above Earth. The communications satellites will then be able to perform their functions from a safe several thousand miles farther out.

After seeing to it that the satellites are delivered properly, some members of Columbia's crew are scheduled to take space walks, before the five-day fifth mission ends on Monday, testing space suits in the process.

Americans everywhere will be wishing the crew a safe space and customer-pleasing voyage.

Wichita, KS, November 12, 1982

The unsuspecting gulls wheeled symbolically away as the great white ship's main cluster of engines came to life Thursday morning. Columbia remained motionless in its mounts a few seconds longer as power built, and then, as the solid fuel booster rockets erupted, it suddenly was away once again.

It rolled lazily onto its back and a dull, crackling roar filled the ears of the onlookers as America's tried-and-true space shuttle traced an arc across the clear blue Florida sky, a powdery-white contrail snaking after it. And once again, all those watching thrilled to the sight of Americans catapulting into space in what is perhaps the most awe-inspiring vehicle yet designed for that purpose.

The magnificent launch of the Columbia on its fifth mission — its first paying commercial assignment — came off precisely on schedule. There were no glitches or holds; the countdown clock never was halted to allow time to make last-minute repairs. Lift-off was slated for 6:19 a.m., CST, and that's exactly when it came.

Thanks to the picture-perfect weather conditions at Cape Canaveral, watchers were treated to perhaps the best view yet of the dramatic separation of the spent boosters from the space plane and its huge liquid fuel tank, only minutes after launch, but many miles above. It was, simply put, a flawless start to another adventure in space.

And, while the airplane-style landing capacity of the shuttle is what makes it unique, there remains an utter fascination with seeing it thunder aloft.

The thousands who watched in person, and the millions looking on through television, felt the thunder that began on the ground in Florida Thursday morning and then streaked heavenward on a fiery tail. Few among them surveying the scene must not have felt their own hopes and aspirations ascend with Columbia and its brave crew.

The Hartford Courant

Hartford, CT, November 18, 1982

The fifth flight of the space shuttle Columbia, which ended Tuesday after five days in space, was significant mainly because it marked the dawn of a new era of competition in space.

The challenge facing world governments is to keep the competition peaceful.

The cost of the latest space shuttle flight was, in small part, paid for by Telesat of Canada and Satellite Business Systems, a joint venture of Aetna Life and Casualty, the Communications Satellite Corp. and International Business Machines Corp. The astronauts deployed two communications satellites for their first customers, and others are on the shuttle waiting list.

There now is competition in the launching services business. Both NASA and the European Space Agency have been using expendable rockets to hoist commercial satellites into space for such uses as the transmission of long-distance telephone calls, high-speed data communications and the distribution of television programs.

Manned shuttles offer solid advantages over expendable rockets. For example, the space shuttle can carry larger, more powerful satellites into orbit; it can launch satellites more cheaply and shuttle crews can repair satellites.

Malfunctions in the $2 million spacesuits, built by United Technologies Corp. in Windsor Locks, that made them impossible to use outside the shuttle during the latest flight, probably will be corrected before the next mission.

After the liftoff last Thursday, President Reagan radioed his support to the astronauts and said that he wished more people could see the Earth from orbit so that "we might realize that there must be a way to make it as united in reality, here on Earth, as it looks from outer space."

But "here on Earth," the Pentagon — with Mr. Reagan's backing — is preparing for a dangerous, non-commercial competition in space. Weapons are being developed for deployment in space on the theory that the Soviets are doing likewise and that the United States should be the first to take the "high ground."

To discourage that competition, one of the first items that the new Congress should take up in January is a resolution, first introduced in September, urging the president to begin immediate talks with the Soviet Union and other nations on a treaty to ban weapons of any kind from space.

The Outer Space Treaty of 1967 contains some weapons prohibitions, but has serious loopholes. A 1979 treaty that prohibits militarization of the moon and other celestial bodies has not yet been ratified by any nation.

The extraordinary difficulties involved in working out agreements to limit or reduce the number of American and Soviet nuclear weapons on Earth should be proof enough to Mr. Reagan that the time to stop a military competition in space is before any weapons are introduced there.

President Reagan also should consider proposing a cooperative project to build and launch a permanent, U.S.-Soviet manned space station. That would serve both as a good will gesture to the new Soviet regime and as a positive step to reduce the danger of militarizing space

Competition for satellite business has some commercial advantages, but competition for military dominance in space would be a net loss for everyone on the planet.

The Dispatch

Columbus, OH, November 18, 1982

THE SPECTACULAR accomplishments of the space shuttle program continued this week with the successful completion of the *Columbia's* fifth flight. Yet while the flight wasn't exactly ignored, it did not generate the same public attention that the earlier flights did.

This is probably to be expected even with such a marvelous machine as the shuttle, the first reusable, revenue-generating space vehicle. The first, second and third flights attracted worldwide interest as the sight of the huge and vastly complex vehicle ascended into space and then descended flawlessly. The fourth flight drew somewhat less attention and the fifth even less.

Part of this trend is due to the fact that the *Columbia* has performed so well. Everything has gone as planned, more or less, and the deployment of the first two commercial satellites was executed without a hitch. The only problem that arose during the most recent trip was the failure of two, $2 million space suits to perform properly, forcing the cancellation of a planned space walk. The suits will be repaired and the space walk rescheduled for a future flight.

The United States, through the space shuttle program, is developing a tremendous asset for future space adventures. By using and reusing the same space vehicles, and by designing them to land on land at easily accessible sites, the costs of space activities will be greatly reduced for decades to come.

The *Columbia* is scheduled for a 10-month overhaul now and the main duties of the shuttle program will be transferred to the *Challenger*. In all, as many as six flights may be undertaken next year.

The success of the shuttle is proving that even the remarkable can become routine — and that is a tribute to the men and women who have made the shuttle program the success it is today.

The Kansas City Times

Kansas City, Mo, December 1, 1982

America's new space shuttle is being billed as a utilitarian space transportation system, but it still contains the research and exploration benefits that have characterized other NASA programs.

One striking and unexpected boon of the shuttle flights to date has been the radar mapping of ancient riverbeds now buried beneath the sands of the Sahara desert in the region where Egypt, Libya and the Sudan join. On the second test flight of Columbia in November 1981, when several experiments had to be abandoned and the mission cut to two days because of a faulty fuel cell, the shuttle's remote-sensing radar yielded clear images of a habitable terrain that first disappeared 2 million years ago.

Aircraft radar has been used for years to map some of the world's remote jungles where cloud cover prevents aerial photography. But whereas radar can penetrate only a few inches into moist ground, it can delve up to 16 feet deep into sand. Excited scientists of the U.S. and Egyptian geological surveys, the University of Arizona and NASA's JPL laboratories say the Columbia radar readings showed the vanished rivers flowed south and west and were not a part of the Nile system. The mapping from space also suggests promising spots at which to drill for water.

The secrets unlocked from the Saharan wasteland provide archeologists a clue as to where to search for remnants of long-ago civilizations that existed during eras of adequate rainfall in the region. Radar mapping from deep space thus becomes an added bonus of the shuttle flights, and there will be more.

New Space Shuttle Completes First Voyage

The U.S. space shuttle Challenger successfully completed a five-day first flight April 9, 1983 that began with a launch April 4 at Cape Canaveral, Fla. Aboard the Challenger were four astronauts: Paul Weitz, the commander; Col. Karl Bobko, the pilot; Story Musgrave, a doctor and mission specialist, and Donald Peterson, another mission specialist. Musgrave and Peterson April 7 took the first "space walk" by Americans in nine years, spending nearly four hours outside the shuttle in a test of their new spacesuits and of their ability to conduct the kind of maneuvers that would be necessary to maintain and repair orbiting satellites. Both men were attached by lines to the Challenger so that they would not drift away. The Challenger completed its first mission in good shape, space officials said. After a preliminary inspection, ground operations manager James Harrington said April 10, "It truly looks like they just rolled it out" of the hanger. "I can't get over just how clean the ship is," he added, saying that it looked as though there was a good possibility of completing the necessary maintenance in time for a next launch in early June. The highly successful first flight occurred despite the fact that the launch date had to be delayed by more than two months because problems were found with the Challenger's main engines. All three of the main engines were found to have flaws allowing hydrogen to leak, a problem that could have proved disastrous had it gone undetected. The problems possibly stemmed from the fact that the engines for the Challenger were designed to produce more thrust, and thus have more lifting power, than the engines for the Columbia.

The main payload item carried by the Challenger was the $100 million Tracking and Data Relay Satellite (TRDS). The 5,000-pound object was described by NASA as the "largest and most advanced communications satellite developed thus far." The satellite was the first of a planned group of three, all built by TRW, Inc. They were intended to provide a space-based communications network that would relay to Earth heavy flows of data from such sources as Landsat, the U.S. survey satellite, and Spacelab, the European-built research facility that was scheduled to be carried by Challenger late in the year.

The Orlando Sentinel

Orlando, FL, April 9, 1983

NASA got its new model truck off the ground this week, hoping to deliver on its promises of a new industrial age. When it lands today, we'll have further evidence to back up those claims.

This lighter, faster and much more powerful shuttle made a flawless launch. The only real flaw has been the problem with a 2½-ton communications satellite, which was nearly lost in space.

The significance of Challenger's flight and two more to follow this summer should not be lost. These tests flights are critical as NASA prepares to move to regularly scheduled launches out of its Cape Canaveral freight terminal next year.

The reusable shuttle's ability to haul large loads more cheaply than one-way rockets should help buy our ticket into the next industrial age. It is important to attain a fast and dependable schedule because hauling cargo brings in money. Cargo destined for space is perishable, largely because satellite batteries have limited lives. And, like a load of Plant City strawberries ticketed for the New York market, if a truck isn't available, there are competitors in the wings. In this case, those competitors are European and Japanese rocket launchers.

Beyond freight hauling and military applications, the shuttle program has two primary goals. One is to tote into space the building materials for laboratories that will become the seed of a new industry. The other is to establish a new threshold for space exploration.

The first of those goals will be the next generation of high-tech industry, turning out such things as perfect ball bearings — impossible in a gravity environment — and heart treatment drugs, which are expected to be on the market in three years. The second goal will be the means to still more discoveries through such instruments as the space telescope.

Challenger may seem like just another truck whizzing down the road. But its mission is to blaze a new trail to better living through such dependability that it becomes routine.

The Cincinnati Post

Cincinnati, OH, April 11, 1983

Six successful space shuttle flights are now in the record book—five by Columbia and one by Challenger. The last one, says NASA, was the most glitch-free so far.

The crew of Challenger made it all look so effortless and routine, from the start to the finish of their five-day mission, that it may be hard to appreciate how far we have come in making space just another extension of the human environment.

It's like watching huge jetliners come and go at an airport. One has to remind oneself that there are people still alive who were born before the Wright brothers flew.

And the space shuttles are, essentially, airplanes, with a direct line of descent from the Wrights' first flimsy craft.

It will become even more difficult to maintain a proper sense of awe as NASA approaches its goal of 70 shuttle flights over the next four years.

Three more are scheduled for 1983. The next time Challenger takes off, it will land back at Kennedy Space Center, a figurative stone's throw from its launch point at Cape Canaveral.

No space mission would be complete without something going wrong, of course. With Challenger, it was an errant communications satellite that was deployed by the crew. Its second stage rockets apparently misfired, sending it into a lopsided orbit.

NASA hopes to nudge the satellite back to where it belongs. But until the problem with the boosters can be identified and corrected, the orbiting later this year of a companion satellite is in jeopardy, as are a European-built Spacelab and other future shuttle missions that will depend on the communications network.

Thus hardware remains the chancy factor, as it has since the beginning of the space age. Men are at home in space (and so will women be on later flights), as Challenger has once again brilliantly demonstrated.

THE TENNESSEAN

Nashville, TN, April 14, 1983

THE space shuttle Challenger, on its first launch into space, performed extremely well and its crew accomplished each mission with great precision.

The one disappointment that marred the Challenger flight happened well away from the shuttle. After a perfect launch from the shuttle's cargo bay, a giant communications satellite shot off course because of the misfiring of a booster rocket.

The National Aeronautics and Space Administration will try to nudge the satellite back into a proper orbit by firing its small booster rockets in a carefully programmed sequence over the next weeks. But at the same time NASA is investigating why the rocket failed to push the satellite into a high-enough orbit.

NASA had planned to use an identical rocket to put into stationary orbit another satellite in the planned communications-control network. But obviously it can't proceed until it finds out what ailed the one that misfired.

In any case, the Challenger worked to perfection. There was a four-hour space walk that demonstrated the astronaut's ability to move freely in space and to perform a variety of mechanical tasks.

The next step will come with Challenger's mission in June. On that flight, the first woman astronaut, Ms. Sally Ride, will use the shuttle's manipulator arm to put a satellite into space and then retrieve it and place it back inside the shuttle's cargo bay.

All in all, the shuttle program has been extremely successful, and considering the complexity of the systems involved, has been remarkable.

Herald News

Fall River, MA, April 6, 1983

The space age proceeds with the new shuttle space ship, Challenger, in the midst of its initial five-day flight.

President Reagan's proposal to defend the nation against possible missile attack by defensive space weapons that would destroy the missiles gave thenation and the world a vision of defensive space wars in the future.

The Challenger offers another vision of peaceful and productive expansion into space.

A new communications satellite will be left in space by the Challenger, new scientific data will be collected during its five-day trip, and the astronauts on board will take a walk in space.

The new satellite can provide the earth with virtually unbroken information about as many as 26 spacecraft. At present, that information can only be gathered from space stations here on earth, and then only at certain intervals.

The Challenger contains a canister with 60 pounds of fruit and vegetable seeds. It is hoped that changes in the seeds during the flight will provide information about how similar seeds should be packaged in the future so thatfrom them, fruit and vegetables can be grown in permanent space stations.

The implications in the inclusion of the seeds on the Challenger's flight are clear. Bit by bit the country is moving toward a future in which space stations will be permanent outposts of humanity, presumably as way stations for inter-planetary travel.

All this is so remote from the way we live now that it still seems like an H.G. Wells novel rather than reality.

But reality is coming closer to converging with science fiction all the time, and the cargo of seeds the Challenger is carrying opens up a vista of the future in which gardens are planted on vast space stations that provide permanent habitations for large numbers of people.

The seeds for these gardens are in a sense being sown on the flight that is going on right now.

The expansion of scientific knowledge that has made possible the breakthrough into the space age will be enormously accelerated by flights such as this one by the Challenger.

To a certain degree these flights still seem to most of us like enormously successful sideshows that are being sponsored by the government for our entertainment here on earth.

But the fact is that they are brilliant coordinated and detailed scientific expeditions that will in a matter of measurable time alter the conditions of life on earth,

How? By providing the first real escape hatch man has had since the discovery of the New World almost 500 years ago.

The space flights going on now will in the future seem analagous to the early, tentative explorations of the western hemisphere by Europeans looking for a place to expand.

Space, however remote and frightening now, will offer the same kind of opportunity to man in the future.

This is ultimately what the Challenger's flight is all about.

Meanwhile, right now, the present space trip rightly engages our attention and our admiration for those responsible for it, not only the crew of the Challenger, but the small army of scientists and technicians its flight requires.

The Challenger is helping make man's dream of the space age a reality.

St. Louis Globe-Democrat

St. Louis, MO, April 12, 1983

The space shuttle Challenger vindicated itself once it managed to get off the ground.

After it finally got off and away, officials of the National Aeronautics and Space Administration reported following the successful five-day, two-million-mile maiden mission that the Challenger performed better than its pioneer sister ship, the Columbia.

That's quite a compliment since the Columbia racked up an enviable record as it traveled 10 million miles in five missions. People tend to overlook earlier failures once success is attained.

Such was the case with Columbia, too. The initial official target date for the first shuttle launching was set for June 1979 but engine problems delayed the Columbia's takeoff for almost two years, until April 1981. Multiple engine problems also plagued the Challenger and delayed the launch from Cape Canaveral, Fla., for 2½ months.

In the latest mission, only 22 minor problems were recorded. That adds up to five fewer than during the fifth flight of the Columbia. On the basis of its performance, the Challenger could be ready for flight No. 2 by the first week in June.

NASA is looking forward eagerly to the next half-dozen space voyages. In the next journey, the seventh, America's first woman astronaut is scheduled to go aloft and on the 12th mission a third shuttle, named, Discovery, will make its first odyssey into space.

With the $2 million space suits in working order and the engine problems cleared, the shuttle program appears to be set to move full speed ahead.

Houston Chronicle

Houston, TX, April 12, 1983

The adjectives used to describe the first flight of the space shuttle Challenger represent a compendium of superlatives, including "great," "super," "stupendous" and "flawless." Even a more down-to-Earth view of the mission leaves little doubt that such praise is justified.

For the "F Troop" of astronauts, Donald Peterson, Story Musgrave, Karol Bobko and Paul Weitz, the mission was "a great time" and "a lot of fun." For the United States it was definitive proof of the concept of developing a fleet of reusable space vehicles. The mission's successes include the launch of one of the largest communications satellites, the first U.S. spacewalk in nine years and the addition of the second shuttle vehicle to the American fleet.

The mission's success and the lack of significant problems on Challenger's maiden voyage bode well for a quick turnaround for other launches in June and August. Challenger's pinpoint landing speaks for itself, and to the astronauts of "F Troop" go a hearty "well done and welcome home to Houston."

The Birmingham News

Birmingham, AL, April 3, 1983

If all goes as planned, the scheduled Monday launch and mission of the space shuttle Challenger will mark a number of firsts in the U.S. space program.

Among others, the mission of course represents the first launch of a shuttle other than the heroic Columbia, which flew five times before temporary retirement. Also of note is the four-man crew of the Challenger, which represents an expansion of the program into its intended role of transporting human as well as equipment cargo — a program which eventually will see non-astronauts participating. Another first of sorts is marked by the shuttle's remarkable payload for this mission — a massive (5,000-pound) communications satellite, the largest ever put into orbit, which will, among other things, take on a great part of the shuttle program's own space-to-earth communications linkage.

Most spectacular, again if all goes according to plan, will be an extended "space walk" by two of the shuttle's crew. In addition to its photogenic qualities, the external movement of crew members will allow them to perform many functions, such as repair of satellites, which would be very difficult to achieve otherwise.

In terms of the overall shuttle program, another important first is scheduled to come on Challenger's planned second flight, when the craft will return from space to land on a 15,000-foot concrete landing strip at Cape Canaveral, rather than at one of the Western desert sites. The turn-around capabilities of the shuttle will thus be enhanced dramatically.

Amid all the "firsts," however, and more important than any one of them singly, is the continuing progress of the shuttle program in specific and the U.S. space program in general. Although budget constraints have impeded NASA in recent years, the agency is here again demonstrating its scientific and technological genius. The value of the program to the United States — indeed to all the people of the world — goes far beyond dollars and cents, for it involves the very advancement of knowledge.

It is a thought worth thinking as Challenger lifts toward its mission Monday, challenging us all in a very real sense to renew our national commitment to space science.

One also trusts that the sentiment against drunken driving grows until a drink-related accident on our highways is a rare event.

The Dispatch

Columbus, OH, April 12, 1983

THE SECOND ship has now joined this country's space shuttle fleet and the just-completed flight of the *Challenger* serves to increase confidence in the ability of the program's engineers to overcome unexpected problems and to mold a better program because of them.

Once the shuttle got off the ground it performed almost flawlessly. There were only 22 "abnormal mechanical events" during the flight as compared to 82 during the maiden flight of the *Columbia*, the first U.S. shuttle. The astronauts conducted a series of scientific experiments, chased a make-believe satellite across the sky in a practice of rendezvous techniques, conducted engineering tests of their new vehicle and embarked on a 3-hour, 47-minute space walk during which new, more flexible spacesuits performed perfectly.

The crew also launched a $100 million communications satellite and while the launch was perfect, problems developed later in the satellite's propulsion system which caused it to go into an incorrect orbit. Scientists are working now to try to correct that orbit.

The *Challenger's* flight was delayed for more than two months by problems in its new-design rockets. These problems were eventually overcome and the lessons learned from this delay will be used to make future rockets even better.

A "well done" to all who played a role in making the flight of the *Challenger* a successful one.

BUFFALO EVENING NEWS

Buffalo, NY, April 13, 1983

The near-perfect performance of the space shuttle Challenger on its inaugural flight amply affirms the stunning progress of the nation's space agency in mastering the concept of winged craft ferried into and out of space for sophisticated missions aloft.

Only two years ago, officials of the National Aeronautics and Space Administration were unsure about the aerodynamic feasibility of a reusable craft that could land like an airplane. After six shuttle flights, however — five by Columbia and the maiden voyage by Challenger — NASA has established beyond doubt that reusable shuttles are not only reliable but also vastly more economical in providing access to space.

That, of course, was one of the original goals in developing shuttles, and Challenger offers the latest evidence that this space enterprise well warrants the national investment in it.

Except for problems with a communications satellite it carried aloft, Challenger performed flawlessly. During their five-day mission, the four crew members experimented with tools and techniques that can be effective in moving objects in space and in repairing faulty satellites. The flight also featured a four-hour spacewalk by two of the astronauts, the first by Americans in nine years.

That Challenger performed with such impressive precision and withstood the flight without significant surface damage to its structure has given NASA engineers confidence in reducing the delay between future shuttle flights, with the next one scheduled for June.

Plans for future shuttle missions include the refinement of techniques for repairing damaged satellites and assembling large structures in space, in preparation for the day when permanent space stations or solar power systems will be constructed in orbit.

In contrast to the spectacular Apollo moon missions of the 1970s, the achievements in shuttle transportation do not command comparable public excitement and suspense. They are nonetheless necessary steps as the United States reaches out to put space research and science to increasingly practical and beneficial uses in a better understanding of the earth, the oceans and the atmosphere.

America's space program has already brought vast spinoff benefits in communications, weather forecasting, computer technology, medicine and defense; many more such benefits surely lie ahead.

Next year's federal budget calls for $6.5 billion for space shuttle programs; that is a relatively small item in the total budget of $850 billion and represents only a modest 4 percent increase over the outlay for the current fiscal year. The knowledge-expanding goals of the nation's space program clearly deserve continued public support and give valid reason for pride in its human and technological successes.

The Kansas City Times

Kansas City, MO, April 11, 1983

Spacewalks are great fun for the Earthbound TV watchers far below. The astronauts are acting out everyone's dream of flying as they float and bounce around, joyously and acrobatically, and no one enjoys it more than the two Challenger shuttle crewmen, Dr. Story Musgrave and Donald H. Peterson, did on Thursday. They became the 28th and 29th Americans to engage in EVA (extra-vehicular activity, in NASA jargon) counting the 12 who landed on the moon.

But there was serious purpose in the 3 hours, 50 minutes the two spent in the cargo bay of the spacecraft, tethered to 50-foot safety lines. The mission was to test tools and techniques for servicing and repairing satellites in a gravity-free garage. If you drop a wrench in space it won't fall 176 miles to the ground, but it might float away if not secured or watched. Such chores as fixing a satellite with frayed solar panels, refilling the fuel tank or giving the little creampuff its 10-million-mile checkup will feature the many shuttle missions on the schedule ahead.

A specific one next April calls for catching up with a sun-study satellite that has been out of order for three years and getting it working again. A simulated rendezvous with a phantom space target was one of Challenger's experiments last week. But there were others of possibly even greater significance to the overall shuttle program. The second of six tests was made on the electrophoresis device developed by McDonnell Douglas Corporation of St. Louis.

The first test last June produced 500 times more biological materials, such as cells, enzymes, hormones and proteins, and the developers expect to achieve purity levels four times higher than possible in Earth's atmosphere. The potential for production of pharmaceuticals, for one example, in space factories can be readily seen. The two astronauts had a great time jumping around in those eerie space suits (which worked fine after a malfunction last November) but some very important other business was taking place aboard Challenger on this latest shuttle flight.

The Honolulu Advertiser

Honolulu, HI, April 5, 1983

Some day space flights will be so routine that they won't make the front pages of newspapers and the lead position in nightly telecasts, but that is not yet the case.

For one thing, yesterday's launch of the space shuttle Challenger was still a dramatic sight, even for veterans who have watched such events for a quarter-century.

And the history of this second shuttle craft has been anything but routine to date. It was delayed half a dozen times over two and a half months by engine failures and lesser problems.

Not only was Challenger itself untried in space, so was the $100-million satellite it carried aloft to launch. That has been called "the most complicated and sophisticated communications satellite ever built."

Space walks are not new. But the one scheduled for Thursday is the first by Americans since the last Skylab flight in 1972. One from the first space shuttle, Columbia, in its last flight was cancelled by malfunctions in the spacesuits.

Then there is the question of space sickness, which has struck 50 percent of those who have flown aboard the shuttle. Under a controversial new policy, information on such sickness will be withheld unless it affects the flight.

Still, while the problem is not expected to change the five-day mission, it amounts to another uncertainty and something that must be dealt with before such missions actually become as routine as ocean voyages in a cargo ship.

The Wichita Eagle-Beacon

Wichita, KS, April 6, 1983

Monday's flawless launching of space shuttle Challenger may have happened 2½ months late — twice delayed since January — and the satellite it put in orbit 10 hours later was having problems. But if the latter can be corrected, Challenger's debut could have set the stage for America's entry into a whole new realm of far-out transportation. The satellite is supposed to function as the first unit of a communications, space-tracking and data-relay network. A second one is to be parked in the heavens in August, and if all continues to go well, a manned space station is to be launched into orbit in September.

Another space first is scheduled for Thursday when astronauts Story Musgrave and Donald Peterson are to put on space suits and take history's first gravity-borne stroll from an orbiting space shuttle. Such a walk was scheduled last November during shuttle Columbia's fifth mission, but was scratched when troubles developed in the regulators that were supposed to keep the $2 million suits properly pressurized.

Now, the United States has two returnable spacecraft built to be rocketed into space but land like airplanes on their return to Earth. The capsules used in all earlier peopled space explorations had to be dropped into the ocean at the conclusions of their missions. The National Aeronautics and Space Administration hopes to have, eventually, a space shuttle fleet of five craft, at least. Orbiter Discovery is expected to make its first flight in the summer of 1984, and Atlantis is expected to fly the following year.

Though the engine defects and other troubles that caused successive delays in Challenger's debut has thrown this year's shuttle program considerably off-schedule, NASA still hopes to complete four more missions in 1983. They hope to have Challenger ready for its second flight in about two months, and ready again to place the second tracking and communications satellite two months after that. Columbia is scheduled to handle the Spacelab delivery mission in September.

At the moment, America's hopes are dedicated to pilots Paul Weitz and Karol Bobko and mission specialists Peterson and Musgrave. Duties of the latter, in addition to Monday night's satellite launching and Thursday's space walk, include conducting a number of experiments, one related to the potential for space horticulture and one involved with testing a snowmaking device.

The world needs a new set of heroes from time to time. It's grateful to the current space team for fulfilling that need.

Challenger Completes 2nd Flight; 1st U.S. Woman Astronaut in Crew

The space shuttle Challenger completed its near perfect six-day mission June 24, 1983 landing safely at Edwards Air Force Base in California. The mission included a number of firsts, but the one that captured the public's imagination was the presence on board of the first American woman astronaut, Sally K. Ride. Ride, a physicist by training, functioned as a mission specialist during the flight. Ride was not the first woman in space. Two Soviet women cosmonauts had preceded her, Valentina Tereshkova in 1963 and Svetlana Savitskaya in 1982. The other members of Challenger's crew were all men: Capt. Robert Crippen, the mission commander (and the first individual to make a second shuttle flight); Norman Thagard, a medical doctor; Col. John Fabian, also a mission specialist, and Capt. Frederic Hauck, the co-pilot for the mission. The flight was the first mission to have a five-person crew.

Almost all of the objectives of the mission were accomplished. On June 18, Ride and Fabian deployed the Anik C communications satellite, a Canadian satellite designed to hover over the Pacific Ocean in an orbit about 22,000 miles high. The next day the two mission specialists repeated the maneuver with the Palapa B communications satellite, a similar piece of equipment built by Hughes Aircraft Co. for Indonesia and the Association of Asian Nations. The early indications were that the two satellites were functioning as planned. One of the key objectives of the mission was accomplished June 22 when the astronauts, particularly Ride and Fabian, successfully employed the shuttle's mechanical arm to catch hold of and retrieve a satellite that had been permitted to drift free in space, unattached to the shuttle. The satellite used for these tests was the so-called Shuttle Pallet Satellite, or SPAS, constructed by West German scientists. Five times the astronaut released the SPAS and then retrieved it with the mechanical arm. The shuttle jets were turned on in brief spurts to move the craft away from and then back to the SPAS, thus mimicking the maneuvers that would be necessary on future missions when the shuttle sought to retrieve orbiting satellites to carry out repairs or for other reasons. Tommy Halloway, the chief flight director, said June 22 that "today is a significant milestone in the evolution of the overall operations of the shuttle. Everything went exceptionally well."

Richmond Times-Dispatch

Richmond, VA, June 26, 1983

Sally Ride and her supporting cast are back from Out There, safe, sound, and celebrated. The second flight of the Challenger space shuttle which they inhabited for nigh a week was a thumping success: systems functioned, satellites were launched, robotic arms proved dexterous. The only hitch came at the end, when soupy weather over Cape Canaveral forced a landing in sunny California — more a setback for the Florida Department of Tourism than for NASA. Unexpectedly, the space program seems to have entered a happy state: It is becoming routine without becoming boring.

To be sure, Miss Ride added an extra dash of color to this carpet ride. More important, her presence may have lowered the resistance of many women to space fever, that jolly disorder characterized by wild flights of the imagination and a stiff neck from too much star-gazing. Neil Armstrong did not pronounce his first trudge on the Moon a giant step for white males over 30, but such is the modern tendency to view events through the prism of dog-tags data that, possibly for the first time, millions of American females feel "included" in space. Coming soon: black astronauts.

But even minus the Ride factor, something is changing. The '70s were the decade of space spectaculars — the Moon landing, Viking's touchdown on Mars, Voyager's kaleidoscopic do-si-do with Saturn. These events were like a pitcher's home runs — astounding when they happen but too rare to generate sustained excitement. The shuttle is a spatial Pete Rose: not a bleachers-banger but always a threat to reach base. America's extra-terrestrial future, like most baseball games, may hinge more on frequent and timely singles than the long ball.

Although over the centuries America has become well settled, her pioneer spirit endures, as antsy as a lamped-up genie. Perhaps other peoples have accommodated themselves to these planetary confines, but it is part and parcel of the American national soul to reach out. If Horace Greeley were around today, his advice surely would be, "Go *Up*, young man."

Greeley, a century ago, knew that a country as factioned and as fractious as ours needs a common vision and a sense of an adventure shared, at least vicariously. Otherwise, people grow self-centered and sulky, eroding citizenship. A rising GNP will enlarge our purses but not our purpose. In the patchwork quilt that is America, space may be a redeeming thread, stitching together our differences with a needle named Challenger.

THE BLADE

Toledo, OH, June 15, 1983

THE seventh shuttle mission is scheduled to soar into orbit Saturday. If it were not for the presence on board of this country's first female astronaut to go into space, Sally Ride, most Americans would give it less than undivided attention. The present-day era of space exploration and research is not always filled with attention-grabbing breakthroughs.

Dr. Ride, an astrophysicist who will travel with four male astronauts on the shuttle Challenger on a six-day mission, has chided reporters for focusing excessive attention on her gender. But it must be noted that her participation does again reflect the fact that in many heretofore all-male bastions women are gaining new authority. If one wishes to downplay Dr. Ride's gender, it might be noted that 88 American astronauts already have preceded her into space.

Overshadowed by Dr. Ride's ride has been the fact that this week an 11-year-old U.S. space satellite, Pioneer 10, passed out of the planet Neptune's orbit and moved beyond the edge of our solar system, thus becoming this planet's first satellite to do so. Ever since March 3, 1972, Pioneer 10 has been speeding along and transmitting information back to earth, moving through the dangerous asteroid belt, swinging close to the planet Jupiter, and moving out beyond the orbits of Pluto and Neptune toward the stars.

Pioneer 10 continues to flawlessly beam data back to earth even though it had an expected life of only 21 months. Its continuing good health is a testimonial to the quality of the technology and the equipment that went into the 570-pound spacecraft.

Steadiness and performance ordinarily are not the stuff that enthralls most Americans. But that takes nothing away from the long list of achievements the space program's men and now its women are producing.

The Hartford Courant

Hartford, CT, July 2, 1983

Just about the same time Sally K. Ride was launched into outer space, the federal government released a more earthbound study.

According to a Department of Labor poll of 77,000 companies, employing more than 20 million people, women and members of minority groups fared better in companies with government contracts. The government demands that these employers engage in affirmative action.

Employment of women increased by more than 15 percent at companies under affirmative action requirements, while companies not subject to the rule added only 2.2 percent more women to the work force between 1974 and 1980. Promotions also occurred more rapidly at companies operating under affirmative action strictures.

Sally Ride and the other female astronauts are a product of the government's own affirmative action efforts. The National Aeronautics and Space Administration did not start accepting women as candidates for space travel until 1978 and until then overlooked the talents of accomplished female scientists.

Space travel today requires technical expertise, and women can be trained to do the job as well as men. It's not as if most astronauts have experience in space before their first mission. Dr. Ride was specially trained to operate the shuttle's complicated arm to retrieve vehicles in space.

As one of the five other women in that NASA class of 1978, Kathryn D. Sullivan, said on a recent visit to Hartford, "I think we tend to be people who are pretty independent and I suppose pretty strong-minded about what we want to do. A somewhat adventuresome sense about going off and trying new things; some amount of tenacity and self-discipline in the recognition that quality counts."

Independence, tenacity, and commitment to *quality*: Those are the attributes of any astronaut. Had NASA not widened the field of candidates, the space program would have missed out on the talents and tenacity of people like Sally Ride.

But all is not equal in outer space: Dr. Ride was paid less than any of her male colleagues on the mission. Her leap into space was still but one small step for womankind.

Post-Tribune

Guarding Your Interests Daily

Gary, IN, June 26, 1983

Our opinions

It is always thrilling to watch a space shuttle land, even on television. For one thing, the shuttle is at its most beautiful as it glides toward earth. For another, it carries with it all the dreams one has of the universe. And it's always hard to believe that the craft landing so smoothly has just returned from outer space.

Thanks to the shuttle program, space is less mysterious and some of those dreams are within reach. Thanks specifically to the space shuttle Challenger's second flight, space will soon have several practical uses.

Although overshadowed by the hoopla surrounding the first space flight by an American woman, this flight opened a new era for space flight.

Among other things, the shuttle proved both its mobility in space and its ability to rendezvous with and retrieve other objects when it deployed the West German SPAS-01 satellite, then maneuvered around it, then plucked it out of orbit and placed it back into its cargo bay. These capabilities are essential for building a permanently manned space station, a high-priority goal for the National Aeronautics and Space Administration.

The five-person crew, the biggest ever on a space flight, conducted about 40 experiments, many of which were geared toward the commercial use of space.

These achievements, along with putting communications satellites into space, should lead the way for many peaceful uses of space. It's even possible now to conceive of passenger travel in space.

All this, and the TV pictures being sent back were even more spectacular than ever.

Another aspect of the shuttle program that gives hope for the future is the cooperation among countries. Communications satellites have been launched for countries throughout the world. The satellite used for the deploying-and-retrieving exercise was West German. The 50-foot robot arm used to grab the satellite out of space was made in Canada; it cost the Canadian government $100 million to develop it but was provided to NASA free as Canada's contribution to the shuttle program. NASA has ordered three more of the arms at $25 million each.

The technological advancements triggered by the shuttle program have been invaluable; if these can contribute to bettering the world, success will be complete.

AKRON BEACON JOURNAL

Akron, OH, June 21, 1983

THE FLIGHT engineer on the space shuttle Challenger is an astrophysicist. The astronaut's steady grip, once used to ignite the amateur tennis circuit, will guide a German scientific satellite into space, then pluck it back from the darkness of infinity.

Glaring television lights may be anathema to Sally Ride, but she will have to get used to them. When Challenger burst from its launching pad Saturday morning, Sally Ride was hurled into history as America's first woman in space.

Her jubilant mother, peering upward as the shuttle evaporated into a cloud, could joke, "Thank God for Gloria Steinem." The feminist leader was one of the VIPs gathered at the Kennedy Space Center.

In truth, Sally Ride would not be part of the shuttle crew now circling the earth if the feminist movement had never existed.

Of course, women were included in the U. S. space program from its beginning: as secretaries, assembly workers, even as cooks preparing zero-gravity meals. Eventually, the National Aeronautics and Space Administration hired a female engineer to work at Houston Mission Control.

But the growing women's movement inspired confidence in thousands of young Sally Rides to become all that they could be. Settling for less was not good enough.

In Astronaut Ride's case, it was a decision to become a scientific expert in laser physics, a subject long dominated by men. And then, in 1977, after getting her doctorate from Stanford, to reach for the stars.

By that time, NASA, through a combination of scientific study and political pressure, had decided that women could play a role in space.

Sally Ride and seven other talented young women scheduled for future space missions had to endure and master the arduous training faced by all astronauts. They earned the respect of their male colleagues.

For now, Ms. Ride is the center of attention. Until women become commonplace on space flights, they will face a flood of media coverage and insatiable public interest.

Sally Ride, who's not happy with all the hoopla, says it's time people realize that women can do any job they want to do.

America's first woman in space will help to kill a lot of stereotypes. And over thousands of little girls' beds, next to the Star Wars poster, will be a picture of Sally Ride.

The Kansas City Times

Kansas City, MO, June 27, 1983

It was such an anticlimactic ending to this most spectacularly successful space shuttle flight yet. On this seventh mission it had been planned to land the Challenger orbiter at Cape Canaveral, Fla., from whence it was launched six days earlier, in an effort to reduce the turnaround time before Challenger's next flight. But clouds and fog at the Cape forced a cautionary wave-off to Edwards Air Force Base in California, where all but one previous shuttle has landed. Now Challenger must again get back to the Kennedy Space Center bolted piggyback atop a 747.

The last-minute change was a disappointment to thousands, including government dignitaries, assembled at the Florida landing strip to welcome back the five-member crew of astronauts, including America's first woman in space, Mission Specialist Sally K. Ride. But it was even more frustrating for NASA officials to whom a shorter turnaround time by the reusable shuttle craft is critical to its eventual financial break-even status between costs and revenues from launching fees. The shortest interval so far has been 70 days and NASA would like to get that down to 14 days.

But at least on Challenger's second time up all else went smoothly, from launch through the orbiting of two satellites from the cargo bay, various on-board scientific experiments and the deployment and recovery of a test satellite using Challenger's 50-foot robotic arm. Ms. Ride proved to be a skilled workhorse on the numerous mission chores.

Now the shuttle flight schedule will be stepped up as planned, but NASA will also be looking ahead to other projects such as the space telescope to expand our view of the universe, radar mapping of Venus and other probes of Mars and Saturn. But the space agency's fondest hope, despite the funding obstacles, is for a permanent orbiting space station to serve, not only as a shuttle depot, but as a research center, manufacturing facility, observation point and a low-energy launch site for missions to the moon and elsewhere. It is the logical next extension of the shuttle space transportation service.

The Dispatch

Columbus, OH, June 16, 1983

HOWEVER MUCH she might wish to be just one of the crew, 32-year-old Sally Ride can't escape a measure of star treatment in her role on the seventh U.S. space shuttle flight.

She is the first U.S. woman to make a trip into space and one of only eight selected for astronaut training by the National Aeronautics and Space Administration.

Second, if only coincidentally, her scheduled blastoff Saturday comes exactly 20 years after the June 16-18 solo orbital flight of Russia's Valentina Tereshkova aboard a *Vostok* spacecraft. Tereshkova held the "only woman" title until last summer when Soviet cosmonaut Svetlana Savitskaya traveled with two male companions to a link-up with the *Salyut 7* space station.

Also, although Savitskaya, a seasoned test pilot, reportedly performed well on her mission, rumors have persisted that Tereshkova, suffering through a highly propagandized flight for which she was poorly prepared, may have slowed the progress of women in space programs. Inevitably, Sally Ride's performance will draw special scrutiny as she carries out tasks assigned to her on the shuttle mission.

Finally, Ride will be aboard the most glamorous vehicle around — the space-tested *Challenger*. The first flight of the giant shuttle and the five logged by its sister craft, *Columbia*, have rekindled the dreams of Americans, men and women, who look to space as the new travel route to both pleasure and profit.

The Commercial Appeal

Memphis, TN, June 22, 1983

IT MAY be no more than a minor footnote in history that at the same time Sally Ride, America's first woman astronaut, was preparing for her space mission, the Southern Baptist Convention was debating whether women were worthy of ordination.

But minor footnotes are the mortar in which history is set, and they frequently tell us more about the inner workings of a society than the carefully crafted tomes they inhabit as stowaways.

Judging from the tone of the publicity generated by NASA about Sally Ride's trip into space — and by the gushing response of the news media — the 32-year-old astrophysicist has herself been depicted as a historical stowaway. It's as if she, among all women in this country, has become the first to *qualify* for a space ride . . . the first to have the *right stuff.*

Is Sally Ride extraordinary?

Yes, as a human being, for she possesses those attributes valued most by society — intelligence, good humor, courage and persistence. But those qualities are extraordinary among both men and women. To single Sally Ride out as a extraordinary woman is to miss the point. There are any number of women in this country who are qualified to participate in the space program. They have been excluded not because they have not excelled in aviation, physics, medicine or any of the other related disciplines. They have been excluded because the space program was originally viewed as an extension of the military, an exclusive man's world of unchartered frontiers.

"The men had it!" goes the book jacket blurb on Tom Wolfe's chronicle of the early years of the space program, "The Right Stuff." "America's heroes . . . the first flyguys in space . . . putting their lives on the line day after day!"

And the women?

"The women had it! While Mr. Wonderful was aloft, it truly lacerated one's heart that the Hero's Wife, down on the ground, had to perform with the whole world watching . . ."

What she had to do, of course, was go before the television cameras and radiate Miss America smiles, give her favorate recipes and swoon convincingly when asked what it was like to be the wife of Mr. Wonderful.

America expected that of her. America was wrong.

AS SALLY RIDE has proved with good-natured ease, women have a place in the space program, other than that of the gushing wife of Mr. Wonderful. Looking back, it is apparent that too much emphasis was put on recruiting astronauts from the ranks of the military, where women, until recently, were not allowed to get flight training, just as too much emphasis was put on recruiting test pilots with the "right stuff."

Women have a long and distinguished history in aviation in this country, a history that has matched the male effort every step of the way in daring, skill and innovation. Women deserve a better break in the space program, and if the public discards the public relations hoopla about Sally Ride long enough to put her contribution in perspective, maybe, just maybe, they'll get it.

The Houston Post

Houston, TX, June 22, 1983

Old cliches never die. They refuse to fade away. And old perceptions cling to their deeply cut grooves. Sally Ride has a Ph.D. in astrophysics and has been in astronaut training since 1978. But when her turn came to board the space shuttle Challenger, she was bombarded by the media with the same silly questions they asked the first WACs almost 40 years ago.

She was asked if she would burst into tears should anything go wrong. Asked what kind underwear she would wear. Asked if she were taking perfume and lipstick with her. Though Ride is childless and only recently married to a fellow astronaut, one syndicated cartoonist drew her suited up, at the shuttle controls, murmuring that she would have to drive the afternoon car pool and see to the kids before the liftoff. This is the persistent "mom" syndrome.

The London Sunday Times referred to her as "America's first but the world's third spaceperson." Men, presumably, are not persons. Only women are.

At 32, Ride has an unusual breadth of interests, from tennis to Shakespeare. She graduated from Stanford University with a B.S. in physics and a B.A. in English. In graduate school she turned to X-ray astronomy and free-electron lasers. She was chosen to be an astronaut because her research interested the National Aeronautics and Space Administration. It is believed that the free-electron laser may prove to be an efficient way of transmitting energy in space.

Sally Ride was chosen in competition against 8,370 other applicants and easily passed the entrance examinations and psychological testing. While in astronaut training, she earned her pilot's license and a commendation from her colleagues as being "very cool — a cool operator." Her reaction to the silly questions was, indeed, cool: "It's too bad our society isn't further along." What did they ask Mme. Curie?

Rockford Register Star

Rockford, IL, June 22, 1983

We're bursting with pride over Sally Ride — just as we are for all the astronauts who are conquering space. But we're not proud of the national media's coverage of the first American woman in space.

Those breathless accounts of what kind of makeup Sally Ride took — or didn't take — into space make us gag. We can't remember anybody reporting the contents of the other astronauts' shaving kits with such enthusiasm.

Good grief, the woman is a physicist. We don't give a darn what perfume she uses, or about the color of her elastic hair bands.

For all the news media likes to think of itself as non-sexist, this kind of drivel is sexism at its worst.

ALBUQUERQUE JOURNAL

Albuquerque, NM, June 18, 1983

Today, the third woman in the brief history of man's Space Age soars into orbit. For the first time, the woman is an American, astronaut Sally Ride.

Ms. Ride is the first of eight American women astronauts trained for space shuttle missions. Judith A. Resnik is scheduled to be the second American woman in space on a flight scheduled next March.

The National Aeronautics and Space Administration has been an equal opportunity employer since 1978 when six women, including Ms. Ride, were chosen from among thousands of candidates who wanted to be astronauts. Earlier NASA requirements for astronauts and the limited availability of seats aboard U.S. spacecraft contributed to the delay in sending a woman into space.

That all ends today with Sally Ride.

Safe journey, Challenger.

Challenger Completes 3rd Flight; Crew Includes First U.S. Black

The space shuttle Challenger completed its third flight Sept. 5, 1983. Lt. Gen. James Abrahamson, associate administrator of NASA, Sept. 5 hailed the flight as "a fabulous mission, the cleanest mission we've had yet. The crew looks good and the spacecraft looks at its very best." Although the launching and the landing took place at night, a first on both counts, the milestone that perhaps drew the most attention was that the mission marked the first time a U.S. black astronaut had gone into space. He was Guion S. Bluford Jr., 40, a lieutenant colonel in the Air Force with a Ph.D. in aerospace engineering. He served as a mission specialist aboard the Challenger. Bluford was not the first black to go into space. He was preceded by Arnaldo Tamayo Mendez, a Cuban cosmonaut who took part in a Soviet space mission in 1980. The other four astronauts in the crew were Navy Capt. Richard Truly, the flight commander, Navy Cmdr. Daniel Brandenstein, the pilot for the mission, Dr. William Thornton, a physician and mission specialist. Thornton, at 54, was the oldest person to fly in space. The astronauts Aug. 31 deployed Insat-1B, a combination weather-communications satellite built for India by the Ford Aerospace and Communications Corp. The satellite was designed to go into an orbit that would keep it stationary at 22,300 miles above the equator at a point in the Indian Ocean. The night launch of the shuttle had been necessitated by the orbital mechanics involved in putting the satellite in this orbit. Insat-1B was equipped to relay television transmission and long-distance telephone calls. On Sept. 4 the satellite's solar gear, designed to power its operations, failed to open fully in response to a radio command. Indian scientists said Sept. 5 that the satellite was obtaining enough power at a minimal level and there was no danger in losing the satellite. Further attempts would be made to unjam the solar gear, they said. The astronauts used a dummy payload weighing 7,640 pounds to carry out further tests of the robot arm. They also conducted communications tests with the TRDS satellite launched in April. But there were problems with these tests, apparently mostly stemming from hitches with the computers at the ground base in New Mexico. Experiments were also carried out in producing pharmaceutical products in the weightless conditions of space.

The Boston Globe

Boston, MA, August 31, 1983

Lt. Col. Guion Bluford, a mission specialist on the space shuttle Challenger, does not want to be known as America's first black astronaut. Unfortunately, in a nation where race, gender and age define rather than serve only as identifying characteristics, he has no choice.

Just as Sally Ride will be remembered as the first woman astronaut on the space shuttle in June, Bluford will hold a similar symbolic place in history.

His mild objections to that status are understandable. Being first carries a special onus, a common and unpleasant burden for most. It prompts intensive scrutiny, eliminates privacy, often isolates and sometimes detracts from the personal achievement.

A generation ago, when being first was much more common for minorities, the pioneers answered the same questions. What did black people want? Would they rather be white? Many answered with a patience that originated from past parental instructions to be on their best behavior. Because they might be the only black experience for many whites, the burden was on them to provide a positive experience for the good of other blacks.

Now, years of achievements should have silenced those questions. But pioneers still encounter the inquiries, the most painful, "Did you get your job because you're black, or a woman?" That attack on credentials is the biggest detriment to being among a small elite, the achievers who pave the way.

Bluford, 40, an Air Force officer and former combat pilot who holds a doctorate in aerospace engineering, is one of four black astronauts among NASA's 78 astronauts. Another black astronaut, Ronald McNair, a civilian physicist, was active in Cambridge while earning a doctorate in laser and molecular research at MIT. Five percent of the professional staff at the Johnson Space Center in Houston are black, and seven percent of NASA's total 22,000 employees.

"This is someone who had earned the mission.... The people who have allowed him to make this mission are the ones that have passed the test," said Bill Cosby, who was at the launch.

By being first, Bluford breaks more than the racial barrier at NASA. He breaks some other stereotypical barriers as well. He grew up in a middle-class family in Philadelphia; his father was a mechanical engineer and his mother a teacher. Also, he is not a typical hero for whom everything came easy, according to an interview with a younger brother. Instead, Bluford had to persevere in tough situations.

He has acknowledged his history-making role but only "in the sense that I'm opening the doors for other people behind me. I recognize the importance of that, but I also recognize I'll be one of the many others who will be flying in space."

Were he not on this shuttle, more attention would have focused on another astronaut, Dr. William E. Thornton, a 54-year-old physician, who is the oldest American to travel in space.

In a more perfect America, this Challenger voyage would be remembered for a different first, the first nocturnal launch, the liftoff that dazzled for miles.

In a nation less conscious of race, gender or age, space travelers Bluford, Ride and Thornton would still stand out as American heroes, but no more so than other astronauts. That is certainly a goal worth pursuing.

Roanoke Times & World-News

Roanoke, VA, August 31, 1983

NASA, to its credit, has a sense of the dramatic. Every time Americans seem on the verge of being bored by the relatively uneventful flights of the shuttle vehicles, it comes up with some variations.

Yesterday's launch will go down in history reading something like footnotes in a sports book: First nighttime launch (which provided a spectacular liftoff), first night landing, oldest American to fly in space (William Thornton, 54), first black American in space, first live-cell samples aboard an American space vehicle, first 7,600-pound barbell in space (for "exercising" Challenger's robot arm).

And if all that doesn't have your heart pounding, Challenger is also carrying 260,000 stamped envelopes — the better for selling them on earth at a tidy profit — and is putting a satellite in orbit for India.

The most interesting aspect of the flight may be the character of the black mission specialist, 40-year-old Guion Bluford, who, like female astronaut Sally Ride, downplays being first.

Bluford, indeed, may have encountered fewer obstacles than Ride to get where he is. Son of a mechanical engineer and a teacher, he was touched only indirectly by the civil rights movement. "If I had any obstacles," he has said, "they were self-made."

There is an economic footnote to this flight that indicates how far the shuttle program has to go before it is a healthy economic enterprise: India is paying NASA $8.3 million to deploy the satellite, but NASA spends an average of $200 million on each flight. One small step . . .

Post-Tribune

Gary, IN, September 2, 1983

It was a spectacular launch of the space shuttle Challenger — with the dazzling light brightening the nighttime sky. It was also a perfect launch, but that is nothing new for the shuttle, which seems to have developed a knack for leaving and returning to Earth with precision.

There also are several "firsts" connected with this mission. Some were too long in coming; others are possible only because of recent technological advancements. Looking forward while being backward in many ways is an irony this country has yet to conquer.

Air Force Lt. Col. Guion S. Bluford Jr. is the first American black to travel into space. As when Sally Ride became the first American woman in space earlier this year, this first produces a bittersweet feeling. The pride is there, but so is the hurt of waiting so long.

But, beyond the shadow of injustice, the spirits of all blacks should be soaring as high as Bluford is. He will be a role model for young blacks for generations to come.

As Detroit lawyer Dennis Archer, president of the black National Bar Association, said: "It gives black children an opportunity to aspire to new horizons."

That may be Bluford's most important legacy.

Another first involved Dr. William E. Thornton, a civilian who at 54 is the oldest American to go into space. That should fire the morale of the middle-age population into an ego orbit.

It was also the first night launch of the shuttle and the trip is scheduled to end in the first night landing. Only 1½ hours after blasting off, the Challenger, for the first time, communicated with Earth via NASA's new $100 million tracking satellite, bypassing ground stations across the country. (Evenually the system will provide worldwide communications for the shuttle, plus serving as a switchboard for orbiting satellites.)

The Challenger's space medicine machine, being used this trip for an experiment in making pure drugs, is being used for the first time on live cells. The experiment involves purifying pancreatic cells from dogs.

Whatever the firsts, let's hope that as this shuttle soars, it will lift Earth's people into closer harmony.

THE MILWAUKEE JOURNAL

Milwaukee, WI, September 2, 1983

How quickly the extraordinary is absorbed into the commonplace. Five Americans tool around in a giant space shuttle 184 miles above the earth, performing scientific experiments. Yet while it's still news, we all pretty much accept it as part of our everyday lives, like telephones and automobiles.

People tend to forget that not many years ago there were no booster rockets, skylabs or shuttles, or any kind of satellites. Now, some scientists worry about debris cluttering the cosmos. Folks almost expect the reality of space to be true to space fiction. "We saw it in 'Star Wars.' Why can't we buy it in stores?"

What the five-man crew of the space shuttle Challenger is doing is mind-boggling enough. By the time the astronauts land before dawn on Labor Day, they will have launched a satellite for India from the cargo bay of the shuttle, tested a new communications system, practiced with a mechanical arm designed to manipulate major payloads, and performed other out-of-this-world chores.

One of the astronauts, Guion Bluford, is the first black to travel in space, a social milestone. Another astronaut (Wisconsin's third man in space) is Daniel Brandenstein, a 40-year-old naval pilot who went to Watertown High School and the University of Wisconsin in Madison. There he is: Once the kid next door; now on TV, munching a carrot in marvelous weightlessness.

Maybe it's inevitable that we earthlings should become accustomed to the miracles of space. But it's also sad that the thrill and the wonderment should fade. From our down-to-earth vantage point, we raise a glass of congratulations to the Challenger crew and wish it happy landings.

THE SUN

Baltimore, MD, August 31, 1983

One of the major jobs of astronauts aboard the current flight of the U.S. shuttle Challenger, launched spectacularly in early morning darkness yesterday, is to check out operation of a Tracking and Data Relay Satellite (TDRS) that NASA orbited earlier this year. When more of these satellites are added to the system, they will put shuttle flights in constant contact with all points on earth — and eliminate cumbersome ground communications stations which result in long "blackouts" in conversations between earth and orbiting vehicles. Even the single TDRS satellite has improved communications, NASA reported yesterday.

President Reagan yesterday noted "the first ascent of a black American into space" on this flight of the Challenger. His reference was to Guion Bluford, the Air Force lieutenant colonel who flew 144 combat missions in Vietnam and who is aboard the space shuttle. But celebration has been muted; the clear consensus is that blacks have so thoroughly established their competence in such a variety of technological and other complex and difficult tasks that for one to fly on a space mission is not particularly startling.

An important feature of this week's flight is further advanced testing of a promising system for electrically separating organic compounds, an activity that is immensely more efficient in gravity-free outer space. If commercial use of this system to purify now-expensive pharmaceutical products is possible — and increasingly it appears it will be — that alone might pay many of the development costs of the shuttle. The delta-winged craft became almost an object of derision during its development because of delays and cost overruns, but now it has begun to prove itself.

The Salt Lake Tribune

Salt Lake City, UT, August 31, 1983

When the late President John F. Kennedy put America seriously into the space business with the objective of placing an American on the moon by the "end of this decade" the nation's space program became geared to "proving something."

With the Apollo program, which did put a man on the moon by the end of "the decade" — July 20, 1969, America's space program was used to prove the United States was technologically more advanced and capable than the Soviet Union. After all, the U.S. space program got its greatest support only after the Soviets launched their sputnik, widely regarded as an affront to American ingenuity and enterprise.

When America's eighth space shuttle mission went into space early Tuesday with the spectacular night time launch of Challenger there still lingered about that mission an aura of "proving something." That aura also attached to the seventh space shuttle mission which concluded about two months ago.

Tuesday's launched included in its five-member crew Guion Bluford, the first black American in space. The seventh space shuttle trip two months ago included in the crew Sally Ride, the first American woman in space.

In these two shuttle missions the fact that these two people were on board became so important that the missions they were on took on a secondary significance. More attention was directed to the fact that a woman was going into space and that a black man was doing likewise than to what was to be accomplished in space by these two people and their fellow crewmen.

In retrospect, and possibly with a bit of cynicism, because it is all too obvious that both Ms. Ride and Col. Bluford are eminently qualified to fulfill their assignments aboard Challenger, it becomes distressingly easy to wonder whether these two people won their trips into space to again "prove" something other than strictly scientific and engineering facts; that they were sent to space as evidence of just how fair, equitable and just the American government can be.

That is probably unjustly attributing dark-side political motives to the selection of both Ms. Ride and Col. Bluford for their Challenger roles. Yet, the fact that their selections generated so much publicity that singularly concentrated on Ms. Ride's gender and Col. Bluford's race underscores one ugly fact of modern American society; it has yet to become a place where someone is judged solely on ability to perform, rather than irrelevant factors like race and sex.

THE DAILY HERALD

Biloxi, MS, September 2, 1983

The nation's eyes fixed on *Challenger* as it blazed into space Tuesday morning, but by the time the evening network televisions news came on, the shuttle had already been downgraded to the final item of the *ABC* news agenda. By the next morning, it had lost front page exposure and was consigned to the 7th page of the second section of *The Sun* and the 12th page of the third section of the afternoon *The Daily Herald.*

News is perishable, but the shuttle's coverage is itself a commentary of the times, of the rapidity with which any new space flight loses its gee-whiz quality, perhaps even of the jaded attitude of Americans, whose attention span continues to shrink even when the subject is of long-term importance, as space exploration is.

Challenger and her crew are pioneers in the sense that the crew includes the oldest astronaut, Dr. William Thornton, 54, and the first black, Lt. Col. Guion Bluford. They were the first to launch in darkness and will be the first to return in darkness. The work they accomplish in their cramped, gravity-free quarters, will generate significant contribution's to man's bank of knowledge. As this is written, *Challenger's* men and machines have performed with almost perfect precision.

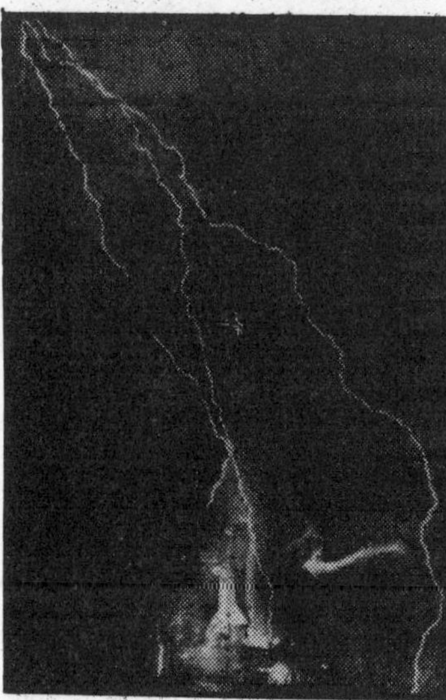

Even the storms which sent lightning bolts dancing menacingly across the pre-launch skies seemed to move obediently away from the launch pad as though directed by a NASA computer. The 17-minute launch delay seemed inconsequential.

The shuttle's prime mission was accomplished uneventfully on the second day with the placement into orbit of a communications satellite that will beam television and telephone calls across a million square miles of India. Other missions include tests of the shuttle's mechanical arm and science and technology experiments.

These latter missions—the tests and experiments, and the experience and knowledge NASA is accumulating with each space flight—comprise the justification for space exploration. They are not the gee-whiz stuff of which newspaper headlines and television bulletins are made, but they are the stuff from which, ultimately, untold benefits will accrue to the United States and to mankind.

The Star-Ledger

Newark, NJ, September 7, 1983

Perhaps the best tribute to the skills our nation has developed in space travel is the fact that missions that a decade or two ago would have stirred the world to cheers and frenzy seem today to many Americans to be almost routine. With attention centered on the horror of the Soviet downing of a passenger plane and on further U.S. casualties in Lebanon, the latest remarkable flight of the space shuttle Challenger failed to attract the normal complement of attention usually reserved for such cosmic feats.

The low-key approach to what was described as the smoothest space shuttle flight yet was accentuated by the decision to have the spaceship leave Earth in the dark and return at the same hour. Even the added feature that, for the first time, a member of the crew was a black man, Guion Bluford Jr., did not add a measurable sense of drama. Black participation in major American achievements has also become routinely accepted.

Uneventful it may have been, but the Challenger mission was nevertheless significant and will add appreciably to our Space Age knowhow. The spaceship traversed the globe 97 times in six days, but came back with less signs of wear and tear than a family automobile that has been taken on a Sunday outing. It was a tribute to the durability of this remarkable spaceship and its ability to operate under virtually any conditions.

In keeping with what has become standard practice, the spaceship crew had some company on its journey—six white rats carried in a special cage. These so-called "astrorats" made the trip with as much ease as their human co-passengers.

The Challenger will get a break now as the veteran spaceship Columbia takes off on a new mission next month, carrying the Spacelab orbiting laboratory and a crew of six in a joint project with the European Space Laboratory. Perhaps the true meaning of the term Space Age is a time when space travel loses its mystery. If so, we are rapidly approaching that happy time.

The Chattanooga Times

Chattanooga, TN, August 27, 1983

The space shuttle Challenger's sorties into space are becoming almost commonplace, but the next shuttle voyage, like the last, will be an historic one from a social, if not a scientific, perspective. Sally Ride was the star of the last Challenger crew, drawing the international spotlight as America's first woman in space. Tuesday night, unless unforeseen difficulties delay the flight, the Challenger will carry Guy Bluford into history as the world's first black astronaut.

Mr. Bluford was selected, along with two other black men, in the 1978 roundup of astronauts which also brought the first women into the NASA program. He was 35 and seeking the experience of working in space, not the limelight. A person who guards his privacy and never an activist in the cause of civil rights, Mr. Bluford has been described as a "reluctant hero" in his role of breaking the racial barrier in space.

But a hero he will be; and an inspiration to young blacks, living proof that, with hard work, they too can come within reach of the stars. And despite his reticence, he seems perfectly cast for the role because Guy Bluford never had an easy time of it. He was just an average student; a former professor said he wouldn't even have remembered him had he not been the only black in the aerospace engineering program. What was not average were the desire and determination that have carried him, now, to historic achievements.

The Houston Post

Houston, TX, September 10, 1983

When the space shuttle orbiter Challenger touched down in the early morning darkness in the California desert, another milestone was marked in the approach to the commercial profitability of the program. The mission also advanced the cause of international cooperation. The main payload was a commercial satellite placed in orbit for the Indian government. The six-day space mission also contributed to future efficiency of shuttle flights, planned to average one a month next year.

NASA officials said they believe the latest flight demonstrated that shuttle orbiters can be routinely launched and landed at night. Another major part of the mission was devoted to testing the Tracking and Data Relay Satellite, which is part of a system designed to improve communications between NASA and spacecraft in orbit. The system will be needed when the ninth shuttle flight is scheduled to place a spacelab in orbit next month. It will be a joint project of NASA and the European Space Agency.

The spacelab mission will be the last shuttle flight this year and is to be performed by the initial orbiter, the Columbia, which has not been aloft in nearly a year. After that, the Columbia will be set aside for two years. It will become temporarily surplus after NASA receives a third shuttle orbiter, the Discovery, next month. Under present plans, Columbia will serve as a back-up orbiter until 1985, when the fourth space ship, the Atlantis, is due to join the fleet.

The growth of the number of orbiters should not lead to complacency. The space shuttle program is one of the few government activities that is being run like a business. Businesses have to grow to survive, and this one has foreign competition. In this instance the United States is far ahead of other nations that are launching satellites with unmanned rockets. To keep it that way, expansion plans need to be made now to meet anticipated demand.

NASA expects commercial and military demands for its space shuttle services to increase its schedule to 40 flights a year by mid-1985. Space officials argue that four orbiters will not be enough to handle business by the end of this decade and ask that a fifth orbiter be authorized soon.

The Birmingham News

Birmingham, AL, September 7, 1983

The spectacular night landing of the space shuttle Challenger was as impressive, in its way, as its light-and-thunder lift-off last week. Gliding in out of the darkness onto its floodlit landing site, the Challenger's pinpoint landing was like some ghost ship out of the night seeking its home.

As stirring as lift-off and landing were, however, what was accomplished during the mission proper was the true measure of the success of this latest shuttle flight. Operating at a near flawless level, the Challenger and its crew accomplished all that was asked of them.

By deploying a satellite for India, the shuttle program impressively demonstrated its launch-for-hire potential. In the further test of its robot-like "arm," it was shown that heavy weights can indeed be handled by the shuttle in space. And the variety of experiments conducted by crew members, including use of the Tracking and Data Relay Satellite deployed on the preceding flight, showed once again the continuing importance of the space program in developing new technologies.

Altogether, this shuttle flight must go down as one of the most successful space ventures ever recorded by the United States. The additional factors of having a black astronaut in space for the first time, and of having a medical doctor as part of the crew, were only so much icing for the cake.

Challenger, dubbed a "wonderful machine" by its crew, now gets a rest as its sister ship, the first shuttle, Columbia, readies for its flight next month. This, too, will be a ground-breaking mission, with a European research facility as the payload.

The continued success and expanding versatility of the Challenger and Columbia are almost enough to make one take the shuttle program for granted. That, however, would be a mistake. What we are witnessing is an epochal development of American space potential, and we should recognize it as such.

Rockford Register Star

Rockford, IL, September 2, 1983

How high may a black man go in America today? In linear terms, Guion "Guy" Bluford, aboard the shuttle Challenger, is orbiting 160 miles over the earth — about the same height most astronauts reach.

But Guy is the first black American to go that high — not just miles in space — but to the status of those in whom Americans place their greatest hopes and trust in a new frontier for humanity.

That puts the American black man's limits, at the least, as far as any American's. If that is tokenism, so be it. That token has now been paid and is as worthy of commemoration as America's first woman astronaut, Sally Ride.

Whatever their impact on minority and women's opportunities in America, these two new memberships in National Aeronautics and Space Administration's (NASA) elite operational ranks signify historic progress. The precedents have been set, the highest NASA opportunities now clearly are available for all — whatever race or sex — willing to work for them.

Before the dazzling, night-time blastoff of Challenger, Air Force flier and engineer Bluford modestly noted he was only one of four blacks in the astronaut program. The "first in space" designation could only go to one of them, and Bluford drew it.

So is was astronaut Bluford who gave the signal whirling into orbit the multimillion-dollar communications satellite for India. And, in the same morning, it was Bluford who received this message from President Ronald Reagan:

"Guy, congratulations. You, I think, are paving the way for many others and you are making it plain we are in an era of brotherhood here in our land, and you will serve as a role model for so many others and be so inspirational."

We suspect astronaut Bluford believes, as we do, that Americans have a long way to travel before reaching such heights as an "era of brotherhood." But we add our thanks to his chief's for showing the determination to get there.

The News and Courier

CHARLESTON EVENING POST

Charleston, SC, September 11, 1983

The U.S. space shuttle program rockets from success to success. The latest flight of the shuttle Challenger produced some varied and tangible achievements. The presence of Guion Bluford demonstrated, as the black astronaut said himself, what a truly integrated society can do. Then there was the problem-free launching of a communications-weather satellite for India; more successful tests of the manipulator arm, this time with an 8,000-pound dummy payload; another advance in communications through the TDRS relay satellite; and the first-ever night landing.

The National Aeronautics and Space Administration has made the shuttle a commercial success yet it is still hurting for funds for further space research. Despite the demand for the shuttle's communications satellite delivery service, NASA finds itself stymied and unable to take the next big step in space — a permanently manned space station.

It is extremely short-sighted to deny NASA the funds to go ahead with space research at a time when the administration has not only demonstrated beyond all reasonable doubt that the next stage in the ascent of man is into space as a permanent residence but has also shown that its programs have commercial possibilities. It is alarming to learn that in order to keep the shuttle program running, Columbia may have to be cannibalized to provide spare parts for Challenger.

Now that it has been shown that the space race offers considerable rewards to those who set the pace, it would be a mistake to allow NASA's first-rate team to be overtaken not only by the Soviet Union but by the Europeans. The next goal, a U.S. space station, must be set now and the funds that are needed to meet the challenge should be forthcoming — on practical grounds and not for prestige.

Columbia Carries Spacelab Aloft; Computer Hitch Delays Landing

The space shuttle Columbia landed safely at Edwards Air Force base in California Dec. 8, 1983 ending a 10-day flight in which the European-built Spacelab was used to carry out around 70 experiments in physics, astronomy and medicine. During the flight a decision was made to extend the mission a day to afford further time for experimentation. Then some hours before the shuttle was due to land problems developed with the craft's computers. A postponement of reentry was ordered while mission controllers on the ground attempted to work out the difficulties. The malfunctions did not readily yield to analysis, but it was decided to go ahead with the landing anyway, with provision made to switch on the backup computers if needed. The reentry went without incident, and the Columbia touched down safely about eight hours later than originally planned. Just as the wheels hit the landing, however, one of the balky computers again switched off.

The mission was notable for a number of firsts. It was the first time that the Spacelab was carried into orbit, and the first time the Tracking and Data Relay Satellite System (TRDS) was used operationally. The six-man crew was the largest lifted into space in a U.S. mission, and it included the first non-U.S. citizen to take part in an American space-flight. This was Ulf Merbold, a West German physicist. Merbold was one of two nonprofessional astronauts on the Columbia; the other was Byron Lichtenberg of the Massachusetts Institute of Technology.

Much of the activity during the mission was concentrated in the Spacelab, the portable science laboratory carried in the shuttle's cargo hold. Spacelab weighed 17 tons, was 23 feet long and had a diameter of 14 feet. It was built in West Germany for the European Space Agency and cost $1 billion. A number of the experiments dealt with materials science, seeking to take advantage of the weightless orbital conditions. Although there were some problems with some of the experimental apparatus most of the experiments proceeded as planned. The astronauts also conducted astronomical observations of stellar X-ray sources, studied the upper atmosphere and photographed the Earth's surface. NASA, in a mission summary said that "the performance of the total facility was considered a success."

Los Angeles Times

Los Angeles, CA, December 9, 1983

Columbia bullied its way back into Earth's atmosphere Thursday with a rich international cargo of knowledge snatched from space.

Astrophysicists, biologists, engineers, cartographers and the merely curious will ponder for months and years the data now secure in Spacelab, nestled in the space shuttle's hold, but the mission was remarkable even before Columbia blasted off 10 days earlier. For one thing, the European Space Agency built Spacelab, which has been in the works for 10 years, and will give it to the National Aeronautics and Space Administration in payment for the flight aboard Columbia. Scientists from several countries could for the first time communicate directly with the astronauts performing experiments that the scientists had designed. And for the first time a mission specialist on a shuttle flight was from another country, West Germany.

Experiments conducted by Columbia's crew gave outsiders a unique glimpse at the kinds of questions that fascinate the scientific descendants of Galileo, Newton and Darwin today. Some experiments provided answers; others failed, leaving secrets for later missions. One gathered information for scientists studying a so-called burster source that emits a tremendous flux of X-rays that can be measured only in space. Another experiment cast doubt on a Nobel Prize-winning theory about the workings of the inner ear; later crews will repeat the experiment in the spirit of scientific confirmation.

Some of the experiments have practical value. One tested space as an environment for growing silicon crystals for computer chips; others might lead to the production of exceptionally strong yet lightweight alloys. One, an attempt to discover why plants grow up instead of, for example, sideways, may have little practical value but would be nice to know.

The mission was not all deep thought and heavy research. Space travel still has its similarities to life on Earth—computers that go down, complicated machines that can be fixed with only the humble screwdriver, jammed cameras, noisy bathrooms, closet doors that warp, and things that go bump in the night.

One such bump shook the space glider and its six astronauts early Thursday when John Young, mission commander, fired Columbia's rocket engines to maneuver the craft into position to descend. When the craft shook, two crucial pieces of equipment failed—one of Columbia's computers and a navigation instrument. The smooth landing at Edwards Air Force Base was eight hours behind schedule.

The failures added to the suspense of the eighth shuttle landing, and provided a reminder that however many ways life in space is similar to life on Earth, it is very different. Weightlessness is the most obvious, with attendant problems such as how to sleep most comfortably in space or how to perform a chore that requires two hands when one hand is needed to hold onto something to keep the performer from floating away.

It also was a reminder that as long as there are questions to ask, practical and theoretical, there will be people seeking answers on Earth and in space, explorers like those working on Spacelab challenging old concepts and defining new ones. As surely as adventurers found America, the space wanderers are finding their own new world.

The Dispatch

Columbus, OH, December 13, 1983

The late-flight computer problems aboard the space shuttle *Columbia* caused a few anxious moments at the conclusion of its 10-day mission, but the delayed landing of the craft enabled many Americans who might have missed the scheduled touchdown to see it live on network newscasts. And it was remarkable how the landing of this amazing vehicle fit in, rather routinely, with the rest of the day's news.

In the past, potentially flight-threatening complications would have caused newspeople to drop whatever else they were doing and focus on the latest space drama. But such a reliable record has been created by the shuttle, its crews and its support personnel that once NASA says a problem is solved people believe it. This confidence is a tribute to all who have brought the space program to the position it enjoys today.

And the future seems even more adventuresome than the past. Next year, NASA is planning 10 shuttle flights that will carry 50 crew members into space. They will deploy more than a dozen satellites and start repairing those already in space that have developed problems. This will be done through space operations requiring that astronauts take space walks during which they will be unattached to the shuttle. The self-propelled space suits that will allow this activity will be tested for the first time in January.

The next flights will be launched long before the scientists in this country and Europe are finished analyzing the data from the many, around-the-clock experiments performed aboard the just-completed mission. In all, 14 countries contributed to 73 experiments that gathered an unprecedented amount of information.

The space program gets more ambitious and sucessful with every flight, and 1984 promises to be a banner year for NASA.

The Virginian-Pilot

Norfolk, VA, December 9, 1983

Even with the computer glitch that delayed Thursday's landing, the voyage of the shuttle Columbia was a triumph for the space program and for science.

And there were enough "firsts" to warrant that this flight wouldn't go down in the books as just another routine pickup-and-deliver mission. The flight:

● Carried the largest space crew ever, six astronauts.

● Included a West German astronaut, the first time a foreign crew member had ridden on an American spaceship. (The flight also marked the first major cooperative space operation between the 11-nation European Space Agency and the United States.).

● Saw retired Navy Capt. John W. Young, the shuttle's commander and pilot, break the record for trips into space. This was his sixth journey.

But the primary objective of the flight was to conduct experiments in the $1-billion Spacelab — the European-built, 23-foot-long laboratory that sat in Columbia's cargo bay. The astronauts ran through more than 70 experiments in Spacelab, and what they achieved in that confined chamber, made the flight a triumph.

Among other things, they demonstrated the potential of space manufacturing. Gravity makes all but impossible the development on earth of some of the extremely strong but light alloys that were whipped together in the Spacelab. Other tests showed that incredibly pure crystals can be grown in minimal gravity; this holds great promise for the nation's microelectronics industry.

The astronauts examined how plants and animals develop in weightlessness. They measured deuterium for the first time in the upper atmosphere — an experiment that will help scientists explain the mixing of gases in the little-understood region 25 to 60 miles above the earth. What happens in this belt affects how much heat reaches the earth, so science is more than a little interested in this zone.

In addition, the Columbia team disproved a venerable theory about the inner ear — the balance organ — that had been used to explain how the human body senses motion. This theory helped its author win a Nobel Prize nearly 70 years ago. If it was right, a common test for defects in the inner ear shouldn't have worked in space because of the effects of low gravity. To everyone's surprise, the test worked on one of the astronauts. Now medical science will have to figure out why.

This is just a sample of the rich trove of findings that will be pored over by scientists from 14 countries who prepared experiments for the Columbia mission.

These days, of course, we've all become somewhat blase about the space program. The 50,000 people who gathered at Cape Canaveral, Fla., to watch the Columbia lift off late last month comprised one of the smallest crowds ever to watch a manned launching. Up to 500,000 people have congregated in the past to watch a single fiery departure. Launches still make the front pages and the evening news. But after the astronauts make it safely into orbit, the shuttle stories tend to get buried under the avalanche of grim news of global conflict and chaos.

But perhaps that ought not to be surprising. Some of the more mundane and commercially oriented shuttle trips had that Buck Rogers-as-truck-driver feel about them, a far cry from the "The Right Stuff" pizazz. NASA tries diligently to build interest in each flight. It has sent the first woman, the first black and now the first European into space. But it's difficult to compete with the Middle East and Cabbage Patch dolls for public attention.

High-tech experiments and amazing scientific breakthroughs, unfortunately for NASA, aren't enough to reignite the public's fancy for space, strongest in the days when men with household names orbited in ridiculously tiny capsules or crunched their boots on lunar soil.

The Saginaw News

Saginaw, MI, December 4, 1983

The greatest danger of space flight may be that it's beginning to seem so routine — and we don't mean in the technical sense.

The human capacities that have enabled us to travel, not merely orbit, in space remain ever marvelous, of course. Each successful space shuttle mission is a reminder of that.

But the current mission illustrates another kind of human capacity: to work together in a common cause — when we can agree on one.

Consider Columbia's latest flight. The co-pilot is Brewster H. Shaw of Cass City in Michigan's Thumb, truly a case of a small-town boy going very, very far. One of his fellow crew members is Ulf Merbold of West Germany, representing the European Space Agency. The ship is carrying the $1 billion Spacelab research station, built by 10 European nations. Aboard are the flags of every United Nations member.

Remember when space was also seen in terms of a race — the alarm over Sputnik, the triumph over Apollo 11? The race is over. Everybody won.

The old cliche about why can't we do this or that if we can put a man on the moon can get tiresome. The point, though, is not really that we know how to go into space, but rather what we are doing with that knowledge. This shuttle is gathering information and conducting experiments for the benefit of all nations, all people.

In light of more earthly concerns, that is not routine at all. We should think more deeply why we can't apply the example to some other kinds of races, so we can hold our world together while we explore the universe.

But then, there are no sacred borders in space — yet.

Buffalo Evening News

Buffalo, NY, December 12, 1983

The six Spacelab scientists and astronauts are back on Earth with some two trillion pieces of scientific data after a successful 10-day mission that was the most ambitious scientific space project yet.

Several hundred scientists from 14 nations were involved in the 70 or more experiments conducted during the intensive mission. The Spacelab, which was carried in the cargo bay of the Columbia space shuttle, was built in Europe at a cost of some $1 billion. The international cooperation that marked the project was symbolized by the presence of a West German scientist in the crew.

Scientists are now eagerly going over the results of the experiments in the fields of astronomy, solar physics, liquid physics, atmospheric studies, biology and materials processing. One remarkable result was the disproving of a theory of the inner ear that won a Nobel Prize in 1914. This new knowledge may assist in the study of motion sickness in space, a problem for many astronauts.

As with past space experimentation, there are likely to be many unforeseen spin-offs for science, technology and industry. Many of the experiments were beyond the layman. One scientist was reported to be studying "a black hole in the process of gobbling up a star rotating around it."

The space-shuttle flights have become so routine that the focus is now on the on-board experiments rather than on the flight itself. But the liftoff is still a remarkable sight, and the view from space was, one scientist said, "spectacular and fantastic." He said "it would have been worth it to work for five more years just to fly for one week."

The success of Spacelab will tend to support the view of those scientists who have pressed for manned, as opposed to unmanned, space experimentation. The scientist-astronauts maintained an intensive schedule of experiments. For the first time, the scientists who developed the experiments were able to communicate directly with the crewmen in setting up the tests, checking results and trouble-shooting. The success of this mission will undoubtedly encourage efforts to establish larger, more permanent scientific laboratories in space.

TULSA WORLD

Tulsa, OK,
December 2, 1983

IT'S a refeshing change of pace to pick up the paper and read about NASA'S earth-orbiting shuttle Columbia.

Its six-member crew seems to be in excellent health and working hard at 12-hour shifts in the new 23-foot long pressurized Spacelab in the shuttle's cargo bay.

The nine-day flight which began Monday opens an exciting new avenue of cooperation between America and Europe. The laboratory was built by the European Space Agency and one of the scientists aboard is from West Germany. European scientists helped devise the more than 70 experiments to be conducted in space, and a team of 100 scientists from Western Europe, Canada, Japan and the United States is supervising the work from a special control room in the Johnson Space Center in Houston.

The flashy, high-drama experiments and space walks belonged to past flights, their lessons learned. Spacelab's sophisticated experiments move into the meat-and-potatoes world of how the human body adapts to weightlessness, precise studies of the physics and chemistry of the earth's upper atmosphere — particularly the density of gases such as carbon dioxide, ozone and water vapor.

Then scientists roll the lab over for some high-altitude astronomy.

As space shuttles move into the realm of methodical science, we broaden our resources immeasureably by including the expertise of international scientists.

And it opens the door for their citizens to feel the excitement of opening the universe — and maybe share the dream of ordinary folks visiting space some day.

The San Diego Union

San Diego, CA, November 29, 1983

The flawless launch yesterday of the space shuttle Columbia may have looked like all the others, but it represented another major step for mankind. The crew of six was the largest ever launched into space. The West German scientist aboard became the first person from a foreign country to orbit the earth in a U.S. spacecraft.

The $1 billion "space lab" in the Columbia's cargo bay symbolizes, more than anything so far, the opportunity for international cooperation on the space frontier. The cylindrical laboratory is a joint project of 11 European countries plus Japan, Canada, and the United States.

NASA clearly has mastered the technology and procedures for launching the space shuttle, maneuvering it in orbit, and flying it back to earth. But this science-oriented mission serves to remind us that we are still in a pioneering stage when it comes to using the shuttle for the purposes for which it was designed.

The four scientists in the Columbia crew will be working alternate 12-hour shifts in the space lab. They have to busy themselves around the clock because nine days in orbit is barely time to get through their assignment — 72 separate experiments employing 38 different scientific instruments. What we need is a permanent station in orbit where scientists need not work under the now-or-never pressures of a short space flight. There are long-term projects in astronomy, earth science, and other fields that await the orbiting of a permanent manned space station.

The space agency has been urging President Reagan to put a first-year appropriation for a space station project in his 1985 budget due to be sent to Congress early next year. We think he should do so, even if new spending programs are anathema around the White House.

NASA has offered a plan which would require an initial appropriation of $200 million for 1985 toward a total cost of about $8 billion by the time the space station goes into operation in 1992. The station would be constructed in space using modules carried aloft aboard the shuttle, which also would carry supplies to the station and provide transportation for its rotating personnel, a crew of six or eight.

The 1992 target for completion of the station is appealing to NASA because it would link the space project to the 500th anniversary of the discovery of America by Christopher Columbus, who is a sort of spiritual ancestor of this century's astronauts.

What recommends the space station project is not its value as a showcase but the fact that it is a logical next step in space. The shuttle, as its name implies, makes it relatively simple to transfer cargo and personnel between the earth and orbit. NASA says the station not only would serve as a base for space research and a workshop for satellites brought in for repair, but would be designed for future expansion to provide "port" services for private commercial spacecraft which are expected to make their appearance in space early in the next century.

The Soviet Union has put small space stations in orbit during the last decade, but has not kept them occupied permanently. The Russians are said to be designing their own space shuttle, which suggests they have in mind the construction of a larger, permanent space station. This simply adds one more reason why a permanent presence in space by the United States should become a national goal.

There is no reason why a decision to go ahead with a manned space station project should be feared by those scientists who are primarily interested in planetary exploration by unmanned spacecraft. The two fields of space activity would not be competing for funds in the near future, and the space station ultimately would make it possible to consider planetary missions now ruled out by problems of cost and logistics.

The scientists aboard the Columbia have a jam-packed schedule in order to make the most of their hours in orbit. Let's offer the next generation of space researchers a little more elbow room and one of those niceties we take for granted here on earth — an eight-hour day.

The Providence Journal

Providence, RI, November 30, 1983

Just when space shuttle launches seem to be getting routine, a new twist is added to remind us that each flight is achieving something new.

On Monday the shuttle Columbia roared into orbit with accustomed precision. For the first time, however, the shuttle carried aboard the compact and efficient Spacelab, equipped to perform the most ambitious schedule of scientific experiments ever tried in space. Spacelab, riding in Columbia's cargo bay, is loaded with enough experiments — 71 — to keep the scientists busy throughout the shuttle's nine-day, 145-orbit mission.

This week's mission can boast several other firsts: the largest crew of astronauts (six); the first foreign scientist aboard (Dr. Ulf Merbold of West Germany); and the first scientist-crew members who have had no astronaut training. Dr. Merbold is one; the other is a 1969 graduate of Brown University, Byron Lichtenberg, now a physics professor at the Massachusetts Institute of Technology. Loyal to his alma mater, Professor Lichtenberg is carrying along copies of rare relics from the Brown library, including a book by the Polish astronomer Copernicus and one of the earliest maps to depict the North American continent.

On this trip the experiments are the big attraction, and earthbound scientists await their results with interest. Spacelab was built by the European Space Agency as its contribution to the U.S. shuttle program. Its 38 instruments will be used to perform experiments on behalf of scientists from 14 countries. They propose to do detailed mapping, study the polar aurora, track distant stars, and study human physiology and the behavior of materials in space. After one day in orbit, four experiments already had been successfully completed.

Can it really be only 25 years since the United States sent its first little space vehicle into orbit?

The Oregonian

Portland, OR, December 9, 1983

It is satisfying and reassuring to learn that a master of spaceships and high technology also knows how to use a screwdriver.

Mission specialist Bob Parker, riding high above the Earth in the Spacelab aboard the shuttleship Columbia, picked up a screwdriver and fixed a high-speed tape machine that had broken down and for 11 hours defied remote efforts of Mission Control to get it started. The broken recorder was delaying some important experiments.

Maybe it should be less surprising that a high-tech astronaut can also run a screwdriver than that the lowly tool is still formidable in the high-flying world of space machines. It is like learning that a woman's hairpin could jump-start a mainframe computer.

SYRACUSE HERALD-JOURNAL

Syracuse, NY, December 12, 1983

The latest space shuttle mission was a scientific bonanza, but our most vivid memory stems from how well American astronauts met problems.

Computer and navigational malfunctions postponed the landing for seven hours Friday. Although astronauts quickly brought the equipment back into operation, ground control delayed the landing until it and the astronauts could make sure everything was in working order.

Flight Commander John Young kiddingly said he "turned to jelly" over the incident. Pilot Brewster Shaw was responsible for correcting the problems, Young said.

The commander's comments not withstanding, this piece of last-minute reprogramming again demonstrated the coolness under pressure Americans have come to expect from their astronauts.

▽ ▽

But, to our mind, more satisfying examples occurred earlier in the flight.

Even with all the 21st-century computer equipment aboard Columbia and supporting the mission from Houston, Yankee ingenuity and an old-fashioned screwdriver proved to be two of the mission's most important elements.

Early in the flight, a high-speed data recorder malfunctioned, jeopardizing the ability to store much of the billions of pieces of scientific information collected during the mission.

Dr. Robert Parker, one of four scientists aboard the shuttle, used a screwdriver to fix a reel on the recorder. Mission officials described him as a "repairman extraordinaire."

A few days later, Parker zipped himself into his sleeping bag — improvising a mini-darkroom to fix a camera used to test the potential for making maps from space. An important experiment was saved by common sense and a steady pair of hands.

▽ ▽

The incidents brought back memories of "The Right Stuff." The current movie describes how, early in the space program, scientists developing rockets considered astronauts nothing more than a "payload." Computers could handle the flying; monkeys could serve just as well as humans and were less trouble, the scientists believed.

Astronauts rebelled and became an integral part of all future missions.

Acknowledgment of man over technology was a stroke of genius, the wisdom of which has been ratified countless times. After all, can you imagine a computerprint on the moon or a robot whipping out a screwdriver from its handy tool kit?

The Evening Gazette

Worcester, MA, December 10, 1983

Spacelab is racking up some spectacular successes in scientific experiments. The payload potential is becoming obvious to laymen as well as the research community.

Oddly, the first companies to indicate a willingness to participate in Spacelab were European. One of the people aboard was Dr. Ulf Merbold, a West German scientist. He was working with British and European scientists on the ground in an unusual cooperation of 14 nations including the United States.

Those aboard Spacelab were enthusiastic about the research possibilities in astronomy, solar physics, atmospheric studies, biology and materials processing. Crystals, liquids and metals were being grown or changed or combined in new ways in the weightlessness of space, with all kinds of industrial applications possible.

The Spacelab travelers needed less sleep because they were not fighting gravity and exerting themselves. So they worked 12-hour shifts tending pots, pans, ovens, computers, electron beams, scopes and "gardens" and observing the Earth, the galaxies and each other, all the while recording data of all kinds. Does a fungus keep the same biological clock it has on Earth? Does a "black hole" gobble up stars? Why do some crew members get motion sickness? All these questions, experiments and data will have relevance to the future of life at home as well as in space.

Spacelab has proven its worth during the past week. Even though it had a few computer problems that delayed its landing for several hours, it was a great scientific success. It has fulfilled all the expectations of its creators, and more.

The Miami Herald

Miami, FL, December 6, 1983

HOW quickly the extraordinary becomes routine; so quickly, in fact, that the two contradictory attitudes seem to co-exist. How else could it be that an otherwise undistinguished Midwestern senator becomes a major Presidential candidate on the strength of his celebrity as an early astronaut, while most voters today could not name the crew of Spacelab?

Yet John Glenn's brief first orbit of the earth is history, while Spacelab's equally historic flight is happening now, today, at speeds and heights unfathomable from the earthbound perspective. The seemingly routine nature of the Spacelab project obscures the importance of the scientific and technological breakthrough that it represents.

Built in Europe, Spacelab is what its name implies: a working scientific laboratory in space. Made possible by the success of the reusable Columbia space shuttle, the 23-foot-long lab carries equipment for more than 70 experiments designed by researchers as far apart as Japan and Belgium.

Spacelab thus can assay gases in the Earth's upper atmosphere with an unprecedented accuracy. It is opening new windows into mankind's understanding of radiation, astronomy, the behavior of gases, celestial light, and other natural phenomena. So many experiments are facilitated by Spacelab's instruments, and so eager are Earth's scientists to conduct them, that the astronauts at times complained of pressure to work too fast and do too many things at once.

The pace aboard Spacelab marks a quantum leap in man's relationship to space. No longer is the space man an aberration, a larger-than-life hero, an object of intense curiosity. That first tentative step off the planet has been surpassed just as the heroic first explorations of Lewis and Clark into the interior of North America were surpassed.

For with Spacelab, man *works* in space — sometimes more than he'd prefer. Like the settlers who followed Lewis and Clark, the Spacelab crew draws on rare courage and preparation in order to expand man's place in the new territory, not just to see the place but to benefit, to profit, and, ultimately, to be at home on the new frontier.

The Hudson's Bay Company, which sent a generation of trappers to explore and mine a new continent, is today an urban chain of department stores and supermarkets. One wonders what new elements will be spawned as the descendants of Spacelab work their way into everyday living.

Astronauts Perform Spacewalk; Challenger Loses Two Satellites

The space shuttle Challenger February 3, 1984 began its fourth flight and the 10th mission of the shuttle program. The highlight of the mission came Feb. 7 when two U.S. astronauts became the first humans to fly freely in space, untethered to their spacecraft and propelled only by backpack jets. The successful spacewalk followed days of mishaps in the shuttle mission, including the loss of two satellites and a target balloon. Prospects for a flawless mission appeared good when the shuttle lifted off on schedule carrying a crew of five astronauts. The crew members were Vance Brand, the mission commander; Navy Cmdr. Robert Gibson, the pilot for the flight, and mission specialists Army Lieut. Col. Ronald McNair, Navy Capt. Bruce McCandless and Army Lieut. Col. Robert Stewart. McNair, a physicist, was the second American black in space.

Eight hours after take-off, when the shuttle was orbiting the Earth at a distance of 190 miles, its crew released Western Union Corp.'s Westar 6 communications satellite, manufactured by Hughes Aircraft Co. The satellite apparently misfired, strewing debris in its wake as it unsuccessfully attempted to reach its intended orbit 22,300 from Earth. A search for the remains of the satellite forced a two-day delay in the launching of a nearly identical second satellite, Indonesia's Palapa-B. The launching of the Indonesian satellite, also made by Hughes Aircraft, had been scheduled for Feb. 4. The Westar 6 was located Feb. 5 in an elliptical orbit not far beyond the shuttle. NASA officials said the satellite was unusable in its current orbit even if it had remained in working condition. Westar 6 could not be retrieved by the shuttle's grappling arm, NASA officials said, because satellites designed for orbits far beyond the shuttle's reach were not equipped with the necessary grappling hardware. The mission's second major setback occurred Feb. 5 when a target balloon launched by the shuttle exploded upon inflation. The balloon had been designed to serve as a target for practicing maneuvers to be used in a repair mission to a disabled satellite planned for April. The mission experienced a third major failure when the shuttle unsuccessfully launched the Palapa satellite Feb. 6. The loss of both satellites was attributed to a failure of the first stage rocket engine that was to have lifted them into their intended orbits. The satellites, valued at approximately $75 million each, had been insured for $100 million to $110 million, which covered replacement costs and lost revenue, as well as the loss of the satellites themselves. NASA officials said the shuttle itself was running smoothly, except for a malfunctioning toilet. The toilet mechanism reportedly had been a problem on every shuttle mission.

The troubled mission took a new turn Feb. 7, when McCandless and Stewart left the shuttle and floated in space without lifelines attaching them to the craft. The astronauts were driven by jet propulsion backpacks designed by McCandless who became the first human to float freely in space when he maneuvered out of the shuttle's cargo bay to a distance of 150 feet from the ship. As McCandless left the craft, the shuttle was orbiting 170 miles above the Pacific Ocean. McCandless flew beside the shuttle at 17,500 miles per hour, the same speed as the spacecraft.

Richmond Times-Dispatch

Richmond, VA, February 5, 1984

Though it was marred by some major disappointments, the latest voyage of the space shuttle Challenger was more a success than a failure. It accomplished the major goal of the mission, which was to test the ability of astronauts to move freely in space, untethered to their ship.

This Bruce McCandless and Robert Stewart did. They moved out on jet-propelled backpacks to prove that astronauts can leave their spacecraft and return safely to it.

This means that it will be possible for astronauts on a later mission to emerge from their spaceship, go out to get a damaged scientific satellite and bring it back to the ship for repairs. It would be far cheaper to work on the satellite there than to return it to Earth for repairs and then re-launch it. And it would be even more economical than building a new satellite at a cost of $280 million. Astronauts will undertake such a repair mission on Challenger's next flight.

The most dramatic disappointment of the latest flight was the loss of two communications satellites the Challenger crew carried aloft. Both satellites — one owned by Western Union, the other by Indonesia — disappeared into space soon after the crew deployed them. Officials emphasize that the astronauts were not at fault, but the experts do not know exactly what did cause the accidents. Obviously, they will have to solve this mystery before they attempt to put another such satellite into orbit.

In contrast to its troubled flight, Challenger's landing was a major triumph. For the first time, a space shuttle put down at its Cape Canaveral launching site. In the past, weather conditions have forced returning space shuttles to land elsewhere, costing the space agency a vast amount of extra money and time. So Challenger's perfect touchdown in Florida did much to assuage the disappointments of its voyage.

The Saginaw News

Saginaw, MI, February 16, 1986

For mankind, another great step...a leap...a...just what is it that people do when they're floating in the void? Free flight is as good as any term, we suppose. But it's apparent that the space program will defy earthly language as well as the highest frontier.

When Challenger shuttle astronauts Bruce McCandless and Robert Stewart left their craft behind, technically they became human satellites. But the freedom afforded by their jetpacks actually turned them into the first true space travelers, no longer bound to mother ship or mother earth.

That marvel was enough to make this shuttle flight another success. Satellites may stray, but they are dumb and inanimate, if expensive, objects. We humans know where we're going. Unless our vision fails us, that's farther and farther.

Can there be any doubt that some day we will travel to the planets? Well, yes. There are feet-on-the-ground realists who question President Reagan's proposal to build a manned space station, and not only for its projected $8 billion cost. They point out that almost anything humans could do in space can be done as well, or better, by remote control.

Scientifically, that is probably right. But science had little to do with the free-flight experiment. That was for our own satisfaction, a comic-book panel come to life. Not everything we do needs practical justification; we travel for pleasure as well as business, romance as well as revenue.

While space offers all sorts of technical potential, fulfilling our human dreams remains the driving motivation behind the program. We are embarked on what Challenger Co-pilot Robert "Hoot" Gibson termed a "terrific adventure." Let's not cut this trip short for lack of traveler's checks.

The Arizona Republic

Phoenix, AZ, February 10, 1984

IF the weather is favorable and the space shuttle *Challenger* lands literally like an airliner at Florida's Kennedy Space Center on Saturday, add another remarkable feat to a space program of remarkable feats.

Americans are justly proud of their space program. But they sometimes lapse into a casualness about the space program that borders on indifference.

Perhaps that's a price of successes that now seem almost routine. The going and coming of space shuttle flights, as well as other space adventures, in time may be treated with the same disregard as the routine arrivals and departures of commercial jetliners.

But surely even the most jaded cannot fail to be impressed when reflecting about how far the U.S. manned space program has gone in the relatively short span of about 23 years.

Rocketry, of course, goes back nearly 60 years in this country. Dr. Robert H. Goddard fired a liquid-fuel rocket on March 16, 1926, at Auburn, Mass., that traveled 184 feet in 2.5 seconds.

It wasn't until April 12, 1961, that manned space flight began, with the one-hour, 48-minute flight of Russian cosmonaut Yuri Gagarin.

For the United States, manned space flight began a month later on May 5, 1961, when astronaut Alan B. Shepard Jr. arced up into space on a suborbital flight from Florida.

The phenomenal achievements are obvious when one considers the contrasts between the primitive, passive nature of Shepard's flight in the cramped Mercury capsule and the extraordinary controlled activities of *Challenger* crewmen who routinely spent five to six hours flitting in and around their mother ship in jetpacks. The *Challenger* astronauts were, for all purposes, individual space vehicles capable of sustained flight away from the shuttle.

Despite the faulty orbits of two communications satellites launched from *Challenger*, other successes of this mission have profound implications for the future.

The astronauts have demonstrated that man now can cut free from orbiting space vehicles to perform specialized tasks — such as repairing damaged space vehicles or satellites, constructing space stations, and even commuting from one orbiting vehicle to another.

But there are literally tens of thousands of other U.S. space achievements, many of them lacking in drama, that have provided exotic advances in medicine, communications and aerospace technologies, photography, geophysical exploration, agriculture and life sciences.

Someday, the routine use of space for colonization may prove to be one of Earth's best defenses against nuclear devastation.

And it also may become a new gravity-free environment for the manufacture of medicines and other chemical-base compounds to spare mankind suffering and disease.

Just as it was difficult for the untutored observer to predict what would follow Alan Shepard's primitive flight 23 years ago, it is also difficult to imagine what more conquests the U.S. space program can make after the shuttle.

But, bet on it. There is drama ahead.

The Sacramento Bee

Sacramento, CA, February 11, 1984

It's not clear who or what should be blamed for the two failed satellite launchings in a row from the space shuttle Challenger. That will require further study. And even then, the conclusion may be only that the misfiring of both satellites' propulsion systems — sending each into a wrong, low, looping orbit that left it fully functioning but useless — was simply a fluke.

That shouldn't, however, be particularly shocking. Too many Americans may have forgotten it in the glow of one space program success after another, but it remains a fact of life that things do go wrong with mechanical equipment, and the fancier the technology the more there is to go wrong. Even in the space program, with its spectacular array of pretesting programs and fail-safe mechanisms — even in a case such as this, where the propulsion system was no longer considered experimental and had already been a success in several previous launchings — there is no such thing as a perfect technological control or predictability.

It is entirely appropriate to accept such risks in the space program, or in any other effort at technological advancement. Error and accident are, indeed, integral to exploration and development.

What is unacceptable is to act as if the risks aren't there, and that's a lesson the Challenger's two unexpected failures should have driven home. What makes no sense is to use inherently fragile technologies as if they were utterly dependable and put in their control the capacity to do such damage that the mistakes one ought to assume will happen can have catastrophic results. The lesson in the Challenger's back-to-back failures is not that there's anything wrong with the space program, but that there's something disastrously wrong with the assumption of technological perfectability that too many people thought we had learned from the space program.

Imagine if it hadn't been the communications satellites for Western Union and for Indonesia that had been misfired, but nuclear missiles, or one or two of the president's "Star Wars" space-launched weapons that inexplicably had been sent — fully functioning — into the wrong, low, looping orbit. Imagine leaving American troops dependent on the perfect functioning of high-tech tanks and weapons that can do so much more than conventional arms, but only when they work. Think, indeed, of any number of "technological fixes" that have been developed to protect us from mechanical and natural disasters, and then relied upon for safety while the extent of the potential disaster forfended is expanded astronomically.

Those are the scenarios that the Challenger's experience ought to conjure up — not of unacceptable technological incompetence, but of unwarranted technological faith.

Buffalo Evening News

Buffalo, NY, February 14, 1984

"Manned maneuvering unit" is what they call the jet backpacks that enabled astronauts to leave the space shuttle Challenger last week and float freely and wondrously in space.

But to any American child who read the "Buck Rogers" comic strips half a century ago, such space maneuvering, with a smaller but similar backpack, was entirely familiar. It has taken science a mere 50 years to catch up with the fantasies of science fiction.

Astronauts have left their spacecraft before, but until now they have always been tethered to them by lifelines. Last week, the world watched on television screens as white-suited astronauts flew away from the spacecraft into the blackness of space, becoming, in effect, tiny celestial objects on their own, traveling in their own orbits at 17,500 mph.

People nowadays have seen such amazing things that it is difficult to stir a sense of wonder, but many felt it on seeing these frail human beings in the hostile atmosphere of space, staring into an eternal darkness. There was that same sense of wonder when Neil Armstrong stepped down onto the surface of the moon and when TV cameras in 1968 first showed our own planet from space — in the words of Archibald MacLeish, "small and blue and beautiful in that eternal silence where it floats."

The untethered spacewalks of the Challenger astronauts are perhaps not that epochal in nature, but they capture the imagination because they starkly demonstrate the power of a lone human being, through courage and ingenuity, to conquer space. It is easy to understand the importance of this new achievement as a step toward working more routinely in space on such projects as building space stations.

The successful spacewalks helped to counteract some of the disappointment over two successive failures to launch communication satellites into orbit. Rocket failures put both into useless orbits and cast some doubt on the reliability of the space shuttle for such commercial operations.

WORCESTER TELEGRAM.
Worcester, MA,
February 9, 1984

Sixty-four years ago, when Dr. Robert Goddard was talking about space flight, The New York Times pompously explained in an editorial why such a notion was scientifically impossible.

What would The Times have said if Goddard had predicted that men in space suits would one day float free in the upper ether?

Well, it happened this week, while the world gaped. It was eerie to see astronauts Bruce McCandless and Robert L. Stewart drifting out of the cargo bay of the space shuttle into the dark void. There they were, two Americans 170 miles above the United States, in the strange realm where the vectors of Newtonian theories meet the principles of Einstein.

That is a realm of no wind, no sound, no blue sky. If they looked in one direction, they saw their mother, Earth. If they looked in another, they saw the limitless chasm of space, dotted with the stars, hundreds, thousands millions of light-years away. Although they were moving at the tremendous speeds necessary to neutralize gravity, they had little sense of motion.

Small wonder that The Times in 1920 found the idea hard to credit. It was difficult enough believing in it, even as one gazed.

Robert Goddard would not have been surprised; he foresaw all this, and probably much else. But for the average earthling who is afraid of heights, the whole thing had a touch of the unreal about it.

How far will man's restless drive and imagination take him? Is there no limit?

FALL RIVER Herald News
Fall River, MA, February 8, 1984

The photographs of two American astronauts floating 169 miles above the earthbrought home to the country more of the truly remarkable implications of its space enterprise.

The astronauts were moving without connective links to thê space shuttle, and were propellling as well as directing themselves with tiny jet thrusts of nitrogen gas.

Their exploit paves the way for future strolls in space that will presumably only be limited by the maximum amount of sustaining fuel that a person can carry in a jetpack.

These astronauts were walking on air, in a fashion that is bound to become commonplace in years to come, but right now has the effect of science fiction come true.

The space flight of which the strolls in space were a major part has been plagued by troubles, including the malfunctioning and loss of two satellites that were the property of Western Union and the Vietnamese government respectively.

Their loss, apart from the expense involved to their owners, may make other prospective users of the space shuttle for similar enterprises unwilling to take the risks entailed.

If so, this would certainly affect the commercial potential of the space ship and might indeed restrict the number of its future flights, depending on the appropriations the government is willing to make for this purpose.

In view of the mishaps to the two satellites, the success of the strolls in space is especially vital.

Not that the whole purpose of the daring feat of the two astronauts is simply public relations. Far from it. They were extremely practical pioneering ventures.

But there is no doubt that their spectacular nature demonstrated in a most graphic way how federal funds are being used to good purpose in the space flights of the shuttle ship.

All of them are paving the way, as the President has clearly indicated, for the future establishment of a manned station in space.

In view of the federal government's enormous deficits, the outlay of billions of dollars for such a purpose at this time is bound to be controversial.

It is thought by many to be basically military in purpose, but in fact, whatever its future use, its establishment would mark a decisive expansion of humanity's capacity to move beyond the limits of this planet.

It would mark another stage in the development of the most important human enterprise since the colonization of this hemisphere by the countries of western Europe.

Nothing more clearly indicates the position of leadership the United States occupies in the world today than its leadership in that enterprise.

The space strolls the astronauts took this week have dramatized that enterprise to millions of Americans whose approval of the appropriations for this purpose is urgently needed.

Over and above their intrinsic value, the walks in space had this vital symbolic importance as well.

They made us all see Buck Rogers as real.

And because they did, the American public may well give its support to the thrust forward of the space enterprise that the President has endorsed.

In a democracy that approval is vital.

The two strolling astronauts this week did a great deal to win it for the space enterprise as a whole.

They have shown us all a vision of the future, and it works.

The News and Courier
Charleston, SC, February 18, 1984

They called themselves "Flash Gordon and Buck Rogers," those two intrepid astronauts Bruce McCandless and Robert Stewart, who took the first, untethered spacewalk, free from that umbilical safety cord tying them to the spacecraft.

What they did in space is difficult to comprehend by those of us jailed here on earth by gravity. While it might have appeared on television that they were barely moving in relationship to the spacecraft, they and Challenger were hurtling along at 17,500 miles an hour in empty space 175 miles above the earth. They accomplished their manuevers with quick blasts from tiny jets built into their backpacks, devices Bruce McCandless has spent years helping to perfect. The spacewalk has paved the way for future astronauts to perform maintenance work and repairs on other satellites.

To cap the success side of the story, Challenger returned on time to make the first landing at Cape Canaveral, thereby cutting six to eight days off the shuttle turnaround time by eliminating the costly, cross-country ferry from California on the back of a 747 jet. Also, Challenger returned in better shape than previous shuttles, meaning it will be ready faster for its return to space.

All was not roses for the Challenger mission. Two communications satellites that were delivered into orbit have apparently just added to space debris because of failure of their rocket motors. Also the launching of a plastic balloon to be used for manuevers by the spacecraft went awry.

But that shouldn't detract from what Bruce McCandless described. Recalling Neil Armstrong's words as the first man on the moon, "That may have been one small step for Neil but it's a heck of a big leap for me." It's also a big leap toward fulfilling the space programs President Reagan has set in motion — missions that would see astronauts working among the stars.

The Star-Ledger

Newark, NJ, February 14, 1984

Between a perfect blastoff and a flawless return, NASA's latest space venture recorded some giddy highs and depressing lows.

Another historic achievement was recorded with man's first free flights in space, the exhilarating demonstration that confirmed the feasibility of using jetpacks to maneuver untethered in the weightlessness of the vast vacuum beyond Earth's atmosphere.

There was, of course, more to the experiment than a footnote in the annals of space exploration that Bruce McCandless and Robert Stewart were the first human beings to jet around without being tied to an orbiting craft. The demonstration showed that astronauts are ready for April's scheduled satellite rescue and repair mission, the next daring milestone for the shuttle program.

Another significant first assured permanent recognition for this eight-day flight by the shuttleship Challenger. That was the landing at Cape Canaveral. Never before had a mission ended at the same base where it had started. What this does is to shorten the turnaround time so the craft can meet the April 4 launching date—nine days faster than the best previous turnabout time.

Everything, however, did not come up rose petals for Flight 10 of the shuttle program. There were some nasty thorns as well.

Two exceedingly expensive communications satellites were launched by Challenger and were lost when they failed to go into their proper orbit—a serious problem that must be corrected if the program is to realize its commercial objective.

A second major setback occurred when a target balloon burst, canceling a planned rendezvous rehearsal. Finally, there was disappointment over the malfunctioning of the wrist on the robot arm of the spacecraft, another problem that will have to be corrected before the next mission.

Once the bugs are diagnosed and corrected, as they surely will be, the pace of shuttle flights is destined to quicken, until the missions become quite frequent and routine. NASA's goal is to operate four shuttle-ships on twice-a-month launchings by 1988—an ambitious schedule that will have to be maintained just to service paying clients who want satellites placed in orbit for commercial or research purposes.

Shuttle astronauts will also take on the worthwhile task of assembling the permanent space station, another Buck Rogers concept that President Reagan wants NASA to tackle as the next major thrust in penetrating the mysteries of the universe, shrinking the boundaries of the unknown and increasing our knowledge of the world around our planet.

The Dispatch

Columbus, OH, February 16, 1984

The space shuttle *Challenger* returned safely to Earth last week but not before a few mishaps shook an admiring world from its overconfidence about the predictability of space travel. Fortunately, none of the mishaps was life-threatening, but they did underscore the almost-forgotten fact that things could go wrong despite the beauty and the precision of man's most wonderful machines.

The shuttle coasted to a stop at the Kennedy Space Center in Florida, just a few miles from where it lifted off eight days earlier. The landing capped a 2.9 million-mile journey that included the first space walks by astronauts unconnected to a base ship. The successful testing of new self-propelled space suits was a highlight of the trip. The suits will be used in the near future to begin the recovery and repair of malfunctioning and non-functioning satellites.

The list of malfunctioning satellites grew by two during the *Challenger's* just-completed flight. The shuttle's crew deployed two $75 million satellites perfectly, but rocket failures on each satellite prevented the instruments from attaining desired orbit. The two are now floating uselessly in space and it is not clear whether they can be salvaged at a cost that would make recovery feasible.

The space-walking crew members were supposed to deploy a balloon in order to practice rendezvous techniques. But the balloon burst soon after deployment making the exercise impractical.

Then the heretofore perfectly operating remote-control robot arm on the shuttle developed a glitch when the mechanism in its "wrist" failed to function as expected. This caused the modification of another rendezvous exercise.

These events awakened space spectators from their complacency — a complacency created by the near-flawless flights of the shuttle in the past. And they underscored the realization that travel in space is always treacherous and potentially life-threatening. Perhaps more so than in previous flights, the safe landing of the shuttle produced a sigh of relief.

The coming months will be busy ones for the nation's space program as no fewer than nine more flights are planned involving 42 astronauts. We salute the crew of the *Challenger* and send our best wishes to the brave women and men who will follow their paths.

Challenger Completes 5th Mission; "Solar Max" Repaired in Orbit

The space shuttle Challenger April 13, 1984 completed a seven-day mission highlighted by the first repair of a damaged satellite in space. Complications in the recovery and repair of the damaged satellite forced the extension of the mission by one day of the six originally planned. The crew for the mission was headed by Capt. Robert L. Crippen of the Navy, who was making his third shuttle flight. Crippen was the the pilot on the first shuttle orbital mission in April, 1981. The other four crewmen were making their first trip: Francis R. Scobee, the pilot who shared the controls with Crippen; Dr. George D. Nelson, an astronomer; Terry J. Hart, an electrical engineer and National Guard pilot, and Dr. James D. Van Hoften, a physicist. The mission set a new altitude record for the shuttle. The initial trajectory, planned to put the craft into position for rendezvous with the damaged Solar Maximum Mission Satellite (Solar Max), established an elliptical orbit ranging between 131 and 289 miles above the Earth. The orbit was the "circularized" at an altitude of about 293 miles. Altitudes on previous shuttle missions had ranged between 150 to 190 miles.

The first task of the mission, the deployment of a module to test the durability of materials in space, went off as planned April 7. The shuttle's robot arm lifted the Long Duration Exposure Facility (LDEF), a 30-foot cylinder containing 57 varieties of space-age materials, out of the craft's cargo bay and placed it in an orbit above the Earth. The module was to be brought back to Earth in February, 1985 by another shuttle crew for testing of the long-term effects on the contents of the heat, cold and cosmic rays of outer space. It would be the first long-term test in space of the durability of a variety of new technological developments ranging from electronic parts and solar cells to space-age fabrics. Among the test materials were Pentagon "Stealth" substances aimed at making spacecraft resistant to radar identification, fiber optics that could be used as laser beam carriers in space and metals for use in a space-based radar surveillance satellite.

The focal point of the mission was the effort to recover and repair the damaged Solar Max satellite. The 5,000 pound satellite had been launched in February, 1980 to study a variety of solar phenomena, including solar flares. However, the failure of an electronics module in November of that year had made it impossible to orient the satellite's sensors accurately. The ability to perform repair work in orbit had been touted by NASA as one of the chief commercial advantages of the manned shuttle over conventional disposable space launchers, which simply delivered their payloads to predetermined orbits. NASA officials said it would take $235 million to replace Solar Max with a new craft, while the the cost of the repairs would be only $50 million. The first attempt to capture the satellite, made by Dr. Nelson April 8 failed. But on April 10, the shuttle snared the satellite with its robot arm and hauled it into the cargo bay. Nelson and Van Hoften April 11 had little trouble performing the repair, which consisted of replacing the 500-pound altitude control module. The complicated electronic package, described as the most sophisticated of its kind, was responsible for keeping the satellite precisely oriented on its solar target.

THE PLAIN DEALER

Cleveland, OH, April 13, 1984

Someday, schoolchildren reading their history texts will marvel at the primitive quality of space flight in the early 1980s. They might ask how routine satellite repairs could be so exciting. They might look at the Challenger as we now look at the Nina, the Pinta and the Santa Maria.

It is the nature of progress to change perception. Ten years ago, the space shuttle was a fantasy. Thirty-five years ago, the idea of a man in space—on the moon!—was sheer science fiction. As technology fulfilled our dreams, we forsook them for grander, more fanciful endeavors. Tomorrow? Space colonies ... interplanetary flight ... star travel!

Our wonderment at the space shuttle, its crew and their successes will be fleeting. They have retrieved, repaired and re-orbited a satellite. By doing so, they have expanded man's ability to probe the universe. The frontier of space has begun to fall with the same rhythm as did the early American wilderness: sorties, then outposts and finally settlement.

An important aspect of the American "personality," if you will, is its perception of frontier and nature. Early Americans viewed the continent's deep forests as malevolent and threatening. Gradually, that changed; nature became benevolent and, ultimately, precious. So too could it be with space. Technology will make it more accommodating. Man's relationship to it will change. Our wonder will decline.

The marvelous accomplishments of Challenger and its crew reflect a curious duality. They give substance to imagination, yet foster insouciance. The shuttle's success should be welcomed and praised, for it has made space more hospitable, and space might someday be our salvation. But soon enough, we will forget the courage and innovation that the Challenger's missions required; soon enough we will remember the flight only as a hallmark of man's technology rather than a measure of man's vision.

The Cincinnati Post

Cincinnati, OH, April 13, 1984

"We deliver," proclaimed the crew of the Columbia spacecraft back in November 1982 after they successfully spun two communications satellites into orbit from the shuttle's cargo bay.

The astronaut corps can now boast a new slogan: "We pick up, repair and deliver."

In the shuttle program's 11th mission, the crew of the Challenger achieved another first and demonstrated the capability to retrieve, repair and restore to orbit an inoperative satellite—the Solar Max sun observatory, which had been dead for more than three years because of blown fuses.

True, the bill for the "service call" was a bit steep—about $48 million. But it was a bargain compared with the $235 million NASA says it would have cost to replace Solar Max with a new satellite.

The only disappointment in the mission was the inability of spacewalking astronaut George Nelson to snag the 21-foot-tall satellite and pull it to the cargo bay. A small stud the designers forgot about apparently protruded far enough to prevent him from locking onto the satellite with a special grappling device.

Nail-biting hours followed as ground controllers worked to stabilize the now wobbling Solar Max with the little power it had left. This done, Challenger carefully maneuvered close to the satellite and snagged it with its robot arm.

The rest was unprecedented but not difficult: replacement of the fuse box and other electronics and return of Solar Max to its 300-mile-high orbit for years more of useful life in photographing and analyzing the violent storms and flares that occur on the sun's surface and directly affect life on earth.

Each shuttle mission has been a milestone in man's progress both in utilizing space and learning more about the cosmos in which we are all space travelers.

Well done, Challenger.

THE ATLANTA CONSTITUTION

Atlanta, GA, April 17, 1984

The Challenger shuttle crew's resurrection of the dormant Solar Max satellite last week was a magnificent achievement, no doubt about it. Terry Hart's manipulation of the shuttle's arm to grasp Solar Max and draw it into Challenger's bay and James van Hoften's and George Nelson's zero-gravity repairs were both products of consummate skill and arduous hours of practice.

It may be some time, though, before NASA's capacity to play Mr. Fixit is standard operating procedure. Last week's mission, in fact, was something of a special case, but an important special case.

For one thing, the Solar Max was within reach of the shuttle; some of America's most important orbiters are not. The shuttle's maximum altitude is 350 miles — close to some scientific data-gathering craft but nowhere near some communications satellites, commercial and military, which fly 22,500 miles above the Earth.

Then there's the cost-effectiveness factor. The price tag for last week's Challenger mission was about $50 million; Solar Max was a $250 million piece of equipment — certainly worth saving. But it is questionable whether NASA would go to the same lengths for last February's wayward Westar-6 and Palapa B-2 satellites, each valued at $75 million.

Finally, Solar Max was one of a new breed of space hardware, easily set aright by the mere replacement of a module that permits it to fix its position against those of selected stars

Earlier satellites tend to be singular contraptions, likened by space authorities to handcrafted customized cars that might be overhauled back in the shop but probably not in space. Sure, the life of some of them could be prolonged simply by refueling them, but as NASA points out, none of them was fitted with gas caps.

The mission clearly confirmed the supposition that the human machine would be able to work effectively in space. In that, the enterprise was genuinely historic.

For the rest, it demonstrated the wisdom of the new direction in satellite engineering: toward a more assembly-line kind of construction, emphasizing interchangeable parts. That development, when matched with the now-proven human ability, will usher in an era when astronaut-handymen can routinely ply their trade in space with the same unruffled ease as back in the garage — but don't expect it next week. Space is not quite prosaic yet.

The Washington Post

Washington, DC, April 16, 1984

WHERE ON EARTH could you find better garage mechanics than the astronauts who towed in that broken-down 1984 Solar Max the other day, worked to all hours to repair it and had it moving smoothly before breakfast Friday morning? True, they didn't call first with an estimate, but what would we all have said, anyway? Never mind, we'll buy a new one? Besides, the crew says the satellite is better than new now—and ready for all sorts of engineering tests to be conducted by flight controllers at the Goddard Space Flight Center in Greenbelt. Parts and labor are guaranteed for 90 days or untold light years, whichever occurs last.

There is the matter of the bill, of course, which runs about $48 million. (We think that is the same as the last one we received for minor repairs to our car here in town; however, they don't take Visa, that being the principal difference.) But talk about payments: list price on a new Solar Max, from factory to delivery in space, runs about $235 million with accessories. The one just out of the shop should do fine. It is for sun-watching, which may someday tell us all a little more about the weather around the world than even Willard Scott does.

So marvelously routine are these space feats now, in fact, that we'll no doubt see Willard doing a remote from the space shuttle some morning. Already the astronauts can hold press conferences there; and they have all sorts of interesting missions in mind for shuttle runs to come.

So quite aside from the gee-whiz aspects of space shots, America's considerable investment in space is starting to yield significant dividends. You'll know it has peaked when all you can find on a Sunday in space is gas-and-go stations that never have mechanics on duty.

Los Angeles Times

Los Angeles, CA, April 15, 1984

It's hard enough to get hold of good repairmen for anything these days—the kind of people who show up on time, take pride in their work and fix things so that in the long run you save money. All the more reason, then, for honoring the crew members of the space shuttle Challenger and the quite elegant bit of work that they did in plucking the 5,000-pound Solar Max satellite deftly out of orbit, replacing its defective parts and easing it back into the task that it was designed to do. They made it look easy, they didn't dawdle and, what's more, they had all the needed tools and parts with them instead of back at the shop. Theirs was a rare accomplishment indeed.

In time it may evolve into a fairly commonplace one. The Challenger crew demonstrated the practicality of overhauling satellites that for one reason or another have malfunctioned or are simply sagging from old age. The promise of the performance is that over the long haul hundreds of millions of dollars might be saved by fixing up satellites in space instead of having to replace them with new ones either launched from the ground or from the shuttle's capacious cargo bay. Later this year a shuttle mission may try to rescue the two stranded satellites put into orbit in February.

Shuttle flights—last week's was the 11th—are getting to be routine, but the wonder of them stays bright because of the achievements that they continue to ring up. And even after the tasks in space of a particular mission are finished and noted, there remain those precious and never-palling moments when the huge plane glides in for a majestic landing for all the world to see, the voyager returning in another quiet demonstration of technological excellence and triumph.

Roanoke Times & World-News

Roanoke, VA, April 16, 1984

THE SHUTTLE program has had its rough moments, so it's a pleasure to hail one of its notable successes: the first-ever retrieval of a broken-down satellite, followed by successful repair and relaunching.

It was nothing like the glamorous, suspenseful exploits that became famous, such as the moon landings and space walks. In fact, an earlier "walk" on this same flight — the untethered venture by astronaut George D. Nelson — became a fiasco: While trying to hook onto the Solar Max research satellite, Nelson instead tipped it into tumbling flight. With Solar Max's on-board battery power almost drained, it was all that ground control could do to right the satellite so that the shuttle's mechanical arm could achieve what Nelson couldn't.

Even so, the repair in orbit harmonized nicely with the shuttle's own concept of reusable, renewable (and money-saving) spacecraft. The repair was millions of dollars cheaper than making and launching a new satellite. All this is a refreshing departure from the economic philosophy that seems to prevail down here: Buy it, use it until it breaks down, then throw it away.

And while this was not a space spectacular, it was arresting in its own way: Imagine repairmen hauling in a 5,000-pound machine, tinkering with its complex innards on the spot, then putting it back to work better than new.

The shuttle Challenger has met an important challenge. All things considered, there seems to be a promising future for bright, handy young men and women to run a roving, radio-dispatched space repair service. And on the side, they might operate a salvage business. We hear there are thousands of pieces of space junk up there.

The Star-Ledger

Newark, NJ, April 17, 1984

During our remarkable era, we have seen space flights and space walks, space stations and even space golf shots. But never until now have we seen the spectacle of space appliance repair.

The feat of the space shuttle Challenger in replacing defective electronic units on the crippled Solar Max satellite while it was in orbit is surely one of the more remarkable accomplishments of our entire space program. It will keep the satellite healthy and doing the job for which it was intended—but it also means a good deal more.

It means that in the future, service calls in space will be possible for any number of purposes—refueling, maintenance and even space construction jobs. This will mean a saving in equipment, in money and possibly even in lives as space travel becomes increasingly commonplace.

There are also benefits to be reaped of a less immediate but perhaps even more significant long-term nature. Solar Max, after three useless years in which it was disabled, will be able to resume its task of studying the sun. This will provide scientists with an understanding of the solar storms that affect our communications and the climate of our planet.

The manner in which the repair mission was accomplished was simplicity itself. After a false start, with Challenger flying close to Solar Max, astronaut Terry Hart reached out with the shuttle's 50-foot arm, snared the satellite and brought it back into the cargo bay. There, repairs were made in what seemed almost routine fashion.

There was a final bit of drama involving the landing of the 98-ton shuttle. When bad weather aborted a scheduled landing in Florida, the landing was diverted to the California desert with less effort than a commercial airliner requires when it is diverted from Kennedy Airport to Newark.

The good news of the landing came on Friday the 13th. Challenger and its crew of five flying space repairmen are heroes who deserve our most fervent admiration.

The State

Columbia, SC, April 12, 1984

THE SUCCESS of *Challenger*'s astronauts in plucking the ailing *Solar Max* satellite out of the sky for repair work is a solid boost to America's space programs. There is no telling what the damage might have been had the historic mission failed.

True, there have been some partial failures in past space shuttle missions — equipment didn't work, and in one case a satellite flew out of its intended orbit. But no shuttle missions have been outright failures.

Challenger's principal purpose was to demonstrate the capability of astronauts and the spacecraft to locate and recover orbiting satellites, to fix them in orbit or bring them back to Earth for repairs.

In this case, the first effort to slow *Solar Max* and to grasp it with *Challenger*'s robot arm failed. But man's ingenuity on the ground and in the spacecraft overcame severe technical problems which were unanticipated in the flight plan, and the recovery was made.

There is another measure of the importance of the mission — dollars and cents. *Solar Max* cost $44 million to build and launch in 1980. It failed to function three years ago. The replacement cost of the scientific satellite today would be $235 million, according to the National Aeronautics and Space Administration. The rescue mission cost about $48 million.

ST. LOUIS POST-DISPATCH

St. Louis, MO, April 14, 1984

"Triumph in Adversity" should have been the official motto for the latest sortie of the space shuttle Challenger. This mission has shown both the flexibility of the space shuttle itself and the professionalism of the ship's crew.

The primary job of the flight was the retrieval and repair of the Solar Max, the out-of-commission, star-gazing research satellite. If the shuttle could carry out the repair, a $235-million satellite would be salvaged and returned to its job for only $50 million. It would also mark the end of an era — the end of the "throwaway" satellite. From now on many communication and research satellites that break down in outer space will be repairable.

But when the initial efforts to capture the 2½-ton satellite by space-walking astronaut George Nelson failed, many feared that it represented a major setback for the shuttle program itself. However, there is more than one way to catch an errant satellite; and with the display of first-rate piloting and operating skill the Challenger crew was able to capture the Solar Max with the shuttle's robot arm. Repairs were quickly made and the satellite was relaunched on its delayed mission.

There was still more adversity, however. Bad weather prevented the Challenger from landing at the Kennedy Space Center in Florida. As it turned out, that was of little consequence to the shuttle and its crew. With minimal effort, the Challenger put down at Edwards Air Force Base in California, thus, again demonstrating the operational flexibility of the space shuttle. For Challenger and its crew, it was a job well done.

RAPID CITY JOURNAL

Rapid City, SD, April 15, 1984

By retrieving, repairing and returning a satellite to its space orbit, Challenger's astronauts refuted the claim that you just can't get good service these days.

After failing to collar the dead satellite during a space walk last Sunday, the Challenger's crew plucked the slowing spinning Solar Max from space with the space shuttle's remote-controlled arm on Tuesday. The satellite was placed in the cargo bay where the defective parts were replaced during a record-breaking space walk by two crew members. Then Super Max was returned to its orbit where it will resume helping scientists better understand the sun and how it affects weather.

Retrieving and repairing the satellite was the primary mission of the latest space shuttle flight. By bringing it to a successful conclusion, the "Ace Satellite Repair Co.," as the Challenger crewmen referred to themselves, demonstrated that satellite servicing can be done.

As the number of commercial satellites increases to meet the demand for state-of-the-art communications, the need for in-space repairs will grow. Capability of the space shuttle to make those repairs will create a market for its services and help defray the cost of the shuttle program while the space vehicles and their crews continue to contribute to scientific knowledge.

The repair bill for Solar Max amounts to a healthy $48 million. But even that's a bargain compared to the $235 million a new satellite would cost.

Failure to retrieve Solar Max on the first attempt did cause the Challenger mission to be extended. And bad weather at the scheduled landing site in Florida forced the landing at the alternate site, Edwards AFB. But these were relatively minor problems and Americans, who like things to work the way they are supposed to, can take pride in the fact that the Challenger crew and its ground support were able to rise to the challenge and accomplish what they set out to do.

The Birmingham News

Birmingham, AL, April 13, 1984

The "entrapment" of the Solar Max satellite by the crew of the Challenger space shuttle marks another triumph in the almost routine progress of the spectacularly successful shuttle program.

The Challenger didn't capture the errant satellite on its first try, it is true, but in a very real way that initial failure only emphasized the remarkable achievement involved in the subsequent success. The failure showed how very difficult and daunting a project it is to rendezvouz in space with a large satellite and maneuver it into the cargo bay of the shuttle.

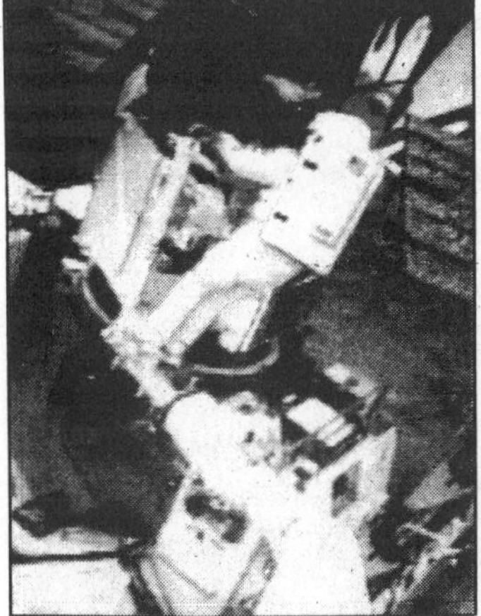

James van Hoften fixes 'flat tire'

Yet, given another shot, the Challenger's astronauts managed the job with real grace under pressure. The effort also dramatized the vital interplay among the shuttle crew, the ground experts (who stabilized Solar Max through remote controls and feverishly calculated the shuttle's fuel supply) and the sometimes balky hardware involved.

The satellite retrieval itself demonstrates a virtual revolution in the way we can think about satellites themselves. Since the launch of Sputnik, satellites have been expendable objects — in orbit for their allotted time, but doomed eventually to run out of fuel or otherwise malfunction and then to burn out in the Earth's atmosphere. The shuttle's mission shows that now, however, satellites can be maintained in orbit indefinitely. One can easily imagine programs to enhance an existing satellite's capacities by having shuttle astronauts add on new instrument packages and the like.

Yet again, the shuttle program has expanded our space cabilities in exponential fashion, and all those connected are due the nation's thanks for their splendid work.

The Houston Post

Houston, TX, April 14, 1984

The news earlier this week that U.S. astronauts aboard the space shuttle Challenger had succeeded in capturing and repairing the ailing Solar Maximum Satellite was welcome indeed. If the mission accomplished nothing else, it saved this country roughly $200 million compared with launching a new Solar Max. At a time of record national deficits, the merit of thrift needs no elaboration.

Besides, the information we were getting from Solar Max was important. It has helped us learn a good deal about solar flares (which affect communications) and the solar constant and sunspots (both of which have a bearing on the Earth's overall temperature and plant growth). In an era when this planet is arguably overpopulated and certain to get more so, this is truly valuable knowledge.

All of which makes this a good time to re-examine a myth about the space program — one that has been around almost since its inception. That is the argument that our expenditures on space are justified by the technological spinoffs which society has reaped from aerospace research. The examples are legion: the miniaturization of electronics; faster, more powerful computers; improvements in medical diagnostics and treatment; high-impact plastics.

It certainly is true that these advances have changed our lives. Because of these discoveries, we are healthier, safer and better cared for than ever before. Perhaps we are even happier. But if high-tech products were all we wanted, how much simpler and cheaper it would have been to develop them in the first place.

NASA's work should not have to lean on the spinoff argument, for two reasons. First, many facets of its work are fundamental scientific inquiry of the highest order. Second, we are headed toward an age in which space endeavors will pay for themselves. Communications satellites have given a hint of this; the latest shuttle triumph illuminates the future even more clearly.

Does this mean the federal government should open its pocketbook unquestioningly to the space program? Certainly not. Times are hard, and the aforementioned deficits are much on everyone's mind. The program still needs to justify its expenditures as much as ever, but can't NASA and Congress be a little more intellectually honest with each other about the rationale for shelling out the bucks?

It's time we gave the tired old spinoff smoke screen a rest. Now, more than ever, what is happening in the space program carries its own justification. It needs no auxiliary defense.

Discovery Completes Maiden Voyage; Perfect Mission Follows 3 Delays

The space shuttle Discovery, the third in a projected fleet of four U.S. space shuttles, made its maiden flight August 30 to September 5, 1984. NASA officials were jubilant over the near-perfect flight, which was highlighted by the successful launch of three communications satellites from the craft's cargo bay. The flight had been twice postponed in June, the second time by a dramatic engine shut-down just four seconds before lift-off. It was postponed again Aug. 28, the night before its scheduled takeoff, by a problem in the electronic system that was to jettison the ship's booster rockets. The six-member crew was unchanged from June's and included Judith A. Resnick, who became the second American woman in space.

Eight hours after lift-off, the crew deployed the first of three satellites to be launched from Discovery: the SBS-4, a communications satellite owned by Satellite Business Systems of McLean, Va. About 45 minutes later, the satellite's booster engine ignited flawlessly propelling it toward an ultimate position 22,000 miles above the Earth. A second satellite, the Navy's Leasat, was launched successfully Aug. 31. The third, American Telephone & Telegraph Co.'s Telstar-3, followed on Sept. 1. Shuttles had successfully launched five satellites on earlier missions, but the failure of two satellite booster rockets in February had cast a pall over NASA's efforts to develop the shuttle as a viable commercial launch vehicle. NASA was technically responsible only for getting the customer's satellite into the shuttle's low-Earth orbit. The customer was responsible for the booster that lifted it into a higher orbit. Practically, however, unreliable booster rockets weakened the competitive position of the shuttle as a launch vehicle. The two commercial satellites launched by Discovery were carried aloft by versions of the booster that had failed in February.

In other highlights of the Discovery mission:

■ One of the crew members was Charles D. Walker, the shuttle's first commercial passenger. On Aug. 31 he began operating a drug-processing machine to test the feasibility of making biological materials of a purity unattainable under gravity. Walker's way had been paid by McDonnell Douglas Corp., the developer of the machine. Its automatic controls failed several times and Walker was forced to operate the machine manually. The drug was later found to be infected.

■ The crew Sept. 1 deployed an experimental solar-powered panel in the first test of an electricity-generating system to be used on subsequent space missions.

The Philadelphia Inquirer

Philadelphia, PA, September 8, 1984

The space shuttle program was A-OK with the six-day, 96-orbit, 2.5-million-mile mission of Discovery and its five-man, one-woman crew. Three communications satellites were launched successfully, a record number for a single flight. Information derived from experiments in drug manufacturing under weightless conditions apparently has advanced the goal of making the U.S. space program more relevant to down-to-earth problems.

In line with that objective, the National Aeronautics and Space Administration is evaluating a proposal from the University City Science Center in West Philadelphia to use that facility as a site for research in the medical, pharmaceutical and biotechnological applications of the space age. Approval would enable NASA to make a sustained effort, as it needs to do, in producing tangible health benefits as a dividend of the nation's investment in space.

If current plans are implemented, the Discovery mission will turn out to be the first of a shuttle-a-month program extending through 1985. Three more shuttle flights are scheduled for the remainder of this year, with 13 more to follow next year. Three spacecraft — the Discovery, the Challenger and the yet-to-be-launched Atlantis — will be used in rotation. While one craft is in space the others will be readied for flight. It won't be quite as routine as the Philadelphia-Washington shuttle, but it's getting close.

Despite some embarrassing setbacks — delayed liftoffs, botched satellite launchings, plumbing that hasn't worked — the shuttle program has been enormously successful in demonstrating that reusable spacecraft, landed like a glider, are practical and capable of performing useful work.

Exercising extreme caution has been NASA's greatest virtue and should continue to be. There's no compelling reason to launch a shuttle every month, and there should be no hesitation to postpone and reschedule when prudence indicates. The shuttle's best record so far is for safety.

Rocky Mountain News

Denver, CO, September 7, 1984

Now that Discovery has landed safely, it's time to review the eligibility list of "citizen passengers" who will be represented on future shuttle flights. (Incidentally, aren't astronauts "citizens"?)

President Reagan has announced that the very first CP, in a launching late next year or early 1986, will be a school teacher. Students are already drawing up lists of candidates to be shot into space.

Busy with the presidential campaign, Mr. Reagan has asked us to finish the sorting-out process. Happy to oblige, sir. The second CP ought to be 11-year-old Meredith Medley of Port Aransas, Texas. She wrote the National Aeronautics and Space Administration, "Children can do a lot of things adults can do, like going to space."

Last December, a NASA official said "We're thinking of journalists, we're thinking of artists, people who can translate the experience into real terms for the public." Did he mention a *journalist*? We know a glib editorial writer who doesn't get airsick and who's great at knocking icicles off toilets. Good choice for No. 3.

Bob Hope, the humorous octogenarian, has said he'd like to be considered. Write down Bob's name as No. 4.

That brings us to the No. 5 civilian voyager. It just so happens that we've met a chef in Johnston named Marco Polo, who has travelled extensively in the Orient . . .

Birmingham Post-Herald

Birmingham, AL, September 7, 1984

A lot was riding on the maiden flight of Discovery, the third ship in the nation's space shuttle fleet. Fuel valve and computer problems had forced three postponements of the mission in two months.

The shuttle program itself was under a cloud as well. In February, two $75-million communications satellites failed to go into orbit when launched from the Challenger shuttle — although this was due to the malfunctioning of their rocket motors and was no fault of the spaceship or its crew.

But now NASA can boast that it is back in the satellite delivery business with Discovery's orbiting of three communications satellites.

Two other important missions were also successfully carried out: the testing of a 100-foot long expandable solar power array and the manufacture of a hormone drug under the weightless conditions of space.

The only significant glitch — ice forming at two discharge valves — was solved through the type of improvisation for which NASA is noted.

The mission ended with the flawless kind of landing the public has come to expect. It was the 12th since the shuttle series began in 1981.

With three proven shuttles now in the fleet, the space agency hopes soon to begin flying a mission a month, orbiting satellites for the telecommunications industry and for foreign governments and carrying out scientific experiments.

The program faces a serious problem, however, because it also depends on the Defense Department for much of its business. Earlier this year the Pentagon began lobbying Congress for funds to develop a new fleet of unmanned rockets to carry military satellites into space to supplement, if not replace, the shuttles.

Discovery was still in orbit when a panel of experts assembled by the National Research Council backed up the Pentagon.

Because of their limited lifting capacity and long turnaround times between launches, the shuttles, it said, cannot provide as much "flexibility and security" for launching military satellites as could a dual system of shuttles and expendable rockets.

Loss of the military as a client could be a severe blow to the shuttle program. NASA's next big challenge, it seems, may not be in space but right on the ground.

WORCESTER TELEGRAM.

Worcester, MA, August 31, 1984

After three unsuccessful attempts, America's newest space shuttle had a picture-perfect launch. Last-minute problems had forced the space agency to scrub Discovery's maiden launch twice in June, and a glitch in the new computerized Master Events Controller Wednesday forced a 24-hour delay.

The delays put increased pressure on NASA to show its commercial payload customers that the shuttle is a reliable space transportation system. NASA officials are well aware that the comparatively primitive European-built Ariane rocket has shown itself to be a reliable satellite launcher, and Ariane's backers are aggressively courting potential shuttle customers.

Even some scientists here suggest unmanned rockets make more sense than the shuttle for simply lifting satellites into orbit. NASA has contended all along that the shuttle program is looking beyond that to the day when man will live and work in space stations, on the moon or beyond.

The disappointment over Discovery's launch delays may be more a matter of unrealistic expectactions than an indictment of the shuttle or of NASA. Since the beginning of the Mercury manned rocket program, NASA has prided itself on making the impossible look easy, and usually has succeeded.

Watching flight after flawless flight into space, it is easy to take for granted the monumental achievement of putting man into space. It is easy to forget that the shuttle is at the cutting edge of technology, using highly sophisticated, experimental equipment often for the first time.

Sometimes absolute perfection is too much to ask, even for the men and women who define the phrase "the right stuff."

DAYTON DAILY NEWS

Dayton, OH, September 2, 1984

Was it really necessary to describe every one-day delay of the shuttle as an "embarrassment" for the space program, or interpret every little glitch according to its supposed effects on this or that pending contract for space travel? That was the line of TV newscasters at the space shuttle launches.

Sending stuff up in space is a complicated undertaking. Last minute problems would seem to be in the nature of the thing. One marvelous aspect of the U.S. space effort is that the double-checking and preparations are so precise.

So long as there is no great rush, no need to get a project launched before some deadline, what's the big deal?

Los Angeles Times

Los Angeles, CA, September 6, 1984

The space shuttle Discovery made one of those routine landings Wednesday. That is to say, it was a beaut. But the first flight of the previously troubled Discovery was not without some anxiety. Before a routine landing was possible, astronauts on the ground and aloft had to dislodge a potentially dangerous ice buildup on the side of the shuttle.

The tricky job of ice chipping was accomplished in the cool professional manner that is a hallmark of American space crews. Hardly noticed was the fact that two of the crew members involved in the delicate operation were the first female American astronauts, Sally Ride and Judith A. Resnik.

At Mission Control in Houston, Ride used a simulator to see if the shuttle's robot arm could brush away the ice without harming the craft. There were fears the ice might fly off during reentry and damage the shuttle at a critical moment.

Ride, who was the first American woman in space and returns next month, determined that the arm could do the job and radioed instructions to Discovery. Resnik, the communicator on the shuttle end, relayed them to mission commander Henry W. Hartsfield Jr., who completed the task. Resnik sent word of success to Mission Control and Ride responded like any pro, man or woman: "Good job, Hank."

No sweat, as an old pilot might say. Quite a contrast from all the hoopla over Ride's flight in June of 1983, when some skeptics wondered if a woman could handle the rigors and dangers of space flight. Any further questions?

THE BLADE

Toledo, OH, September 3, 1984

THE summer of 1984 may be remembered as the season of the space shuttle's greatest slump.

As the season opened in June, the flight of the Discovery was canceled following an unprecedented shutdown of its engines after ignition on the launching pad at Cape Canaveral. It required three more attempts before. Discovery finally left the launching pad in a smooth lift-off early last Thursday.

Now, as the summer season enters its final month, clear across the continent air force officials face allegations of widespread problems in the construction of the nation's second space-shuttle launch facility at Vandenberg Air Force Base in California.

A hastily initiated investigation may prove that alleged welding defects in the steel launch tower are correctable, an outcome that would permit officials of the joint air force-NASA program at Vanderberg to maintain the present schedule. But if that optimistic result is not achieved, any necessity for massive reconstruction of the launch facility could hurt the already lagging competitive position of the space-shuttle program.

It is ironic that these misfortunes could cast a shadow over NASA. The Vandenberg facility was originally conceived as the launch site for space shuttles devoted almost exclusively to military missions, while the Cape Canaveral launch site would be used primarily for civilian space missions.

That division was made because of Vandenberg's favorable position for launching flights southward over the Pacific Ocean into polar orbits that permit satellite surveillance of the Soviet Union. But the military has scaled back plans for Vandenberg while expanding demands for space on shuttles launched from Cape Canaveral. NASA meanwhile increased its demand for Vandenberg flights, where it is now scheduled for more flights than the air force.

As a result, additional serious delay in the operational date of the Vandenberg facility would be a stroke of particular bad luck for NASA. Setbacks for missions at Cape Canaveral already have raised the cost of shuttle operations and strengthened the position of competitors, such as the European Space Agency's Ariane program, which employs conventional expendable rockets.

NASA seems to have recognized that it needs to eliminate the bugs in its program at Cape Canaveral. But the agency and the space-shuttle program also depend on solving its problems on the other side of the continent.

Lincoln Journal

Lincoln, NE, September 9, 1984

One of our prominent neighbors, James Van Allen of Iowa, marked his 70th birthday last week. Van Allen, a scientist long associated with the University of Iowa, discovered the belts of radiation around the earth that are named for him.

At three score and 10, when Van Allen speaks, other scientists still listen, though those of the National Aeronautics and Space Administration may bridle a bit.

Earlier this year, the Iowan criticized NASA, saying its shuttle and space station programs cost money that could be better used for satellite exploration of the solar system. Allen is not alone taking such shots.

The shuttle, he said, has fallen far short of 1972 projections for commercial usefulness in carrying for-hire payloads, and the space station, planned for the 1990s, is not likely to live up to expectations either.

Undeniably the shuttle has been plagued with delays and failures. But the way Discovery performed on its first mission, putting three communications satellites in orbit without a hitch, may restore consumer confidence.

Van Allen would prefer to see unmanned satellites exploring other planets, sending back information that would advance space science. He thinks that would be more useful in the long run.

It also could be more exciting, depending on what was learned. With the shuttles, the U.S. space program has become routine and lacking in the romance that characterized its infancy and peaked with the first trips to the moon.

Under the present administration, we shall see a teacher launched into space — and that is a worthwhile step, one that may reawaken some of the early excitement — but exploration for the sake of knowledge will get insufficient emphasis. Still, it is good to have people like Van Allen reminding us there are opportunities out there for things other than making money.

The News and Courier

Charleston, SC, September 7, 1984

The success of the Discovery space shuttle mission, which concluded with a perfect touchdown at Edwards Air Base in California on Wednesday, has been justly hailed by NASA as a lifesaver for its satellite launching program. Coming in the wake of two failures, which cost insurance companies millions of dollars, the latest mission dispelled doubts about the future of the shuttle. Discovery placed two satellites in orbit and in so doing established a monthly schedule for shuttle flights from October onwards. This is all good news, because NASA's shuttle operations depend on continued commercial success.

The hard fact of life that NASA's future depends on its profitability was emphasized by a report released by the National Academy of Sciences' National Research Council on the eve of Discovery's return to earth. The report revealed that many Air Force officials support a Department of Defense proposal to build a fleet of expendable rocket boosters to launch military satellites. There are sound military reasons for setting up an alternative means of launching satellites. The report mentions one of them — avoidance of the risk of flying a manned shuttle over the Soviet Union when launching satellites in a polar orbit. Most of the space "watchdogs" sent up by the military are on a polar path which would take a manned shuttle over Soviet air space.

A decision to build expendable vehicles, like the Atlas II-Centaur, the Titan 34D7 or the SRB-X, to replace 10 of the shuttle missions would cost NASA dearly. It is estimated that some $10 billion would be drained from NASA's shuttle program.

Regardless of any future decision about its defense contracts, Discovery's latest mission has flashed a "go" signal for the shuttle program that looks good for another decade of development in space. It has also reminded the world of the importance of good plumbing, something that has been associated with success since the days of the Roman Empire

The Commercial Appeal

Memphis, TN, September 7, 1984

A LOT WAS riding on the maiden flight of Discovery, the third ship in the nation's space shuttle fleet. Fuel valve and computer problems had forced three postponements of the mission in two months.

The shuttle program itself was under a cloud as well. In February, two $75 million communications satellites failed to go into orbit from the Challenger shuttle — although this was due to the malfunctioning of their rocket motors and was no fault of the spaceship or its crew.

But now NASA can boast that it is back in the satellite delivery business with Discovery's orbiting of three communications satellites.

Two other important missions were also successfully carried out: the testing of a 100-foot long expandable solar power array and the manufacture of a hormone drug under the weightless conditions of space.

The mission ended with the flawless kind of landing the public has come to expect. "We've got a good bird there," said Henry W. Hartsfield Jr., Discovery's commander. It was the 12th flight since the shuttle series began in 1981.

With three proven shuttles now in the fleet and a fourth, Atlantis, on the way, the space agency hopes soon to begin flying a mission a month, orbiting satellites for the telecommunications industry and for foreign governments and carrying out scientific experiments.

The next flight, scheduled for Oct. 1, is to carry a crew of seven, the largest number to be placed in orbit at one time. Plans call for the first spacewalk by a woman, Kathryn Sullivan, and a mock refueling of satellites in orbit, a maneuver in which the United States lags behind the Soviet Union.

The program faces a serious problem, however, because it also depends on the Defense Department for much of its business. Earlier this year the Pentagon began lobbying Congress for funds to develop a new fleet of unmanned rockets to carry military satellites into space to supplement, if not replace, the shuttles. Discovery was still in orbit when a panel of experts assembled by the National Research Council backed up the Pentagon.

Because of their limited lifting capacity and long turnaround times between launches, the shuttles, it said, cannot provide as much "flexibility and security" for launching military satellites as could a dual system of shuttles and expendable rockets.

To help counter that criticism, the space agency says that starting next month it will land most shuttles near the launching pad at Cape Canaveral to avoid the cross-country trips of the shuttles from Edwards Air Force Base.

NASA'S STRONGEST argument is that loss of the military as a client would increase the cost of launching a satellite by shuttle to the point that NASA's commercial business would be undermined. A major consideration for Congress and the administration has to be the agency's ability to remain competitive with other satellite programs in Europe and Asia, including satellite service offered by the Soviet Union.

With NASA apparently closing in on its goal of developing regular, reliable space transportation, it would seem self-defeating for the United States to put any crimp in the shuttle program. If the once-a-month flights work out as planned, the technical achievement will be hard to dispute.

Even so, NASA's next biggest challenge may not be in space but right on the ground — in Congress.

Los Angeles, CA, September 5, 1984

Despite the striking successes of the space shuttle Discovery's long-awaited maiden flight, scheduled to end this morning, folks on the ground talked most about the tribulations of a most prosaic piece of equipment: the toilet.

On this flight, as on nine of the previous 11, the toilet conked out. This time, liquid wastes froze before they could be expelled into the limitless beyond, thus bedecking Discovery with ice that threatened to break off and possibly cause re-entry damage. Yesterday, the crew managed to knock the ice free with the shuttle's high-tech robot arm. But for most of the flight, the astronauts had to use what were delicately referred to as "special chemically treated bags" for certain vital functions.

Designing a toilet for use in weightlessness is tricky enough. Using it is even trickier. Flushing is out, because there's no gravity to drain the bowl. Accounts of how space toilets work have been blessedly oblique, but we understand suction is somehow involved. Anyway, the contraption is still far from foolproof.

But why all the interest in the problem? Maybe it's because, although most of us know beans about the shuttle's abstruse tasks, we all know about toilets. And it's refreshing, if that's the right word, to realize that they can be just as troublesome in space as they are on the ground.

Someday, no doubt, NASA will invent a space toilet that works at least half the time. But we're sure that, no matter how advanced the technology, astronauts will have to jiggle the handle afterwards in a futile attempt to make it stop running at night. ■

The Wichita Eagle-Beacon

Wichita, KS, August 31, 1984

Space shuttle Discovery is at last in orbit, but Americans should keep their fingers crossed. A lot rides on the successful completion of NASA's 12th shuttle, which is the inaugural flight of the Discovery, the third shuttle service entry. Not only is national prestige at stake but also the need to demonstrate the shuttle program can deliver dependably.

Thursday's launch was doubly gratifying because it followed three delays — the most recent one only this week — in Discovery's debut, originally scheduled in June. Similar delays to solve technical problems preceded the initial flights of Discovery's older sisters, the Columbia and the Challenger. There were more frustrations in February when two communications satellites launched from the Challenger failed to reach proper orbits and were lost. Since then, some of NASA's commercial customers have switched their business to the European Space Agency's Ariane program.

But three satellite launchings were made part of Discovery's maiden mission, the first one scheduled for late yesterday. Each of three communications companies has paid NASA in excess of $10 million to do the job. The net effect of this mission will be to prove either that the shuttle program can do the job or it can't.

So far, NASA has scored some impressive firsts in space research and commerce. The current flight carries the shuttle program's first paying passenger — a space industry engineer operating an orbiting pharmaceutical lab. Here, it is hoped a new type of hormone of potential benefit to millions can be produced to a degree of purity impossible on Earth. During the mission, a new type of solar sail also will be tried out; its function is to convert sunlight into electrical power for space stations.

It's clear that, despite earlier problems, the shuttle program's promise remains bright. Let us hope Discovery's voyage reaffirms that promise.

Teacher to Get Shuttle Ride; Citizens In Space Weighed

The relative comfort of shuttle flight meant that space was no longer the preserve of highly trained astronauts. Only brief training periods of two or three months would be needed before scientists, engineers, doctors and even artists, journalists and politicians could make flights into space. NASA voiced hope that as the shuttle evolved in years to come that it could carry large numbers of people into orbit, and return them safely to the Earth. It has been suggested that the shuttle's airplane-like design could be modified so that some of its 65,000 pound cargo capacity be transformed into a high-seating density and a no-frills passenger cabin for about 350 people. Spaceliner shuttles could dock at a space station built as an educational-recreational resort where the passengers could experience a once-in-a-lifetime revel in space. At the shuttle's present operating expense, which was expected to drop lower in the 1990s if it was operated commercially, the cost would be around $10,000 a person. Aside from education and tourism, the shuttle-derived space liner would open the space frontier to entrepreneurial development.

President Ronald Reagan announced August 27, 1984 that an elementary or secondary school teacher was to be chosen as the first "citizen passenger" to ride into space aboard the shuttle vehicle. All of America would be reminded that day "of the crucial role teachers and education play in the life of our nation," he said. "I can't think of a better lesson for our children or our country."

The Kansas City Times

Kansas City, MO, January 3, 1984

The National Aeronautics and Space Administration plans to issue regulations soon for carrying three or four civilian passengers a year on space shuttle flights. This idea serves a lot of useful purposes. For one thing, it will give the American public—or at least a few of its more adventurous members—a chance to identify more closely with the space program. And that should generate stronger popular support. Such backing is needed because, as NASA administrator James M. Beggs says forthrightly, "the public is paying for this."

The act of taking along a few ordinary folks on future space rides will demonstrate that space travel is not only for superbly conditioned and trained individuals. Most people in good physical and psychological shape can withstand the pressures of a space trip. With a crowded future schedule of shuttle flights on a variety of missions such as orbiting, repairing or retrieving satellites, NASA wants to reinforce the image of the reusable spacecraft as a dependable bus and truck operation that is here and now—not some futuristic Buck Rogers fantasy.

NASA is thinking in terms of priority for writers, journalists and artists, who can translate their space experience into more meaningful terms for their fellow Americans than can the astronauts, busy with their technical tasks during most of the flight. But the civilian passengers can expect to lend a hand occasionally.

The space agency expects to begin taking applications from prospective civilian passengers before too long, with the first such flight as early as 1985. The chance to fly in space, even though one doesn't possess the superior attributes required of astronauts, can become a new American dream of being part of the modern version of conquering the frontier. And a fortunate few will realize that dream before too much longer.

Fort Worth Star-Telegram

Fort Worth, TX, August 31, 1984

The teaching profession is getting a deserved lift.

A teacher will become the nation's first private citizen to soar into space.

President Reagan, appropriately, has decided that a teacher should be the first citizen passenger to join a space shuttle mission.

Journalists, novelists, artists and those in other professions have been vying for the honor. They will get their chance later. Fair enough.

For, as the National Aeronautics and Space Administration Administrator James Beggs said, "A good teacher can have an impact on a person, not only in his or her formative years, but throughout life. It will give an opportunity to our children to look up to someone who has flown in space."

The first private citizen to go into space will be chosen from America's 2 million elementary and secondary teachers. NASA expects as many as 80,000 teachers to apply for the position. A "reasonably healthy" individual can go into space, officials said, and the work for the teacher chosen will not be stringent.

The teacher selected will work for NASA at civil servant pay during eight weeks of prelaunch training and for a year after the flight when he or she will make appearances in the nation's school rooms.

The teacher-astronaut will be under no restrictions after completing the NASA work.

A citizen-astronaut will bring special insight to the space program, which is why officials have wanted for some time to permit a private citizen to go into space. And what better private citizen to begin with than a teacher, who is especially trained to not only get information but to impart it to others?

If space exploration has lasting value for the nation — and it most certainly does — then today's younger generations will be the ones carrying that effort to new heights. A teacher, by becoming the first citizen-astronaut, can have a profound effect upon encouraging them to do that.

In recent decades, the teaching profession has slipped a bit from the stature in which it once was held. It is time some of that lost stature were restored, and this assignment into space should help do that.

In fact, the mere fact that the announcement has been made that a teacher will be first into space surely has already had a positive effect upon the teaching profession.

And sooner or later, that positive effect will rub off on all those who come the teaching profession's way.

Which, in short, means everybody.

Richmond Times-Dispatch

Richmond, VA, September 5, 1984

Should a teacher be the first workaday American to ride into space aboard the shuttle? Yes, concluded President Reagan, because when the craft lifts off, the nation's esteem for teachers, traditionally high, will soar. No, replied National Education Association president Mary Hatwood Futrell, her sights fixed on astronomical dollar signs. Instead of spending "a lot of money to send a teacher into space," she groused, "the president might better put some resources into the nation's classrooms."

This being an election year, no one should ignore political content in either the action or reaction. It can be assumed that Mr. Reagan is not unaware that there are 2.1 million teachers in this country, that they tend to vote, and that his image among them could stand some polishing. But it also can be safely assumed that anything President Reagan does between now and Nov. 6 will elicit a knee-jerk, negative reaction from the NEA, whose officialdom endorsed Walter Mondale even before he won the Democratic nomination.

Politics aside, directing the space agency to search the nation's elementary and secondary classrooms for the right person to achieve this historic "first" makes sense. The president didn't select a lawyer. He didn't select an architect, an artist, a veterinarian, a butcher, a baker, or a candlestick maker — or a journalist. The designation of a teacher says a lot about the central place education has in American life, and, in fact, in such U.S. technological triumphs as the space shuttle itself. The teacher who turns astronaut for that 1986 mission will have the adulation not only of one school's pupils but of a whole nation of students.

Big Government advocates like the NEA believe that no president can be pro-education unless he vastly increases federal spending for education. But President Reagan recognizes that primary responsibility for financing public schools constitutionally rests with the state and local governments, and that initiative-sapping federal control inevitably accompanies any big increase in federal outlays. Erosion of standards occurred, he noted the other day, "during the very period when spending on education was up by over 600 percent." A 15-year decline in test scores during a period when Democratic-controlled Congresses lavished federal dollars on education supports that contention.

Though the NEA may deride them as empty, gestures from the presidency in support of excellence in education do appear to make a difference. The "nation at risk" alarm that a presidential commission sounded last year has accelerated the pace of school reform in most states. Some 350 secondary schools, including Hermitage High School in Henrico County, earned plaudits from the U.S. Department of Education for goal-setting, administrative leadership and inspired teaching that make them exemplary. And, know what? As a new school year opens, the heartening news is that test scores are rising again, and so is public confidence. In a recent Gallup Poll, 42 percent of Americans gave local public schools an A or B; only 31 percent did so a year ago.

President Reagan deserves his share of credit, though he will receive none from the NEA, for raising the nation's educational sights. By the time a teacher-astronaut lifts off, Mission Excellence should be, if not complete, well on its way.

St. Petersburg Times

St. Petersburg, FL, November 9, 1984

They used to say the sky was the limit on congressional perquisites, which include free parking, postage and foreign travel, cut-rate haircuts, subsidized lunches and a pension plan second to none. But the sky is the limit no longer. Behold space, the final frontier!

The National Aeronautics and Space Administration has chosen Sen. Jake Garn, R-Utah, to fly in a space shuttle, possibly as early as next year. NASA explained that it is "appropriate for those with congressional oversight to have flight opportunities to gain a personal awareness and familiarity" with the shuttle program.

WHAT THAT gobbledygook really means is that Garn holds NASA by the pursestrings and NASA wants to butter him up. As chairman of the Senate appropriations subcommittee in charge of NASA's budget, Garn has been angling openly for such an invitation ever since 1981. He also happens to be giving the agency some trouble over its impending request for $8-billion to develop a permanent, manned space station. Perhaps to make Garn's invitation look less craven, the space agency also invited flight applications from three other key subcommittee chairmen, one of whom is 73 years old.

Garn is a relatively youthful 52, and, NASA pointed out, he has logged more than 10,000 hours of pilot time in the Navy and the Air National Guard. His background, while plausible, is beside the point: For each politician who serves as supercargo on one of those costly shuttle flights, some highly trained scientist or astronaut will be left behind. And besides, what ever happened to President Reagan's promise that a teacher would be the first non-astronaut into space?

GARN WOULDN'T be the first in any case. That honor belonged to Charles D. Walker, an engineer for the McDonnell Douglas Corp., who flew on a shuttle last August to conduct a company experiment. But Walker was a paying passenger ($80,000); he had been trained intensively for his scientific mission. One doubts that the "relatively short period of training" NASA has promised Garn will qualify the senator to do anything but stay out of harm's way.

With Garn's precedent, it will probably become routine to have some junketing member of Congress aboard every flight. This will make grist for the national humor mills; one can already imagine the jokes about spacy senators and far-out congressmen. In all seriousness, the space program should not be politicized this way. It is more likely to erode than enhance public support and it will interfere with the space agency's scientific and defense missions.

Birmingham Post-Herald

Birmingham, AL, August 30, 1984

The National Aeronautics and Space Administration has been talking for a couple years about making room for "citizen passengers" on its space shuttle flights.

A lot of people assumed a reporter would be the first such observer and journalists were, in fact high, on NASA's list, which also includes poets, artists, scholars and even show business people.

But educators were ranked highest of all among the options submitted to President Reagan, says NASA administrator James M. Beggs — and election-year pressures had nothing to do with it, he claims.

The president won't be politically damaged, of course, by his announcement that he has directed NASA to choose an elementary or secondary school teacher — "one of America's finest" — to be the nation's first citizen astronaut.

Leaders of teacher organizations, who are working to defeat Reagan's re-election, may grumble that they would rather have more money spent on the schools, but they can hardly complain about his tribute to their profession.

"When that shuttle lifts off," said the president, "all of America will be reminded of the crucial role teachers and education play in the life of our nation. I can't think of a better lesson for our children, and our country."

NASA officials say they expect as many as 80,000 applications from the nation's two million schoolteachers between Nov. 1 and Jan. 1, the period during which applications will be accepted. From this number two will be selected — one to go into space, the other to serve as a backup.

Any teacher in reasonably good health is eligible, and he or she need not be a science or mathematics teacher. The winner will have to be able to get a lengthy leave of absence — for astronaut training, the flight itself (in late 1985 or 1986) and then a year of traveling around the country in "communications and education missions" on behalf of NASA.

But what school administrator would dare refuse the chosen teacher that chance of a lifetime?

In future years, other ordinary Americans will voyage into space aboard the shuttles. The first, however, will have been a teacher, and there couldn't have been a better choice.

Discovery Rescues Lost Satellites

The space shuttle Discovery completed an eight-day mission November 16, 1984 that included the launching of two satellites and the first-ever salvage of lost satellites. It was the second flight for Discovery. Discovery carried a crew of five. The mission commander was Navy commander Capt. Frederick Huack, a pilot and physicist who was making his second shuttle flight. The pilot was Navy Cmdr. David M. Walker. The other three were Dr. Anna L. Fisher, a physician, who became the third woman and the first American mother to fly into space, and Joseph P. Allen and Navy Lt. Cmdr. Dale A Gardner, both of who were making their second shuttle flights. The crew Nov. 9 successfully deployed a communications satellite, the Canadian Anik D-2, which was eventually to join five other Canadian satellites already in service. Because demand was lagging behind projections, however, the owner, Telesat Canada, did not intend to put the satellite into service immediately. Telesat went ahead with the launch because it was less costly to store the satellite in space than on Earth and because NASA's launching costs were scheduled to double at the end of 1985. The shuttle deployed the Leasat 1, a military communications satellite, Nov. 10. A sister satellite, the Leasat 2, had been deployed in August. The two launches cleared the shuttle's cargo bay for the satellite salvage operation that was the centerpiece of the mission.

In a pair of demanding and unprecedented space maneuvers, astronauts Joseph Allen and Dale Gardner, assisted by Anna Fisher, successfully snared the first of the two lost communications satellites, the Indonesian Palapa-B2, on Nov. 12. Using the same procedures they retrieved the second, Western Union Corp.'s Westar 6, two days later. NASA flight director Jay Greene Nov. 8 called the mission the "most challenging" since the first Apollo moon landing more than a decade earlier. "We've deployed satellites before, we've picked up satellites before and we've repaired a satellite before," he said. "But we've never before done all of these things together on one flight." Both satellites had been floating in erratic orbits since February. Controlled firing of their thrusters over the previous two weeks had brought them into more circular orbits that intersected that of the shuttle, and the shuttle had begun closing in on them Nov. 9.

The Miami Herald

Miami, FL, November 16, 1984

SCARCELY 23 years ago, as the first Americans ventured into space, the men who rode the rockets wondered if they had any role beyond serving as test flesh for sensors to monitor. Monkeys could do it just as well, the skeptics whispered. Space was the province of high-tech machinery and computer-meisters in mission control, they said. Astronauts were sent along for image purposes only, to hook the public imagination so the taxpayers would pay. Astronauts, one scornful phrase put it, were little more than "Spam in a can."

Such limited vision was the product of little minds lacking imagination, as the subsequent stream of stunning successes in manned space flights attests. If any doubts about that possibly remained, surely they were dissipated by the latest adventure this week by America's spacemen.

For six hours on Monday two American astronauts in Buck Rogers jet-pack space suits grunted and groaned and wrestled a 1,200 pound, nine-foot long bum satellite out of dead orbit and into the cargo hold of the space shuttle *Discovery*. With this feat, space was no longer reserved as the testing ground of futuristic engineering, and man's role there no longer was confined to the poetic experience of adventure and the dry realm of scientific inquiry. As of Monday, space also became another place where humans perform sweaty muscle work to get necessary but mundane things done.

What astronauts Joe Allen and Dale Gardner did was retrieve a $35-million communications satellite that had gone off course. They labored like tow-truck operators pulling a car from a ditch. They worked with their hands like good mechanics, dismantling a brace, bolting on another with a power wrench, and muscling the massive satellite into position so it could be brought home for repairs. Two days later, they did it again — salvaging a second satellite.

To be sure, there remained an aura of grandeur. How could there not be, with TV cameras recording the planets whizzing by as the mechanics strained, 224 miles above Earth? But where once American adventures in space seemed all wonder, majesty, challenge, and potential, now they are that and something more. Now they are practical. Now they extend the routine functions of the human workplace into the heavens. Now an astronaut salvaging a broken communications satellite is the space-age lineman up a pole.

The last frontier is being tamed. How far we've come, and how fast.

The Virginian-Pilot

Norfolk, VA, November 15, 1984

Inspiring pictures and words depicting the space shuttle Discovery's daring and resourceful salvage deeds contrasted this week with a report from the Congressional Office of Technology Assessment that tossed cold water on high-flying NASA at its apogee of public appreciation.

The space administration's high-frontier goals "for advancing U.S. interests into the second quarter-century of the Space Age" are not justified on scientific, economic or military grounds, the report asserted. It asked: "How can the U.S. people and government justify, today, continuing such truly great and continuing public expenditures on space-related matters perceived by most of our general public as lying well outside of the mainstream of their personal interests and concerns?"

The answer can be found in a couple of questions for the authors of the OTA report: Where do they get the idea that most Americans are uninterested and unconcerned about space research? Not from public opinion polls. A Harris survey, for example, reported a year ago that 70 percent of those sampled saw positive benefits in establishing a permanent space station, NASA's next steppingstone in space.

And what are those "truly great and continuing public expenditures" the OTA report writers criticize? The $2 billion spent annually on space shuttle development is less than half of 1 percent of the total federal budget.

And what does the nation — indeed, the planet — receive for that comparatively small investment? The easiest to quantify are space spinoffs like pocket calculators, which were on the market by the time the first man, an American, walked on the moon, a feat made possible by the number-crunching power of navigational computers. Now, desk-top versions store for the average office worker what 16,000 brains can remember. Meanwhile, top-of-the-line, state-of-the-art computers can perform 100 million operations a second, and in a decade the average computer will be performing 8 *billion* operations a second.

Less easy to quantify are the space firsts that shuttle missions routinely achieve — first woman in space, first black in space and, soon, the first teacher or congressman in space (a sort of space race within a space race). Who can know how such symbols inspire young citizens of the world? Or who can know how modern satellite communications benefit primitive cultures? Surely they are among our best foreign-policy instruments.

The authors of the OTA study do allow that a solid case can be made for NASA plans. But their conclusions come off sounding like the latest screed from The Flat Earth Society. They should be reminded that President John F. Kennedy was advised just before the history-making 1961 blastoff of Cmdr. Alan B. Shepard Jr., the first American in space, that "manned space flight will be man's most expensive funeral." The president gambled and won. Now, we know more and gamble less — and benefit more. The OTA ignores the fact that history — especially space history — has a trajectory all its own.

THE TAMPA TRIBUNE

Tampa, FL, November 13, 1984

SPACE SCIENCE can do wonders, and so can man's perseverance and foresight, the crew of the space shuttle Discovery proved yesterday.

Both traits played key roles as the astronauts rescued an errant communications satellite, the Palapa B2, a task they will seek to duplicate today with the similar and also errant Westar.

It was science that brought the Discovery within 35 feet of the Palapa and enabled astronaut Joe Allen, flying free with a rocket pack for propulsion 224 miles above Earth, to snare the satellite and bring it to the Discovery's cargo hatch.

But it was human ingenuity and foresight which enabled Allen and astronaut Dale Gardner to maneuver the 21-foot, 1,500-pound space vehicle by hand through the hatch and into the cargo bay for its return to Earth.

They had to fall back on the backup manual method they had rehearsed under water to approximate weightlessness during training, because a protruding section of metal on the Palapa prevented attachment of a grapple fixture to permit the Discovery's robot arm to bring the satellite into the shuttle. The glitch was not specifically expected, but the pre-flight training included plans for contingency procedures just in case.

Completion of the task had Lloyd's of London sighing with relief. It had insured the Palapa, owned by the Indonesian government, and the Westar, owned by Western Union, against malfunction. The satellites were intended to orbit 22,500 miles from Earth, but rocket misfires instead put them into elliptical paths 161 to 700 miles high, useless for their communications missions.

Lloyd's had settled the owners' insurance claims for $180 million. The firm paid the National Aeronautics and Space Agency $5.5 million to attempt the retrieval. That money was augmented by $5 million in pre-flight costs paid by Hughes Aircraft Co., which made the two satellites.

Lloyd's hopes to refurbish both (built at a price of $35 million each) and resell them for as much as $25 million each, thereby recouping at least a part of what it paid on the claims.

This Discovery mission, the 14th shuttle operation, is being partly financed by other private interests as well. Hughes Aircraft paid another $17 million for the Discovery launch of another communications satellite it will lease to the U.S. Navy, and the Canadian company Telesat paid $10 million for the launch of its eighth satellite.

Those who spend millions for the communications and other advantages of space doubtless share Lloyd's enthusiasm for the success astronauts Allen and Gardner achieved yesterday. The ability to retrieve and repair malfunctioning satellites makes the use of space more commercially attractive.

Beyond the commercial, however, the Discovery crew's reliance on manpower when things went wrong lifts the spirit. The Palapa retrieval reminds us that however marvelous the missions on which we send the space age's marvelous machines, mankind has some marvelous assets of its own.

ST. LOUIS POST-DISPATCH

St. Louis, MO, November 19, 1984

After a 3.2 million-mile voyage, the 104-ton Space Shuttle Discovery has returned safely in a brilliant dawn landing at Cape Canaveral. The mission was a complete success. The highlight of the flight was of both historic and practical significance. It was the recovery of the wayward Westar and Palapa communication satellites, which had failed to go into their proper orbits during their initial launchings. Two other communication satellites were also orbited.

The double satellite rescue operations were the first such recapturing of useless satellites in space and they proved that the space shuttle and its crew are capable of doing one of the major jobs the shuttle was designed to do. The recovery should allow the insurance company that paid for the $10.5 million salvage operation to recoup about $70 million once the two satellites are refitted and resold to other users.

This latest shuttle mission came amid reports that the U.S. and the Soviet Union were considering a joint space mission that would involve a U.S. space shuttle docking with an orbiting Soviet space station. The humanitarian purpose of such an operation would be to test in-space rescue techniques in case either U.S. or Soviet astronauts face an in-orbit space emergency.

Although U.S. space officials say the Soviets are cool to the proposition, such a mission is a good idea — both practically and symbolically. Sooner or later, there will be a serious emergency in space; preparations now could provide the two space powers with the ability to keep an emergency from becoming a tragedy. Just as important, it would show that Washington and Moscow can work in common.

THE CHRISTIAN SCIENCE MONITOR

Boston, MA, November 16, 1984

INITIALLY the United States space program seemed a dramatic novelty, like Magellan's global circumnavigation, Fulton's steamboat, and the maiden flight by the Wright brothers. But with the latest successful mission of Discovery, the space shuttle program has advanced firmly into the workaday world, as did its predecessors in other fields. The shuttle this week showed its versatility and utility as an orbital workhorse.

The retrieval of two satellites for earthly repair was extraordinary drama; at one point, 130-pound astronaut Joe Allen "carried" the 1,200-pound Palapa satellite once around the globe. The televised sight of astronauts physically shoving the two satellites around was one with which the average working person could identify: It humanized a program that had often seemed esoteric. A result is likely to be increased public support for the space program and, one hopes, greater government support.

Like the circling astronauts, the shuttle ground team performed well in getting off two shuttle missions in a little over a month. It has shown encouraging progress toward the kind of turnaround capacity needed to support the goal of launching up to 24 missions a year.

Further, the Discovery mission gave evidence that it is millions of dollars cheaper to retrieve — and, presumably, to repair and relaunch — defunct satellites than to leave them untouched and launch all-new equivalents.

Yet Discovery's achievements will probably do little to allay the concern the US Defense Department has expressed about the shuttle's reliability for carrying out crucial military missions. The Challenger was to perform a secret Pentagon mission early next month. Now its launch is postponed until late January because of serious problems with its heat-shielding tiles.

Had that mission involved orbiting a critical communications or spy satellite, the delay would be intolerable. Therefore the department plans to keep open the option of launching its payloads on expendable rockets. This is bad news for shuttle managers who had counted on a near-monopoly of Pentagon business to help the shuttle pay for itself.

That operation already faces stiff commercial competition from Western Europe's Ariane launcher. The Europeans underscored this by orbiting two satellites even as Discovery's crew was delivering payloads for two customers. Competitive pressure will increase now that private US companies and countries such as China and the Soviet Union are beginning to offer launch services.

Furthermore, the National Aeronautics and Space Administration is under orders to raise its prices to cover costs. As one step it will double shuttle prices next year.

The space shuttle team has amply demonstrated the potential of its orbital transport system as a key to further development of US spaceflight capability. Now it has to convince would-be customers that it can provide reliable service at a realistic, yet competitive, price.

THE SUN

Baltimore, MD, November 17, 1984

It is hard to contain the superlatives in describing this week's mission of the U.S. space shuttle Discovery, which retrieved two communications satellites that had gone into improper orbits at launch. The "learning curve" of the astronauts between retrieving Satellite No. 1 on Monday and No. 2 on Wednesday was magnificent. Once again human improvization proved even more reliable than the most sophisticated technology.

One writer suggested that the Monday maneuver, though successful, was so awkward as to resemble a Laurel and Hardy comedy. What made it difficult was an unanticipated glitch, an improperly designed "grapple frame" which could not be attached to the satellite as planned. This required astronauts to jockey the huge barrel-shaped satellite into the shuttle's cargo bay manually — at a high cost in time and dignity.

By Wednesday, however, when they rescued satellite No. 2, they had turned a liability into an asset: They discarded the idea of the grapple frame altogether and used manpower, plus the shuttle's mechanical arm, entirely. Their planning, based on Monday's experience, was so superb that Wednesday's execution was flawless. The astronauts even had an hour to spare to rest and take pictures. They completed their mission, with rescued satellites aboard, on Friday morning.

This sort of activity was precisely the reason the shuttle was invented. It is man's first spacecraft with a major cargo-carrying capacity which also can enter space, perform work there and then return to earth. The rescue of the two satellites was praised by insurers (who had paid owners of the two malfunctioning satellites $180 million because of the failure) as "an enormous feat of flight." It will now be possible to refurbish the old satellites and put them back in orbit, instead of losing them altogether as would have been the case in pre-shuttle days.

Shuttles have launched more than a dozen satellites, repaired one in space and retrieved two. Some they cannot repair or rescue because the satellites are in orbits too high for shuttles to reach. Although the Congressional Office of Technology Assessment has suggested that a current NASA proposal for a permanent manned space station is half-baked, it seems obvious that the United States will soon have some sort of manned station, even if not precisely the one NASA now visualizes. Why? Well, among many other things, it will be needed as a base for vehicles which will be able to reach higher orbits than the shuttle can now reach. As the shuttle's successes mount, there is less and less reason to place limits on our vision of man's future in space.

The Providence Journal

Providence, RI, November 14, 1984

Flights of the space shuttle have long since passed the "gee whiz" stage, but the spectacle of two astronauts capturing a sulking satellite and wrestling it into their own cargo bay was one to marvel at. The Discovery thus added a new and practical use to the shuttle's resume, providing a kind of back-up for satellites that malfunction or go astray.

Jaded Americans already take for granted the blast-offs and landings of the space shuttles, from either Vandenburg Air Force Base in California or Cape Canaveral in Florida. They even yawn as television photos of astronauts performing intricate maneuvers (or hamming it up for diversion) appear on the nightly news.

But behind the facade of nonchalant space walking and sure-footed (or back-packed) flight, lies the boldness of the explorer into new modes of existence and the painstaking preparations of dozens of scientists who back up the space flights at NASA headquarters in Houston. The combined skills of astronauts and scientists have now retrieved an abandoned satellite — and may retrieve another, if their luck holds out. On the practical level, this exercise in space recycling holds promise for the sponsors of future satellites that may malfunction. No longer must such "duds" be given up for lost. They can be recaptured, returned to earth, repaired and presumably launched anew to do the work they were meant to do.

That, at least, is the objective for the Indonesian satellite that was picked from orbit and hauled aboard Discovery, and for the Western Union satellite that is next on the rescue list. The feat means a great deal more to insurance companies that have insured such space vehicles. For slightly more than $10 million, insurers who paid $180 million to the owners of those two "lost" satellites will get them back; and the chances are good that after repairs the satellites can be sent back up into orbit to perform their original functions.

For this opportunity, the owners and insurers must be grateful to Dr. Joseph P. Allen, who held on to the captive satellite for an hour and a half while his colleague, Comdr. Dale A. Gardner, put clamps on it that enabled the Discovery crew to secure it firmly in the cargo bay. These modern explorers in their shining space suits, controlling their flight with jets from back-pack motors, showed again the amazing results that can be achieved with the ingenious and concentrated efforts of NASA.

If there was heartening affirmation of NASA's program in the performance of the current mission, NASA was creating nothing but trouble for itself on another front. By promising Sen. Jake Garn that he could have a ride on an early flight of the shuttle, the agency opened its arms to junketeering by congressmen and other officials on whom it must depend for funds and support.

Senator Garn is not likely to gain any knowledge about usefulness of the shuttle program by going on one of its missions that he couldn't have learned by talking with the NASA scientists. He and other members of Senate and House who have been offered rides would do better to decline and leave the berths on future flights to technical people who can perform useful functions on the still experimental spacecraft.

THE LOUISVILLE TIMES

Louisville, KY, November 20, 1984

Those who grew up watching the first, tentative wonders of space travel probably thought its magic would wear thin eventually. Not so, as the excitement created by last week's successful retrieval of two wayward satellites by the shuttle Discovery attested.

The days are gone forever when the projects of the National Aeronautics and Space Administration are so spellbinding that virtually every school in America stops classes to watch on television. But the flawless performance of the astronauts who walked out into space to grab the malfunctioning satellites and haul them aboard for return to earth was still great stuff.

Further, astronauts Joseph P. Allen and Dale A. Gardner demonstrated yet again that the world needs human ingenuity — even in the rarified realm of space flight where every contingency is supposedly provided for. When the device the astronauts were to use to latch the wayward satellites onto the shuttle's mechancial arm failed, they used Dr. Allen instead. He stood on the shuttle arm and grabbed the satellite, then held onto it while a colleague inside, Dr. Anna L. Fisher, guided it into the cargo bay.

As if all that weren't enough, Drs. Allen and Gardner completed their recovery of the second satellite in less time than the schedule allotted.

Now *that's* truly amazing.

The Dispatch

Columbus, OH, November 6, 1984

The space shuttle *Discovery* is set to take off on an exciting mission tomorrow that will see the launch of two satellites and the hoped-for recovery of those two super-satellites that went into useless orbits last February.

The shuttle will lift off at 8:18 a.m. The crew will include Anna Fisher, America's fourth woman in space and its first mother. She and her husband, Bill — also an astronaut — became parents of a daughter last year.

The flight will last for eight days. Early on, the crew will launch communications satellites owned by Telesat of Canada and Hughes Communications Services.

Then, with its cargo bay empty, the shuttle will close in on the two large communication satellites that need to be retrieved and repaired. The satellites were set loose by the shuttle early this year, but faulty rockets on the satellites caused both to go into errant orbits.

Insurance underwriters, who shelled out $180 million when the satellites were lost, are paying NASA $5.5 million to bring the satellites back to Earth. They will then be repaired and sold.

"I guarantee you, we're going to have a lot of fun," exclaimed commander Rick Hauck.

It sure sounds that way.

The Evening Gazette

Worcester, MA, November 13, 1984

There's something refreshingly American about the mission of Discovery to reclaim stray satellites from space. It's sort of a space version of kids hanging around the edges of the golf course to retrieve lost golf balls.

Discovery was assigned to pick up the two communications satellites which misfired into useless orbits last February. One has now been stowed away and the other is scheduled to be recaptured tomorrow.

A combine of British and American underwriters has obtained title to Westar, formerly owned by Western Union, and Palapa, which had belonged to the government of Indonesia, after paying off insurance claims. The claims cost Merrett Syndicates of London and International Technology Underwriters of Washington, D.C., a total of $180 million. To those costs will be added a $5.5 million charge for Discovery to pick up the errant satellites and $5 million paid Hughes Aircraft for building equipment and nudging the satellites into orbits from which they are to be rescued.

Post-salvage costs are still to be reckoned. Obviously, the underwriters are convinced that this expensive effort is worth all the technological tricks it takes to bring Telestar and Palapa back to Earth for re-use.

There are a number of exciting facets to the Discovery mission, including that of growing crystals in microgravity by using six football-sized chemical reactors aboard ship. One is the pre-rescue task of launching two other satellites, Anik D-2, which the Canadian government is having placed in "storage" for a back-up to five other communication satellites in service, and the Leasat-1, which the Navy is leasing from Hughes Aircraft for communications. It may sound routine, but the Discovery mission is science and technology at its best.

Meanwhile on Earth, hundreds of scientists are preparing for future missions. The Canadians, for example, will send up Anik C-1 in February, and once it is established they will sell it to a private company or another nation. Capitalism in space!

The Arizona Republic

Phoenix, AZ, November 14, 1984

THOSE uncommon men with the right stuff who first ventured into space were heroes.

Alas, mastery of space has become so commonplace that those who venture forth now are little more than high-tech mechanics.

Most Americans with any interest in the U.S. space program can remember the names of the first seven astronauts, whose flights, actually, were primitive to the point of requiring less flight skill than derring-do.

But who can remember the names of the dozens of U.S. astronauts who have followed, including the first American women in space, those who've walked on the moon and those who have taken death-defying, untethered extravehicular trips outside the space shuttle?

Just how prosaic space flight has become, and how extraordinarily proficient astronauts have become, is evidenced by the latest mission of the the space shuttle *Discovery.*

The crew aboard *Discovery* is to recapture two communications satellites — Palapa and Westar — that have been orbiting uselessly since February.

Palapa was snared and stored in *Discovery*'s cargo bay for return to Earth, and Westar is to be snared and stored today.

In snatching Palapa from useless orbit and storing it, the Discovery crew used its wits as well as modern grappling devices designed for space work to do the job.

However, this is more than an exercise in space acrobatics.

Big money is at stake. Insurers of the two commercial satellites paid out $180 million in losses to the government of Indonesia and Western Union who were owners of the misfired satellites.

Now the insurance firms hope to repair the $35 milion satellites and sell them as "flight tested" comunications systems to other owners.

Surely, with such routine work in space being performed, it won't be much longer before some irreverent comic is making jokes about buying a used satellite from some political sharpie.

DESERET NEWS

Salt Lake City, UT, November 14, 1984

By snaring a nearly one-and-a-quarter-ton satellite out of wayward orbit and hauling it back into the space shuttle this week, the U.S. managed to give some dramatic justification for its expensive space program.

One of the reasons advanced for the shuttle's existence was the ability to retrieve or repair objects in space. The attempted rescue of a second wayward satellite is scheduled later in the week.

Another historic precedent during the satellite retrieval was the nature of the space walk by astronaut Joseph Allen. Using a jet-powered backpack, he launched himself from the shuttle to the nearby satellite — the first time an astronaut has sailed into the void without a tether tying him to safety.

Yet despite these successes, something seems to be missing from the space program — the lack of a transcendent goal beckoning into the future. The excitement of early space flights was not just because of their newness and risk; they also were part of a larger plan, namely, getting to the moon. That was real exploration that people could appreciate.

While orbital space flights offer some possible advances in technology, science, industrial techniques, and even medicine, they essentially aren't going anywhere except around and around the earth — an exotic environment, to be sure, but still growing somewhat routine and mechanical.

The U.S. space program needs a more definite goal of just where it is headed and why. The justifications don't need to be linked to finances or certain earth-bound benefits any more than when President Kennedy electrified the nation by pledging to put Americans on the moon in less than a decade. In fact, since the moon-landing days, the space program seems to have gone into something of a retreat.

The acquisition of more orbital expertise may be enough for the moment, but the goals need to be farther out — a base on the moon, for example, followed by an expedition to Mars or the asteroids. Whatever the targets, they should be clearly stated and within certain timetables.

Without that kind of vision of the future, the space program may find itself relegated to the sidelines, a form of tinkering in orbit.

Sen. Jake Garn Picked For Space Flight

NASA announced November 7, 1984 that Sen. Jake Garn (R, Utah) had accepted an invitation to take a ride on the space shuttle. No specific flight had yet been chosen. Garn, 52, was an experienced pilot and chairman of the Senate Appropriations Committee's subcommittee on independent agencies, which oversaw the NASA budget. President Ronald Reagan August 27 had promised that a schoolteacher would be the first civilian passenger on a shuttle mission. In response to criticism over its offer to Garn, NASA Nov. 9 said the senator would not be aboard as a civilian observer but rather would be flying "on an inspection visit in a management role."

After being bumped from one mission, Garn finally traveled on the space shuttle Discovery mission of April 12-16, 1985. But a stillborn satellite marred the mission, the 16th of the U.S.'s space shuttle program. The launching of two communications satellites was the mission's primary objective, and a Canadian satellite, Anik C-1, was deployed successfully April 12. The second satellite to be released, Leasat 3, was built by the Hughes Aircraft Co. for leasing to the Defense Department for operation in its worldwide communication network. It was to be put in orbit 22,300 miles above the equator above the Indian Ocean. But when it was deployed from the shuttle April 13, the barrel-shaped satellite—14 feet wide, 20 feet high and 7,900 pounds in weight—went into a slow rotation of about two revolutions a minute. The $40 million machine just slowly rotated. The remainder of the mission, which was prolonged two days because of the problem, was spent in an unsuccessful effort to trigger the inactive satellite into operation. The rest of the flight's tasks were decidely low-key. Garn had volunteered for medical tests related to motion sickness, which normally struck at least half of a shuttle mission's crew, usually in the first two days in orbit. Garn was equipped with stethoscope microphones around his waist to record sounds from his intestines. Appearing in a brief telecast from the shuttle April 14, Garn corroborated that he was rigged for monitoring "good old bowel sounds."

Lincoln Journal

Lincoln, NE,
November 13, 1984

Some government agencies will go to extreme lengths to impress members of Congress. Count the National Aeronautics and Space Administration as one. Going more than 100 miles high ought to strike anybody as an extreme in lengths.

The Journal expressed modulated approval two months ago when NASA said it would select a schoolteacher as the first citizen passenger on a space shuttle flight. If nothing else, at least there would be several levels of positive symbolism for education and teachers in that.

But the enthusiasm is substantially tempered by NASA's invitation to Sen. Jake Garn to be the first licensed politician making a sky ride.

The Utah Republican just happens to be chairman of the Senate appropriations panel that has jurisdiction over NASA's budget. Having a buddy in a strategic position to wrestle powerfully for agency askings is a kind of bureaucratic-legislative marriage made in . . . well, the heavens.

Currently the space shuttle program has the nation transfixed. Monday's salvage of a misplaced satellite is an example of super technology and human daring which spell drama. But all of that is surely possible without having a member of Congress aboard, witnessing. And once one of the 535 members is lofted (and presumably returned), the pressure will be on to give some of the remaining 534 a comparable ride — at a cost unparalleled by previous congressional junkets.

Is NASA's agenda for shuttle tasks so limited that room has to be made for a politician?

[Forget not, too, that the next shuttle flight is to be entirely military in character. Secret equipment going up. It may not have occurred to most Americans, but it surely has to the Kremlin, that a U.S. space shuttle operation which can snatch back a satellite made in the U.S.A. can do likewise with a satellite made in the U.S.S.R.]

All the congressional oversight needed for NASA can be obtained, it seems to us, on the surface of the planet.

The Wichita Eagle-Beacon

Wichita, KS, November 14, 1984

Sen. Jake Garn, R-Utah, has hinted more than once at his willingness to "kick the tire" (figuratively speaking) of a space shuttle, because "it is a necessity that congressmen check things out that they vote for." Considering that Mr. Garn chairs the subcommittee overseeing the National Aeronautics and Space Administration's budget, it's no surprise NASA last week announced he will be the first politician ever to leave the earth's atmosphere (literally speaking).

NASA officials insist Mr. Garn's not-so-subtle hints — voiced during subcommittee hearings to determine the magnitude of various NASA budgets — had nothing to do with their decision to put him on the crew of an as-yet-unspecified shuttle flight. Considering that Mr. Garn not only is in excellent physical condition, but a pilot, that could be true.

What's really bothering us is what Mr. Garn will do after the first hour of the mission — once he satisfies himself NASA's tax dollars are being spent sensibly. As Mr. Garn no doubt would be the first to agree, that portion of those funds devoted to putting him into space shouldn't go to waste. Accordingly, there follow some suggestions on how he might occupy his time aloft:

- Determining the effects of weightlessness on filibustering. (For the sake of the rest of the crew, Mr. Garn should do this alone, in the payload bay.)
- Forcing an errant satellite back into orbit by threatening to cut off its funds. (This may prove the best repair technique yet.)
- Space campaigning. (NASA could make this more than an exercise in futility by including another Utahan in the crew.)

No matter how Mr. Garn spends his time in space, we admire him for this imaginative venture in congressional fact-finding (euphemistically speaking).

Detroit Free Press

Detroit, MI, November 12, 1984

SEN. JAKE Garn, R-Utah, has solicited and been offered an invitation to fly aboard the space shuttle. By no small coincidence, Sen. Garn is a power on the Senate Appropriations Committee, where he takes a strong interest in the budget of the National Aeronautics and Space Administration. For that conflict of interest alone, the senator ought to be grounded.

Sen. Garn, an old Navy pilot, would not be much practical use aboard the shuttle. NASA administrator James Beggs says the senator needs to check things out in orbit, the better to oversee the financing of the space program. Mr. Beggs, we think, is underestimating the public's intelligence. You don't need to be an elephant to teach zoology.

Nor does NASA need to work hard to convert Sen. Garn to its cause. The senator is a staunch supporter of the space exploration program and his home state of Utah holds $1.5 billion worth of shuttle rocket-building contracts.

For NASA to squeeze a deadweight senator into the flight schedule only arouses the suspicion that the agency doesn't have a better use for its expensive hardware. A junket is a junket, after all, even if it's out into space. And orbiting impresses folks on earth much less than it used to. If NASA and Sen. Garn have any doubts about that, they should check the terms of endearment that Senate candidate Jack Lousma, former Sen. Harrison Schmitt, or presidential candidate John Glenn heard from the voters recently.

Los Angeles, CA, November 11, 1984

We hate to be cynical about something as all-American as the space shuttle, but we can just picture the meeting at which NASA decided to give Sen. Jake Garn a ticket for a ride on the space shuttle. NASA is still looking for a teacher to make the trip, but Garn may be the first non-astronaut to soar into orbit, possibly as early as next spring.

"Who should it be?" we imagine a NASA bigwig asking his colleagues. "A poet? An editorial writer? A housewife from Peoria? Or the chairman of the Senate subcommittee that approves our budget?"

It must have been a tough choice.

In all fairness, though, the selection of Garn makes sense, even if the ultra-conservative Utah Republican isn't our favorite politician. As a former Navy pilot, he actually has more flight time than all but one of America's astronauts. As chairman of the Senate panel that oversees NASA appropriations, he's an important figure in the U.S. space program, of which he is a strong (but hardly uncritical) supporter. On top of all that, he really wants to make the trip.

Alas, so do we. Color us jealous. ■

THE BLADE

Toledo, OH, November 22, 1984

THE Blade has been as critical as anybody of congressional junketeering, the proclivity of congressmen and senators to take expensive foreign trips — at taxpayers' expense — which really have no legitimate basis or justification.

So we would be abandoning our responsibility if we overlooked the ultimate junket: the scheduled space shuttle flight next year of Sen. Jake Garn, Republican of Utah. Senator Garn is courted by the National Aeronautics and Space Administration because he is chairman of the Senate committee which oversees federal funding of space exploration.

In other words, unless he gives the word and provides the bucks, the birds don't fly.

Senator Garn's flight, as one might expect, is billed as an educational experience for the senator, the better to weigh the merits and many advantages of a suitable space program. But we imagine that this political space pioneer will blaze the trail for a whole parade of lawmakers from Capitol Hill. First thing you know, an engineer will be bumped off a flight to make way for the head of some obscure subcommittee.

We wish that NASA had stuck to President Reagan's plea that the first non-NASA astronaut be a teacher. When it comes to symbolism, sending educators into orbit makes more sense than sending politicians. A teacher-astronaut will not be selected until next summer, however, and his or her flight will not come until 1986.

On the other hand, how nice it would be if we could just persuade Senator Garn to take the federal deficit with him into space aboard the Columbia, and simply jettison the darned thing into oblivion.

Pentagon Asks Shuttle Secrecy; Washington Post Reveals Payload

Defense Department officials December 17, 1984 announced that extraordinary measures would be taken to keep secret the next mission by the space shuttle Discovery, scheduled for January 23, 1985. The launch was to be the first all-military mission for the spacecraft fleet. Brig. Gen. Richard F. Abel, Air Force director of public affairs, told the press Dec. 17: "The more mission information they [the Soviets] have, the easier it is for them to counter the capabilities of those payloads." Thus, Abel said, the press would not be allowed, as it customarily would, to follow the launch countdown, and news articles that "speculated" on the shuttle's secret payload would be investigated as breaches of national security. To disguise the launch time, foreknowledge of which would allow easier tracking, the shuttle's schedule would not be announced in advance, and duration of the mission would not be announced until 16 hours before landing. In response to a reporter's question about possible penalties for press coverage of the shuttle mission, Abel declined to speculate but said it "would depend on the story that was written" whether to launch a probe.

The Pentagon's stricture concerning news articles about the Discovery mission was tested the following day, when the Washington Post published a story describing the shuttle's January payload as a "signit," or signals intelligence, satellite for use over the Soviet Union. Although the story contained little technical detail, it brought an immediate blast from Defense Secretary Caspar Weinberger. "It's the height of journalistic irresponsibility," Weinberger said, "to violate requests that are made...There are certain things that we have to do that we should do, that when they're published can only give aid and comfort to the enemy." Post Executive Editor Benjamin C. Bradlee responded that Weinberger's reaction was "not justified." He said the paper had published the story only after "careful review" and that virtually all the information in the story was already public. A story in the paper Dec. 19 detailed more of the public domain information that had gone into the original story.

The original Post story said that the $300 million satellite that the shuttle would be carrying was a successor to an existing series used to intercept Soviet electronic communications and data transfer. It would join four or five "signit" satellites already orbiting above the Soviet Union, but would be the most massive yet—at least 30,000 pounds—making it too large to launch by conventional rockets. The story said the new satellite would be deployed in a geosynchronous, Earth-stationary orbit, somewhere over the Western Soviet Union, where it could be used to monitor missile tests and verify arms control compliance. The Post cited several military experts who explained why it was easy to identify the payload from the public record.

THE SUN

Batimore, MD, December 21, 1984

Secretary of Defense Caspar W. Weinberger says the Washington *Post* story on the forthcoming use of the space shuttle to put an intelligence-gathering satellite into orbit "can only give aid and comfort to the enemy." Those are harsh words. The Constitution invokes "aid and comfort" in its definition of treason.

This extreme and irresponsible language is not surprising. He and his president have been more sensitive to leaks and more insensitive to the public's right to know than their predecessors. The Reagan administration has at one time or another championed widespread lie detector testing, lifetime censorship of federal employees and the banishment of reporters from war zones. The policy of trying to shroud the space program in secrecy is hardly a surprise coming from this mindset.

The case of the shuttle mission is a classic case of insiders wanting to keep information inside. We don't mean keeping it secret or classified. We mean just keeping in the community. According to one insider, Senate Intelligence Committee member Daniel P. Moynihan, "I saw nothing in the article that you wouldn't just naturally know if you knew anything at all about this subject."

Probably all the *Post* did was cull the air and space journals, other technical writings and non-classified congressional testimony, and ask some informed questions of some informed officials and former officials. Then it presented to the general public a story that every expert (and Soviet intelligence agents are experts) already had figured out. We absolutely agree with Richard Smyser, the Oak Ridge, Tenn., editor who is president of the American Society of Newspaper Editors: "The only significant group that did not have the information was the American public."

It is, of course, the media's job to provide the public with such information. Newspaper editors no less than military officers and their civilian superiors are aware that there are things which must remain secret. Editors and reporters keep secrets every day. Newspaper editors are also aware that much censorship is self-serving, unjustified, or a coverup for mistakes or waste or potentially unpopular activities. Security is not always a prime motivation. Because of that, newspapers often decide to go with a story rather than withhold it. Reputations may be at stake, or policies, but not national security. Mr. Weinberger's prattle about aiding and comforting the nation's enemies is poppycock.

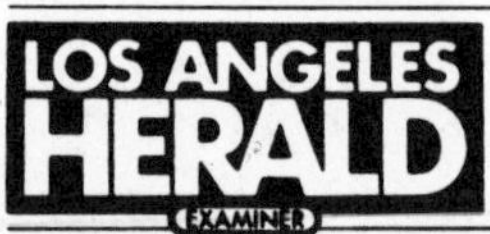

Los Angeles, CA, December 21, 1984

The Washington Post, followed by several other news organizations, reported this week that next month's secret space shuttle mission will carry a satellite capable of intercepting a variety of Soviet transmissions — whereupon Defense Secretary Caspar Weinberger flew into a rage, charging that the paper had endangered national security. All this occurred after defense officials had asked several news outlets to withhold, in the interest of national security, any details they had uncovered about the flight's military payload. (Ben Bradlee, the editor of the Post, claims no such request came to him, although he says he nonetheless did withhold some sensitive information about the launch.) For some reason, the Pentagon's sensitivity even extended to when the shuttle would lift off — as if that wouldn't be immediately apparent to a Russian trawler just off the coast or, for that matter, anyone within sight of the launch. The Pentagon went so far as to threaten investigations if stories merely *speculating* about the load were published. Rarely, if ever, has the government made such a public to-do about a secret it wanted to keep, which, of course, fueled the very speculation that the Pentagon claimed it didn't want.

Were Weinberger & Co. simply trying to pick a fight with the press and the networks to show them who's boss? It's hard to imagine any other rationale for the Pentagon's advance warnings or for Weinberger's tantrum. The information the Post published was no secret: Aviation Week, the Associated Press, NBC and CBS News and other news outlets already had information about the shuttle's mission, and, if they could get it, we imagine the KGB could, too.

What's most troubling about this case is not the Pentagon's rather lame attempts at pre-publication censorship or that its secret was so quickly blown. What's troubling is that the military is now using NASA, instead of its own space program, to launch spy satellites. This role for the shuttle has been in the cards all along. NASA, however, has always been a civilian agency wide open to public scrutiny, which helps explain why the U.S. space program is so admired around the world. We will sacrifice that openness at our peril. What's more, we'll bet the Pentagon would find it a lot easier to keep its spy-satellite secrets if it used its own rocketry and technology instead of NASA's. ■

The Pittsburgh PRESS

Pittsburgh, PA, December 20, 1984

According to The Washington Post, the "secret" military payload the space shuttle Discovery will launch next month is a satellite that will remain in stationary orbit over the Soviet Union, intercepting radio, telephone and satellite signals and data from missile tests.

Did the Post, in printing that information Wednesday, violate national security?

Defense Secretary Caspar Weinberger, calling the action the "height of journalistic irresponsibility," says security was breached — and he's right. Only the violation was committed in Mr. Weinberger's own Defense Department.

In addition to the Post, at least five news organizations — the Associated Press, ABC, CBS, NBC and Aviation Week — knew about the satellite. Most of them got their information from Pentagon sources and had more details than the Post published.

So the Pentagon's attempts at unprecedented secrecy for the Jan. 23 launch apparently had been doomed from the start. And Mr. Weinberger's unhappiness with the Post was only one sign of the brass hats' larger frustration in trying to clamp an impossible blackout on a program that, until now, had been a model of openness and public involvement.

In fact, that openness had helped assure the funding for — and the success of — this country's space explorations.

Surely Mr. Weinberger does not expect Americans to believe that when newspaper, television and magazine reporters know about shuttle cargo, the Soviet Union, with its vast intelligence apparatus, can be in the dark.

But let's assume that (a) Moscow's spies are incompetent, (b) they don't read congressional testimony about satellite launches and (c) the blast off occurs amid the secrecy the Pentagon desires. What next?

First, the Soviet trawler sitting three miles off the Kennedy Space Center tracks Discovery with its powerful telescopes and radar. Then space stations in the Soviet Union observe the satellite launching. They study the device's distinctive configuration, and one scientist says to his comrade: "Grigori, it's the new American sigint (signals intelligence) satellite. They've got five up there now."

This doesn't mean that Mr. Weinberger has nothing to worrry about. In fact, as he was denouncing the Post, a Northrup Corp. technician was being booked for trying to sell Russia "Stealth" technology, advanced methods of hiding bombers from radar.

In recent years there has been a hemorrhage of sensitive information from defense contractors, military officers and even an FBI agent to communist spies. Mr. Weinberger would do well to thwart such serious espionage and worry less about shielding the American public from things that the Russians already know or inevitably will learn.

THE SACRAMENTO BEE

Sacramento, CA, December 21, 1984

To hear Defense Secretary Caspar Weinberger tell it, you'd think that the Washington Post's decision to publish details about next month's space shuttle cargo had imperiled the security of the nation. This is "the height of journalistic irresponsibility," Weinberger declared, which "can only give aid and comfort to the enemy."

That's a grave charge but the facts do not support it. What's going on here has more to do with Pentagon politicking and media-bashing than with keeping vital secrets.

There's no doubt that the Pentagon has a legitimate — and powerful — interest in not disclosing matters that it regards as critical to the nation's defense. But the broad description of the forthcoming shuttle mission as published in the Post, including its orbit and general purpose but not including sensitive detailed information, poses no threat to national security. Once the Air Force announced earlier this year that it would be sending a payload beyond the space shuttle's low orbit, it was easy enough for the scientifically informed to puzzle out the nature of that payload.

Only three categories of military satellites operate at such high orbits, and only one, electronic intelligence satellites, have ever generated such high-level security concern. Those facts — and considerably more — have already been published in the technical literature. And as Sen. Daniel P. Moynihan, the former vice chairman of the Senate Intelligence Committee, pointed out, Pentagon spokesmen have said at least as much in open testimony.

All that makes the cause of Secretary Weinberger's pique the topic of considerable speculation. There's informed conjecture that the Air Force may have promoted this fight, in order to secure the funding it needs to launch this new generation of spy satellites on its own — independent, that is, of NASA's space shuttle program.

Some Washington insiders have also surmised that Weinberger may have seized the occasion to take on the press, in hopes of winning new popularity for himself and his budget. Evidence for this hunch comes from the comment of Fred Ikle, undersecretary of defense for policy and the most vigorous promoter of new secrecy measures within the department. Weinberger's real concern, Ikle said, is "procedures," — meaning the Washington Post's defiance of a general administration admonition not to publish — and not the substance of the disclosures.

All of this is of more than passing concern. For if the satellite is part of the administration's Star Wars planning, as some people suspect it is, then its launching is a matter of the highest public interest. But even if old-fashioned space spying is all the Air Force means to accomplish, there's no reason to keep that fact out of the press. Such information is hardly likely to surprise the Soviets, who have been known to read technical journals, follow congressional hearings — even keep a watchful eye on the doings at Cape Canaveral. Had this information remained unpublished, the only group that would have been left uninformed would have been the American people.

The Augusta Chronicle

Augusta, GA, February 2, 1985

It is impossible to keep the launch of an American space shuttle a secret for the obvious reasons — you just can't hide the spectacle, nor has the United States ever attempted to do so before.

But the U.S. Air Force would have liked to have kept secret the just-completed Mission C-51. The spacecraft *Discovery* was launched Jan. 24 from Kennedy Space Center in Florida to place into orbit a very large satellite to monitor the Soviet Union's communications. *Discovery* returned last Sunday, and the mission was reported successful.

This was the nation's first use of the space shuttle for a military mission after 14 flights for scientific and commercial purposes with civilian crews. We must remember, however, that the spacecraft's payload bay was built to the Pentagon's specifications and intended for the military's use.

In the future, one-third of the space shuttles' missions will be for military purposes; the National Aeronautics and Space Administration owns the space vehicles and controls the launches and operations, but the Defense Department will pay for the missions. The one just completed cost the Pentagon $31.2 million (of the estimated total flight cost of $125 million.)

Secrecy, which is instinctive with the military, is going to be a complex matter in the future, as it was with the mission last weekend. Both the United States and the Soviet Union have spy satellites constantly monitoring for missile launches as well as intercepting each other's communications.

The Kremlin's intelligence operations were certainly capable instantly of monitoring the recent launch, and we do not doubt they also know precisely where *Discovery* placed the new spy satellite in space. What they may not know, however, are the satellite's technology and range of capabilities, which we are led to believe are much more sophisticated than any the Soviets now have.

The predicament of trying to conceal the launch of such a huge missile with its piggy-backed space shuttle gave humorists and cartoonists an exceptional opportunity for creativity.

Just the same, the mission was extremely important to our national security, for it adds significantly to our "national technical means" of monitoring military and other activities in the Soviet Union and its allies in Eastern Europe.

When the debate comes again over verification of the Soviets' compliance with any nuclear arms agreements, our means of verification are substantially better than what they were before Mission 51-C. And, our capability of "listening in" will improve our assessment of the Soviets' intentions, which is a chief goal of intelligence.

Maybe in future military missions for the space shuttles the Air Force can accomplish more secrecy than it has with this first mission. But, frankly, we are heartened by what we have learned. The satellite is a benign instrument of defense which should help keep peace on earth.

The Wichita Eagle-Beacon

Wichita, KS, December 21, 1984

Defense Secretary Caspar Weinberger's outrage at the Washington Post for publishing a story about the nature of next month's NASA-Air Force secret shuttle mission purportedly is so immense he would have Congress pass laws to prevent future such "journalistic irresponsibility." Had the Post truly revealed secrets, and had the revelations truly damaged national security, Mr. Weinberger's outrage might be understandable.

It's a safe bet Mr. Weinberger, deep down, isn't so much outraged as frustrated that his latest attempt to intimidate the press has ended in failure. In the end, it was news executive themselves, not the Defense Department, who decided what to publish. And it's clear the Post acted responsibly in making that decision: The story, which ran in the Post's Wednesday editions, was reviewed carefully before publication, and technical details that conceivably could have been of interest to the Soviet Union were withheld.

The gist of the Post's story — that the shuttle payload would be the first of a new generation of military intelligence satellites and would be deployed over Western Siberia — wasn't quite as "secret" as Mr. Weinberger insists. In fact, that information already had been bandied about in Washington in a variety of places, most notably in open congressional hearings and in reports of other newspapers that dealt with the Air Force's shuttle program. All the Post reporter did was find Defense sources willing to confirm what already was known.

One suspects this incident, which Mr. Weinberger has blown far out of proportion in relation either to its news value or to its effect on national security, really is the latest offshoot of the Reagan administration's war on "leaking" — the process by which administration insiders, for a variety of motives, convey newsworthy unpublished information to journalists. Leaking always has gone on in Washington and other centers of government, and is one of the ways by which readers, viewers and listeners learn the full effect of governmental decisions and actions.

Mr. Weinberger apparently either has a self-serving view of this process, or doesn't understand that government only can work as the framers of the Constitution intended — as a participatory entity in which citizens themselves ultimately make critical decisions — when information flows to the people unimpeded. That's why it's a blessing that he only can suggest press controls, not enact them.

WORCESTER TELEGRAM.

Worcester, MA, December 22, 1984

The Defense Department's efforts to stamp "TOP SECRET" on the Jan. 23 shuttle flight has backfired. The first shuttle flight to be fully devoted to military operations has generated more public interest and news coverage than the Pentagon wanted. In the shuttle's cargo is a new generation of intelligence satellite with a capability of monitoring radio signals on earth.

The Pentagon triggered the controversy when it announced a new policy of secrecy applying to all shuttle flights with military cargoes. Citing national security, Pentagon spokesmen said restrictions would even omit exact launch times and nature of cargo. Some ominously suggested that speculation in the news media as to the nature of the payload might be investigated.

That was a guaranty of more attention, not less. In no time, the restrictions led to a full-scale debate. The policy's effectiveness was questionable from the start. Details about the spy satellite were widely published.

It was all predictable. The National Aeronautics and Space Administration was established as a civilian agency to conduct open and peaceful missions in space. Now with the first military cargo to be taken aloft, the administration has fallen on its face.

Some things should be kept secret, and it is up to the Defense Department to keep them secret. The government shouldn't expect newspapers and television stations to keep its secrets for it. Their job is to disseminate the news, not restrict it.

But why was the administration so set on keeping from the American public information that obviously is available to the Russians? Much of it already is in the public domain, either in technical literature or through Congressional testimony.

Spy satellites are really nothing new. They have been whirling around the earth for more than a decade. They eavesdrop on earth communications, take photographs and pick up all sorts of other information.

The Russians know a lot about the subject. Their intelligence gathering ships closely monitor American space shots. From intercepted radio signals they can get a pretty good idea as to the satellite's mission and payload. Information that the NASA space shuttle will carry a spy satellite in its cargo bay is hardly news.

No doubt it's annoying to the administration when government officials talk out of turn. But that's not the fault of the media.

Far more worrisome are the spies for foreign countries. Only a few days ago, an American was caught trying to pass information about the stealth bomber to Soviet agents. That's the sort of thing our intelligence agencies should be concerned about, not news stories about matters that already are widely known.

DESERET NEWS

Salt Lake City, UT, December 21, 1984

Did The Washington Post violate national security?

Secretary of Defense Caspar Weinberger thinks so. But he's wrong. Worse, by mis-diagnosing the problem, he is making it even more difficult to cure.

This week Weinberger said it was "the height of journalistic irresponsibility" for the Post to report the supposedly secret military payload that the space shuttle Discovery will launch next month.

It's a satellite that will remain in stationary orbit over Russia, intercepting radio, telephone, and satellite signals and data from missile tests.

America would have been better off, all right, if this information about the Discovery satellite had been kept from Russia. But don't blame the Post or other reporters for spilling the beans. Instead, the buck stops at Weinberger's own Pentagon.

Before the Post story came out Tuesday, at least five other news organizations — the Associated Press, ABC, CBS, NBC, and Aviation Week — knew about the satellite. Some of them had more details than the Post published. Most of them got their information from sources within the Pentagon and from the Reagan administration's public testimony on Capitol Hill.

It can be argued that the Post still should have withheld the story, as some news media did, even though it was not among the outlets that the Pentagon asked to withhold the information.

But such an argument is not convincing — unless one happens to believe that the Russians, with their vast intelligence apparatus, are somehow unable to unearth information that the American press can dig up.

Or would Secretary Weinberger have us believe that the American public shouldn't be told what the Russians likely already know?

This doesn't mean that Weinberger has nothing to worry about. On the contrary, there has been a hemorrhage of sensitive information lately from defense contractors and military officers to communist spies. If the secretary wants to prevent security leaks, he has some plumbing to do much closer to home than the offices of The Washington Post

The Salt Lake Tribune

Salt Lake City, UT, December 21, 1984

At a recent conference in Washington, D.C., dealing with the economic implications of various proposed acid rain control strategies, we were introduced to the term "popular journals."

This was how conference panelists and speakers, mostly economists of various stripe, differentiated between those publications readily available to the general public — ordinary newspapers and magazines — and the technical publications that are narrowly oriented toward a very specific audience.

The term comes to mind because of Defense Secretary Caspar Weinberger's unthinking condemnation of the Washington Post account which identified the payload to be carried in the January space shuttle mission.

What was ostensibly billed as a highly secret military mission has now been revealed as a $300 million Air Force effort to launch a satellite that can spy on the western reaches of the Soviet Union and "listen" to all sorts of electronic communieations.

The defense chief protests excessively and unreasonably. What the Washington Post published was, essentially, information long available to the American public, along with the Soviet Union, in all sorts of "popular journals," as well as from congressional testimony and more than just a few non-classified, technical journals.

From those sources anyone else could have foreseen that the United States would be likely to attempt an upgrading of the capabilities of communication intelligence satellites that have been orbiting around the Earth for more than a decade, eavesdropping on Soviet radio communications.

"The congressional intelligence committees have made not the least secret of the fact that we've provided funds for verification methods in space," Sen. Daniel Patrick Moynihan, D-N.Y., former vice chairman of the Senate Intelligence Committee, told the New York Times.

"The details are confidential and should be kept confidential," he said. "But I saw nothing in [the Post] article that you wouldn't just naturally know if you knew anything at all about this subject."

The Pentagon, along with Mr. Weinberger, must heap most of the blame on themselves for the Washington Post's effort.

With their wholesale, pre-launch publicity effort, they, in essence, were like the parent who tells a child that its Christmas present is in a certain closet and then telling the child not to look in the closet. That sort of temptation, along with Pentagon's misbegotten "hope" that some element of the American press would not look in the closet, is too much to resist.

It is appropriate to recall that the Air Force and Central Intelligence Agency have managed to launch military and intelligence satellites through the years in relative secrecy, in an operational pattern that won public and press acceptance. But in this case, the administration dramatically changed both the public routine and open nature of the National Aeronautics and Space Administration.

"If the Defense Department wanted to keep this particular mission classified, it chose the worst possible approach," James R. Schlesinger, a former defense secretary and director of central intelligence, said.

"By throwing the spotlight on this mission, it produced an enticement for people to go after what the mission was about and then to publish what they found out.

"If the objective was re-establishing the capability of the Air Force in the long run to have a classified satellite launch, that's understandable. But if the objective was to keep this particular mission classified, it was almost inevitable that something would leak. Breaking the [public] routine of the shuttle flights and the NASA tradition of 20 years was bound to call attention to this mission."

Mr. Weinberger's Defense Department has, in short, been hoisted on its own petard. Instead of making hollow-sounding accusations about giving "aid and comfort to the enemy," they ought to muster enough dignity to crawl quietly back into some dark corner to lick their self-inflicted wounds.

Teacher Named for Space Flight

Sharon Christa McAuliffe, 36, was selected July 19, 1985 as "the first private citizen passenger in the history of space flight." McAuliffe, a high school teacher in Concord, N.H., was chosen from more than 11,000 applicants received since President Ronald Reagan announced during the election campaign in August, 1984 that he wanted to send a teacher into space to highlight the importance of the profession. Her flight aboard the shuttle was scheduled for late January, 1986. The primary mission of the shuttle would be the launching of a tracking and relay satellite and a satellite to photograph Halley's Comet. The association with NASA would last a year beyond that, according to current plans, when McAuliffe would go on a speaking tour across the country on the experience of space flight.

NASA announced October 24, 1985 that it was seeking a journalist to become a civilian passenger on the space shuttle. NASA administrator James Beggs said the agency wanted to give free shuttle rides to people who would "tell their own story in the best way they know how." Other possible candidates for future flights included artists and aerospace industry workers and even a man-on-the-street "with no particular professional qualifications." Applicants to become the first journalist-astronaut would be required to be U.S. citizens, to have at least five years' professional experience in broadcast or print journalism and to pass a "minimum" NASA physical examination. The screening process would be performed by a professional peer group. NASA would select the winner and a backup candidate from a list of five finalists.

The Times-Picayune
The States-Item

New Orleans, LA, July 8, 1985

The depths of space make a spectacular blackboard and the planets, stars, nebulae and galaxies bright chalk marks to be learned by earthly students. The National Aeronautics and Space Administration has now picked ten finalists for the teacher who will be sent up in the space shuttle Challenger on Jan. 22, but explaining the cosmos is too vast a lesson plan to assign them. Indeed, no specific assignment has been given, and that makes the venture a particularly personal, creative one for the teacher chosen.

The project had all the earmarks of public relations puffery when President Reagan announced last August that he was directing NASA to find an elementary or secondary school teacher to send into space. Critical eyes have been turned on American public education, and the president clearly wanted to spotlight exceptional teachers and attract the attention of every student. But judging from the list of finalists and some of their comments, the project holds definite promise.

The ten are from nine states — Idaho has two entries. Selected from more than 10,000 applicants were six women and four men who among them teach English, romance languages, math, science, computers, business, social studies, government and geography and general grade-school subjects.

A single teacher returning from space cannot be expected to teach the nation's children what he or she has learned, experienced or been moved to propound. She — let us say she, since women are in the majority and the odds with them — will be primarily a public-figure teacher who will travel, make speeches and give interviews to focus attention on education, which, like space exploration itself, is oriented toward the future.

One finalist commented, for example, "I want to go up there and inspire some other students to take an interest in math, science and in their future." Another said, "I plan on taking the whole country with me," and wants to collect questions via a toll-free long-distance number.

Another said she wants "to bring back the wonder of it all," and an articulate teacher should be able to do that with a productive purposefulness. This teacher, Sharon Christa McAuliffe, 36, a high school social studies teacher in Concord, N.H., also remarked that she would be like "a woman on a Conestoga wagon pioneering the West," twin comments that encapsulate the real substance of the venture.

The space shuttle and the permanently manned space station now being blueprinted are indeed wagon trains into space, and the teacher will be the first true civilian, on purely civilian pursuits, to accompany the professional trailblazers.

There have been proposals to send writers, poets and artists into space to benefit from their particular sensibilities and points of view. The time will come for them, and it should be soon, for their work will be valuable. But it is proper that the first true passengers to the new frontier should be teachers, for the accumulation and transmission of knowledge and thought is the distinguishing characteristic of our species.

The Salt Lake Tribune

Salt Lake City, UT, July 29, 1985

Selection of the United States' first non-astronaut, civilian passenger for a space shuttle flight seems to be as precise as a rocket launching. Concord, N.H., high school teacher Susan Christa McAuliffe qualifies in every important regard.

That she is an educator was preordained when NASA decided this is the profession from which the first shuttle civilian passenger would be chosen. Nevertheless, Mrs. McAuliffe is the perfect sort of teacher for the honor and opportunity.

Space shuttle personnel tend to be not only intelligent and intrepid, they usually reflect more than the ordinary amount of idealism, amplified by a natural ability to articulate their convictions. Mrs. McAuliffe is certainly cast in that mold.

By all accounts, she is a teacher who so loves her work she devotes boundless energy to making it as relevant and meaningful as possible. Personally designing courses in law, economics and business, she introduces practical experience to theory in ways that enhance the impact of learning. Rather than moan and grouse about the underappreciated nature of teaching, she retains a selfless enthusiasm for its purposes and potential.

For her part, Mrs. McAuliffe wants her historic participation in the space program to elevate the standing of both teaching and teachers. NASA will expect her to join its other veteran personnel in the ancillary process of strengthening a popular resolve to continue this nation's space exploration and exploitation. It's doubtful either she or the team will be disappointed.

THE KANSAS CITY STAR

Kansas City, MO, May 20, 1985

Undoubtedly, the 114 teachers chosen as nominees for NASA's planned civilian flight are excited about the possibility of going into space in 1986. There is probably much interest at Shawnee Mission Northwest and Goddard High Schools in Kansas, and McCluer North Senior High and Park Hill R-5 schools in Missouri, where four of the state nominees work.

But teachers belong in the classroom, not as spectators on board the space shuttle. The National Aeronautics and Space Administration should tell President Reagan, whose suggestion it was to include a teacher as part of civilian travel, "Thanks, but no thanks."

Training a teacher for space travel is a waste of taxpayers' money. So was sending up U.S. Sen. Jake Garn, who was given little more than busywork to justify his participation aboard the Discovery shuttle last month. Congress should be embarrassed that a senator whose responsibilities on Earth include monitoring the NASA budget was hurling dollar bills at meteors.

NASA should spend its money (yours) wisely and invest in the men and women who have trained long and hard to be a part of the space program. Even the monkeys and sacrificial rats aboard Challenger had a definite scientific purpose. Teachers who want a piece of the action would be better off participating in more realistic activities between NASA and educators.

For six years, NASA and the National Science Teachers Association have been involved in the Space Shuttle Student Involvement Program. The program helps to develop an awareness of space by giving secondary school students a chance to submit proposals for space science experiments. The students also qualify for scholarships.

Teacher participation in a space shuttle mission will not stimulate students' interest in space science and technology. All it will do is create a "king" or "queen" for the day. Besides, most parents enroll their students in schools which have good education programs, not which boast of being the home of America's first teacher in space.

THE TENNESSEAN

Nashville, TN, August 4, 1985

A history teacher from New Hampshire will make history herself next year when she becomes the first educator to ride on a space shuttle mission.

Ms. Sharon Christa McAuliffe, who also teaches law, economics, and women's studies to 10th and 11th graders in Concord, N.H., has been selected by NASA to ride on the Challenger mission scheduled for January.

Ms. McAuliffe's selection made good President Ronald Reagan's vow last year that a teacher would be the first "citizen passenger" launched into space. She will not, however, be the first non-astronaut on a space mission. Two scientists — one German and one American — flew on a Spacelab mission in 1983, and Sen. Jake Garn from Utah was onboard a Challenger mission earlier this year.

The thousands of teachers who vied for the space shot were narrowed down to 114 semifinalists and then 10 finalists. The teachers were asked to submit a proposal on how they planned to use the space mission experience when they returned to their less-than-heavenly classrooms. Ms. McAuliffe plans to keep a daily journal of the flight. She said that a diary would be simple to keep, wouldn't interfere with the mission, and would help "demystify" space flight.

The "teacher-in-space" program was obviously designed to boost citizen support of NASA, which in turn will keep Congress generous with the space agency. If NASA wanted the first teacher in space to project an all-American image, it got just that with Ms. McAuliffe. In addition to being a teacher, wife of an attorney and mother of two, she is a Girl Scout leader, and a volunteer at a family planning clinic. She is also articulate and enthusiastic, qualities that will serve NASA well when she goes on her post-flight six-month speaking tour.

But the program might have more difficulty in achieving its other goal — that of sending a message to American teachers that this administration understands the concerns of teachers and cares about education.

The teachers who applied for the space flight probably benefited by learning more about NASA and by designing their flight projects — so to some extent the "teacher in space" program benefits education. But most teachers are far too busy with earthly problems of crowded classrooms, lack of programs, and low teacher pay to be impressed with or assuaged by Ms. McAuliffe's mission.

The administration hoped that blasting-off a teacher would make teachers feel more professional. What it doesn't seem to grasp is that teachers already are professionals, which is exactly why their morale won't be launched just by sending a teacher into orbit.

The Union Leader

Manchester, NH, July 23, 1985

We proudly dub New Hampshire's Christa McAuliffe the Granite State's — indeed, America's — 1985 *Teacher of the Sphere.*

Since last Friday, when she was singled out from among 11,416 applicants vying for the honor of being the first teacher in space, Christa has given new meaning to the word "vicarious" as millions of her fellow citizens experienced, through her, the indescribable joy of anticipation preparatory to the actual experience of the wonders of space flight. When she returns from her historic flight next January aboard the space shuttle Challenger, the Concord High School social studies teacher may even succeed in updating that time-worn definition of education to read: a teacher at one end of a rocket and a student at the other.

The class exhibited by Christa in accepting the distinction of being the first "teachernaut" and handling its attendant publicity, which surely has already established some kind of record for volume and intensity, affirms the wisdom of NASA's public relations-conscious choice. One need not wonder why 36-year-old Christa was selected for the distinction of being the first private citizen passenger in the history of space flight. She herself has given the answer, given it casually and unpretentiously, through the exemplary manner in which she has conducted herself since Friday's announcement by Vice President George Bush at a ceremony in the Roosevelt Room of the White House.

It is clear that Christa was chosen not simply because she is articulate and attractive. Those qualities go with the territory: the former Christa Corrigan has a Celtic background. Nor was she chosen solely for her intelligence and poise.

Rather, it seems obvious, at least to us, that the decisive determinant in NASA's selection was her delightfully refreshing and unrehearsed *enthusiasm*, essential qualities for any effective communicator of facts and concepts.

Space agency officials have good cause to be smug, for they have pulled off a major public relations coup. It's a lead pipe cinch that Christa McAuliffe, through the sense of wonderment she will bring to her post-flight public appearances and the pages of the journal she intends to keep of her experiences, will prove to be of inestimable value to the entire space exploration program.

Challenger Explodes After Takeoff; All Seven Crew Members Killed

The space shuttle Challenger exploded shortly after launching January 28, 1986 killing all seven crew members. A stunned nation mourned the lost astronauts and pondered the future of its winged craft space program. NASA immediately grounded its remaining fleet of three shuttle orbiters. The disaster, the worst in the history of the American space program, occurred only 74 seconds after takeoff at 11:38 a.m., from Cape Canaveral, Fla.

The craft was flying 1,977 miles per hour, 10 miles up and eight miles down range when an orange ball of fire appeared at the base of Challenger's external fuel tank. The tank held at liftoff a half million gallons of liquid hydrogen and oxygen for powering the spacecraft's three main engines during ascent. Another show of fire flickered around the tank. Then, a few seconds after the first flame appeared, an intense fireball burst out amidship and the craft was instantly engulfed in a giant cloud of fire and smoke. The solid-fuel rocket boosters veered off, trailing white plumes of smoke that began to coil erratically downward. Debris rained into the ocean for an hour after the explosion, making the sight of the fallout, about 18 miles offshore, unsafe to search. The long fallout also testified to the force of the explosion. By nightfall, no evidence had been found that the crew had survived.

The shuttle crew members lost were Francis R. Scobee, mission commander; Navy Cmdr. Michael J. Smith, the pilot; Dr. Judith Resnick, an electrical engineer and mission specialist; Air Force Lt. Col. Ellison S. Onizuka; Gregory B. Jarvis, a Hughes Aircraft Co. engineer; and Christa McAuliffe, a Concord, N.H. high-school teacher selected by NASA to be the first "citizen observer" to ride the space shuttle. President Ronald Reagan addressed the nation from the Oval Office that afternoon, calling the crew members "pioneers" and "heroes."

U.S. space shuttles had flown 24 times without a major mishap since the first flight of the orbiter Columbia in April, 1981. The tragic flight was the first of 15 missions scheduled for 1986, the most ambitious year yet in the program. NASA suspended the shuttle program and named an interim review panel to collect and identify flight data from the mission, pending appointment of a formal investigating board.

ARGUS-LEADER

Sioux Falls, SD, January 29, 1986

Editorial

The shocking spectacle was witnessed Tuesday by millions of television viewers.

A minute after liftoff, space shuttle Challenger exploded into a fireball, killing seven crew members and shattering some of America's hopes for space exploration.

Among the victims was New Hampshire schoolteacher Christa McAuliffe, the first citizen passenger to ride a shuttle. Her death adds to the devastating setback suffered by the National Aeronautics and Space Administration.

Tuesday's unsuccessful flight followed 24 successful space shuttle missions, but it understandably raises questions about the future of the U.S. space program.

President Reagan canceled his State of the Union speech Tuesday and supplied the right answers during his alternative speech to the nation.

Reagan appropriately called Tuesday a day for mourning and remembering. But he also pledged to continue America's space program. U.S. journeys into space will not end, he said.

Victims of the flight should be mourned. And the nation's scientists must determine what went wrong and how to prevent it from happening again. But our journeys into space should not end.

The journey did not end 19 years ago, when three astronauts died in a launching pad fire while preparing for an Apollo flight. Our future and that of our children demands that we remain strong and committed in our hopes for the future.

As Reagan said, sometimes painful things happen along the way.

"The future doesn't belong to the fainthearted," Reagan said. "It belongs to the brave."

McAuliffe, the 37-year-old teacher selected from 11,146 teacher applicants to be the first to fly in NASA's citizen-in-space program, was to have beamed two lessons for live broadcast over the Public Broadcasting System into classrooms across the nation.

Instead, she and the other crew members — Francis R. Scobee, Michael J. Smith, Judith Resnik, Ronald E. McNair, Ellison S. Onizuka and Gregory B. Jarvis — taught us a lesson in purpose and dedication.

That lesson should not be wasted.

The Washington Post

Washington, DC, January 29, 1986

"OH, THE humanity!" Those words, spoken by a weeping radio announcer as he witnessed the explosion and fire that consumed the dirigible Hindenburg nearly 50 years ago, must have come to the minds of some people yesterday as they watched the terrible short flight of the space shuttle Challenger. For a few moments that announcer in New Jersey in 1937 was doing his best as a journalist, describing a disaster that was to prove a turning point in the history of aviation. But suddenly he was overcome by the sight of fellow human beings dying.

The radio and television journalists who brought the first word of yesterday's loss were similarly affected. For a few moments they were, like the rest of us, shaken and horrified by what they had seen, by what was the last thing they or any of the rest of us had expected to see: the deaths of seven people we thought were beginning another routine voyage into space.

So routine, in fact, had these shuttle takeoffs and landings become that many of us didn't bother to turn on the television anymore. All had gone so well so many times that we tended to forget what a combustible combination of fuel and rocket engines is needed to lift a 100-ton craft into orbit—to move it in 10 minutes from Florida to a place high over the Indian Ocean. Remember the cartoons and columns making fun of a U.S. senator's shuttle trip—and printed *before* the flight? They were the work of people certain that nothing could go wrong.

Now we have seen how far wrong things can go. This disaster will undoubtedly have its impact on decisions to be made about the future of American space exploration. But that debate is for another day. There can be no questioning the spirit of the people who have gone aloft in this country's 55 space trips. We were reminded yesterday of the courage it takes to board these outlandish craft and head off beyond the atmosphere, and of the quality of the people who devote their lives to getting a chance to do so.

It was painful to see the reruns of the explosion, and even more painful to see once again the seven crew members and passengers boarding the ship, a cheerful, varied and interesting lot looking forward to a great adventure. The disaster that befell them occurred, horrifyingly, before the eyes of their loved ones and of schoolchildren across the country who were watching this launch as part of an educational project. It was awful to see, but if such things are to be done they should be done in the open. We need to know the people involved and to see their humanity—good people such as Christa McAuliffe, Gregory Jarvis, Michael Smith, Francis R. Scobee, Ellison Onizuka, Ronald McNair and Judith Resnik.

The Orlando Sentinel

Orlando, FL, January 29, 1986

Throughout America, throughout the world, untold millions were by their radios and in front of television sets, waiting for the moment of liftoff, wondering as always: My God, what goes through a man's mind at a moment like this!

That was Tom Wolfe in *The Right Stuff*, a book on the early years of the manned space program. He was referring to the reaction of the American public right before a 1963 televised liftoff of a Mercury capsule. Spaceflight seemed so dangerous in those days. Indeed, it was; four years later three astronauts died as fire swept through their Apollo capsule.

Yet in the 19 years since then, spaceflight has seemed so safe, so routine that few people sitting in front of their televisions Tuesday were wondering what was going through the minds of seven astronauts at a moment like that. Who expected anything other than a safe liftoff?

The seven Challenger astronauts, however, knew the risks and took them anyway. They too had that rare "right stuff" quality of courage necessary to ride a rocket to the stars.

But the shock so many people felt in seeing the shuttle explode on television was followed by a gut-wrenching sense of loss: seven lives wiped out in pursuit of science and exploration.

That a teacher was aboard made it even more tragic. No one was better equipped to tell schoolchildren across the nation what space exploration meant than the personable, enthusiastic teacher from New Hampshire. For their part, her six colleagues represented the maturity of the shuttle program — they, too, were professionals intent on doing their jobs.

Today the questions as to why this happened are many and probably will not be fully answered for months. Nevertheless, NASA must be as forthright as possible in explaining to America what went wrong.

Many people already are starting to question the shuttle program, whether it is worth lives and billions of dollars.

But the manned space program should go on. Its benefits to those on Earth are too great for this tragic shuttle mission to be the last. Just think if the world had stopped making airplanes after the first few accidents.

Indeed, just think if the United States had ended the manned space program because of the fatal 1967 accident. The nation never would have put a man on the moon and never would have had a shuttle program with all its practical benefits, including producing useful drugs. Instead of abandoning manned spaceflight, NASA made it safer as a result of that 1967 accident. For instance, they gave astronauts and the ground crew a quick way out in case of a fire on the launch pad.

That same attention to improving safety has to happen now. Find out what went wrong and correct the problems. Meanwhile other issues swirl around Tuesday's tragedy:

■ **Should NASA be sending civilians, such as schoolteachers, into space?**

Though it was tempting Tuesday to say NASA should never have embarked on a program of sending teachers, congressmen or journalists into space, that hindsight overlooks what someone such as Christa McAuliffe brings to the understanding of space. A teacher in space can motivate youngsters to become interested in a subject that otherwise might have been ignored. That in itself is worth far more than the $100,000 cost of training. The same might be said of writers in space. Through their talent to communicate, they can teach the world.

■ **Should the space program place more emphasis on unmanned military missions rather than on manned commercial ventures?**

Already the tragedy is being used to boost the argument that the space program ought to be shifted toward unmanned military ventures. But this accident is no justification for turning the program into a military tool and abandoning manned commercial ventures. That might be an appropriate emotional response but not something in the best interest of the nation, which is just beginning to use the shuttle for practical commercial applications.

■ **Should there be an independent inquiry in addition to NASA's own investigation?**

Definitely yes. Congress has a perfect right to order an investigation of what happened to complement NASA's own inquiry. It's in NASA's best interest for an independent investigation to be undertaken. Then there won't be lingering suspicions about cover-ups.

✓ ✓ ✓

The horror of Tuesday's tragedy was best reflected in the disbelieving faces of Christa McAuliffe's own students. It didn't take long for them to realize that something had gone wrong, terribly wrong, in what should have been a magnificent day for their teacher and her six fellow crew members.

America mourns with them.

The Boston Herald

Boston, MA, January 29, 1986

WE HAD all become so blase about this space flight business. The few die-hard office fans who turned on the TV to watch the lift-off, found instead the usual array of game shows and soaps. We have taken this most dangerous of missions for granted — until now.

Now seven human beings have perished and the horror of that moment will live with us forever. Lost in the mid-air explosion were pilot Michael J. Smith, Judith Resnik, Ronald E. McNair, Ellison S. Onizuka, Gregory Jarvis, and the person who in life and now in death has touched us more than any other person to don a space suit, school teacher Christa McAuliffe.

Her trip was to be a public relations triumph for the space program — the first "real person" to go into space was a source of great fascination. Youngsters watched film of the teacher-astronaut in training with a new interest. Christa McAuliffe herself brought to the program a freshness and a sense of wonder that made this not just another shuttle flight. She wanted during her on-board lessons from space to "humanize the technology of the space age."

And so, in the most tragic of ways, she has. For as we watched the reruns of a space ship exploding into a fiery ball our thoughts were not of this usually flawless technology, but of a husband, a son, a daughter, and two parents who were at Cape Canaveral to watch the lift-off. "There has been a major malfunction" — NASA's official jargon for the explosion — took on a human dimension, a tragic human dimension.

At Concord (N.H.) High School where Christa McAuliffe's students were cheering the flight, stunned youngsters had the same reaction as usually cynical journalists — "This can't be real. We can't be watching this."

We grieve for the families who have suffered the most direct loss of all. And on this morning we now feel, amid the wonder of man's technological triumphs, the fragility of our own humanity.

Portland Press Herald

Portland, ME, January 29, 1986

It was, of course, bound to happen sometime. It's just that in the relatively short time since the world entered the space age, we had become used to the routine of space flights — even a bit bored by them.

The record has prepared us for the routineness — 56 successful U.S. space missions overall, 24 successful shuttle missions in the last five years alone. "We've become accustomed to success," said Sen. John Glenn, one of America's pioneer astronauts.

The addition of a Concord, N. H., schoolteacher to the crew of the space shuttle Challenger made the routine seem all the more ordinary.

Christa McAuliffe, the first private citizen to make a space flight, was scheduled to perform a most ordinary task for her — teaching — in the most extraordinary classroom imaginable. Her lessons from space were to be relayed to millions of pupils in the Classroom Earth project.

"If we don't prepare kids for the future," she said, "we are not doing our job as teachers."

She hoped, she said, to "humanize the technology of the space age." In a tragic sense, she did that.

McAuliffe never spoke of the risks involved in the flight, only of the challenge of her coming adventure.

But if the mission itself ended in stunning tragedy, the goal this 37-year-old schoolteacher set for herself — indeed, for American education — remains valid.

In death, as in life, McAuliffe taught us a lot about ourselves and our world.

The Toronto Star

Toronto, Ont., January 25, 1986

We had become used to the delays and troubles with the U.S. space shuttle program. But most of us never lost the thrill of watching lift-off. Until yesterday.

All seemed well as Challenger thundered into a clear blue sky from the chilly shores of Cape Canaveral on Florida's east coast. A minute and seconds later, however, Challenger and its seven crew members had disappeared in a colossal ball of flame. Immediately, we knew they were gone.

Christa McAuliffe, a 37-year-old high school social studies teacher and the first private citizen chosen to go into space, was among Challenger's crew. She caught our imagination, and later this week we were looking forward to the first lessons from the Teacher in Space.

The ill-fated flight of the Challenger shocked the world. We've known of disasters much worse, but seeing this one on TV in our own homes or offices was different. Once again, we were witnesses to history. And we wished it hadn't been so.

There have been other deaths. Four Soviet cosmonauts have lost their lives in space, and three U.S. astronauts died on the launching pad almost 19 years ago to the day. It was inevitable that more would die as we pursued our dreams into space. Perhaps this is why we find the deaths of Challenger's crew members so poignant.

No doubt the ambitious U.S. shuttle program, plagued as it's been by technical problems, will be set back. But that shall be overcome. We mourn the deaths of Challenger's crew. And we look to the future.

THE CHRISTIAN SCIENCE MONITOR

Boston, MA, January 29, 1986

IT is perhaps of the schoolchildren, millions in America alone, waiting expectantly for the Challenger space mission this week, that the tragic explosion after liftoff most makes one think. New Hampshire teacher Christa McAuliffe was among the passengers. She was to teach two lessons from space, which were to be viewed in hundreds of schools via the Public Broadcasting Service, during the six-day mission. The loss of Mrs. McAuliffe and the members of the Columbia crew makes us all reach for words of compassion and comfort.

Spaceflight needs to be viewed in the broadest of frameworks — as more than a personal adventure. It is an aspect of the human race's efforts to comprehend and master its universe.

From ocean voyages and the discovery of continents on Earth a few centuries ago, to the first steps by man on the moon's surface in 1969, to the hurtling of the Voyager II spacecraft past the planet Uranus's moons at the outer reaches of our solar system just this week, human familiarity with this magnificent universe has been accelerating. It is in the context of this larger adventure that this week's tragic event should be viewed.

Pioneers endure risks.

Columbus and other navigators lost whole crews. The first Apollo flight team, astronauts Grissom, White, and Chaffee, perished in a simulated launch in 1967. The disabled Apollo 13 had to be swung in a special arc around the moon in 1970 to be returned safely. A special honor is reserved for those who undertake such adventures.

On balance, the manned space program has been relatively free of serious mishap. There have been failures of equipment to function. In the current Challenger shuttle series, delays were frequent, often caused by misreadings of the weather. Disputes have arisen over the mix of military and commercial missions for the Challenger craft. The space program's top administrator has come under a legal cloud for private transactions. An air of impatience about the space program, and competition for federal funding for space exploration at a time of budget constraint, have been evident. But all this must be set against reasonable expectations for an experimental program that to this point has had remarkable success.

"I am a teacher first," Mrs. McAuliffe said of her role in the space mission.

Surely all are heartened and instructed by this sense of purpose which embraced her and her fellow space voyagers.

Chicago Tribune

Chicago, IL, January 30, 1986

In the first aftershock of the explosion of the space shuttle Challenger came the question: What will this mean to the children? How should it be explained? The President spoke to them directly in his painful national address. "It's all part of the process of exploration and discovery," he said. "It's all part of taking a chance and expanding man's horizons."

And the best that can come of this is a recognition of the real weight of those words, the way they should speak to all of us, including the President himself, reawakening the true spirit of the endeavor, its depth and seriousness.

The instinct to embrace the children at a moment like this is very powerful. They are innocent of the tragic sense that prepares people to contend with the horror. To them, space flight is pure adventure, a form of romance, in which death has no part.

Unusually large numbers of youngsters were watching the doomed launch Tuesday because there was a teacher along. This was to be a learning experience. They could identify with Christa McAuliffe because she was an extension of their own narrow world. Later in the mission, she was to teach a televised class period from space. But a little more than a minute after liftoff, the lesson became mortal.

Searing as it was for the children, for the adults it should have been more. Watching the awful image of the blast ought to have brought home a vivid, painful recognition that somewhere along the line we had begun to trivialize the exploration of space. The long, hard and dangerous work of reaching beyond our grasp had become so ordinary that network television had even stopped broadcasting it live.

Locked in its own political struggle—to get funding, to retain civilian control—the National Aeronautics and Space Administration had begun to try to stimulate the public imagination by public relations tricks. This was why a senator and a congressman and finally a teacher went aboard the shuttle. When the attention of the media strayed, in response to a flagging interest by the public, the program lost its political edge.

Mrs. McAuliffe's reasons for going were as fine and simple as the yearning to extend beyond the normal confines of experience and return to tell the tale. But the reasons she was invited were less perfectly noble, less important, more expedient. And if her death weighs specially on the mind, this must be why.

It would be a terrible mistake to let the sacrifice of seven lives destroy our confidence that exploration of space by human beings is worth the risk or the resources. Of course machines can do many things. But it is men and women who truly extend the human reach. No inanimate thing can have the intimate relation with wonderment that is the fundamental motive and animating spirit of all discovery.

But the shattering reality of force and fire 10 miles above the Earth, and in particular the death of Mrs. McAuliffe, should re-educate us all about the seriousness of what it means to investigate the heavens, and the stakes.

As adults, we had to explain to the children the meaning of what they saw Tuesday morning. But if we are to be adults ourselves, we must go on, chastened by the knowledge that we may have become jaded and indifferent, that we may have expected new thrills with every new mission, and that when the hard, slogging work of technological development and preparation for the next bold leap grew stale, the people in charge of the program tried to provide artificial substitutes.

From now on, the commitment should be as serious as the nature of the enterprise. It should be taken in the full knowledge that it is dangerous to the men and women who go aloft and that it may not provide immediate gratification to the appetite for novelty.

Reaffirming the more profound appetite—in President Reagan's words the yearning to explore and discover, to expand mankind's horizons—should be the consequence of the tragedy of Challenger.

We should no longer expect space crews to reflect any social or political purpose beyond the work of expanding knowledge. There need be no journalist, no mayor, no clergyman, no public representative. Anyone who reaches out into space belongs to the people and is the extension of the human species into the realm of the unknown. The astronauts are our hands, outstretched to the sky in a position that is at once vulnerable and humble and aspiring: the attitude of hope and fear, of awe.

WORCESTER TELEGRAM

Worcester, MA, January 29, 1986

The space shuttle had become routine. Since the first spectacular test mission in 1981, the sleek Challenger, Columbia and Discovery had gone up and beyond with what seemed to be humdrum efficiency. Fifty-two million miles without a major mishap. An amazing record and one we came to take for granted.

Then came Tuesday's terrible tragedy: An explosion about 10 miles above the Florida launch pad.

All seven aboard are lost, including the first private citizen in space, Christa McAuliffe of Concord, N.H., who went to high school in nearby Framingham.

Mrs. McAuliffe, 37, mother, vibrant teacher and caring citizen, was an example to her profession, her community and her pupils. She was high on learning, optimistic about people.

Over the last few weeks, her smiling, almost playful expression was much in the news and her spirit was obvious — and infectious.

America and much of the world weep for all of the lost crew members.

Now, NASA will sift through reams of computerized evidence and tapes. It will pore over what scraps of metal it can find as it relentlessly pursues the cause of this tragedy.

There is no worse setback than one that takes life. Our space program must now pause, take a hard look at its methods, procedures and goals, and then decide on the direction our manned-flight space program is to take.

Were Christa McAuliffe back in her New Hampshire classroom, she would tell her students that life goes on no matter what the setbacks. She would say unhesitatingly that there are new challenges to meet and new worlds to explore if ignorance is to be conquered.

That is the American spirit the brave little band aboard Challenger so clearly exemplified.

The Hartford Courant

Hartford, CT, January 29, 1986

The tragedy of the shuttle explosion, and the death of the Challenger crew, remind us that human life remains our real national treasure. Technology advances, limitations diminish, but life — its potential, quality, surely its loss — remains the rule by which change evolves and achievement is measured.

The shock, sadness and regret that attend the death of the seven astronauts only emphasize this: We expect things to function or fail, but we comprehend the machinery through human experience. The shuttle is extraordinary, but the pioneers on board were all too mortal.

It is especially poignant that Christa McAuliffe, the first "teacher in space," should be among those killed. Her public experience had reflected many private aspirations. But her death dramatizes the truth that space flight remains a perilous adventure, and that its risks are as grave as its rewards are high. The apparent ease with which shuttles have flown is an illusion; the success of the program has lulled us into complacency about its abundant dangers. Now, those dangers are evident to all who have seen the lurid spectacle of the shuttle's consumption by fire.

Of course, the space program will go on, and the shuttle series will continue. As President Reagan said Tuesday, "Nothing ends here. Our hopes continue." One catastrophe cannot take away what knowledge and experience have given us in the past. It can, however, add to that knowledge and broaden the experience for the future. There is some consolation in the thought that Challenger's loss may teach some lessons that were not known before, and might prevent such tragedies from recurring.

Yet it is worth noting that the American manned space flight program has compiled an admirable record of safety. The first astronaut went up in 1961, and 56 missions have been flown without incident in the subsequent quarter-century. (In 1967, three astronauts died in a launch-pad fire, and in 1970, the Apollo 13 mission was aborted in flight and safely returned to Earth.) Until yesterday, no astronaut had been killed or injured while in flight.

Indeed, the delays and postponements that have recently bedeviled the shuttle program are an emblem of its wisdom. No one begrudges the highest standards of safety, or the exercise of caution. No one supposes that schedules and timetables are somehow supreme, or that technology demands unnecessary hazards to life. Like soldiers, policemen or diplomats, astronauts assume a calculated risk. But life is not casually expended; manned space flight is a gamble in which we have chosen to influence the odds.

May it ever be so. But when the odds overwhelm the chances for success, the words of the Senate chaplain ring mournfully true: "Our hearts are smitten and we are reduced to silence." The drama of space exploration has produced its first tragedy. It has taken the pain of human loss to drive home an essential lesson in courage.

Wisconsin State Journal

Madison, WI, January 29, 1986

BULLETIN 10:42 a.m. CAPE CANAVERAL, Fla. (AP) — Shuttle Challenger rocketed away from an icicle-laden launch pad today, overcoming finicky weather and faulty equipment to carry aloft a New Hampshire schoolteacher as NASA's first citizen-in-space.

BULLETIN 10:42 a.m. CAPE CANAVERAL, Fla. (AP) — Space shuttle Challenger exploded today as it carried schoolteacher Christa McAuliffe and six crew members into space today.

URGENT 10:44 a.m. CAPE CANAVERAL, Fla. (AP) — There was no indication of the fate of the crew but it appeared there was no way they could survive.

One after another, the wire-service bulletins, urgents and advisories kept moving Tuesday.

The first, announcing what appeared to be a successful liftoff for the irrepressible Mrs. McAuliffe and her fellow space voyagers, carried with it a sense of human victory over the inhuman factors of weather and hardware.

The second brought terse words of shock and possible tragedy.

The third confirmed the nation's worst fears — that six dedicated astronauts and Mrs. McAuliffe, 37, a teacher whose stated goal was to "humanize" the exploration of space, had almost certainly perished.

First pride. Next shock. Then horror.

In the days and weeks that follow, it will be important to keep each of those emotions fresh in our minds. The shock of watching Challenger's fiery disintegration should not be forgotten by those who will be asked to investigate this catastrophe.

The horror felt by the crew's families and friends, including those students of Mrs. McAuliffe who watched the liftoff and explosion will be felt by all who read, hear or watch news accounts of the day's events.

But may it be pride — silent remembrance of those who gave their lives to cross into mankind's greatest frontier — that lingers the longest.

There will be calls for a dramatic curtailment of the nation's manned space program, as there were nearly 19 years ago to the day when three Apollo astronauts died in the searing heat of a launch-pad fire.

Fortunately, those calls went unheeded. Two and one-half years after Gus Grissom, Roger Chaffee and Edward White died on Jan. 27, 1967, fellow astronaut Neil Armstrong set foot on the moon. Their deaths were not in vain.

"If we die," Grissom once said, "we want people to accept it. We are in a risky business and we hope that if anything happens to us, it will not delay the program. The conquest of space is worth the risk of life."

So should it be with the deaths of Christa McAuliffe, Challenger Commander Francis Scobee, pilot Michael Smith, and crew members Judith Resnik, Ronald McNair, Ellison Onizuka and Gregory Jarvis. May their spirit and purpose live on.

The Birmingham News

Birmingham, AL, January 29, 1986

"Vehicle has exploded. ... We are awaiting word from any recovery forces downrange."

With those matter-of-fact words in the almost emotionless tone that we have come to expect, NASA's launch commentator confirmed what nobody wanted to believe.

The space shuttle Challenger had exploded 75 seconds after liftoff from Cape Canaveral yesterday killing seven crewmembers, including teacher Christa McAuliffe, pilot Francis Scobee, co-pilot Michael Smith and crew members Judith Resnik, Ellison Onizuka, Arnold McNair and Gregory Jarvis.

They are the first U.S. astronauts to die during an actual space mission and the first fatalities directly related to NASA's space program since Virgil I. Grissom, Edward White and Roger Chaffee died during a launch pad fire 19 years ago. The last known death in the Soviet program was in 1971.

Although there have been other close calls and many minor glitches, our space safety record has been so good that most of us have been lulled into forgetting how dangerous the technology of putting men and women into space actually is.

Yesterday's tragedy reminds us once again that life is fragile, that our technology is only partially under control.

It will be weeks, months, years or perhaps never before we know exactly what went wrong as Challenger's main engines were given full throttle. Decisions on the future of the space shuttle program must wait until we know more.

For now all we can do is mourn with their families and friends the loss of seven brave individuals who were seeking to move mankind and our frontiers of knowledge outward from this planet.

Newsday

Long Island, NY, January 29, 1986

The ghastly explosion of the space shuttle Challenger little more than a minute after it lifted off its launch pad yesterday was a devastating reminder that space exploration is still far from routine or free of risk.

After 24 successful shuttle flights during the past five years — and 31 earlier manned U.S. space flights with only one fatal accident — many Americans had begun to think of these journeys as hardly more dangerous or uncertain than a trip to the local supermarket.

That illusion was heightened by the space program's evident effort to break the routine quality of the shuttle flights by adding a new ingredient: passengers who were neither military personnel nor trained scientist-astronauts. Sen. Jake Garn (R-Utah) and Rep. Bill Nelson (D-Fla.) were the first two of these to go into orbit. The third, Christa McAuliffe, a schoolteacher from Concord, N.H., was aboard Challenger yesterday — and died along with the other six crew members.

No one who watched Challenger rise briefly into the sky and blow apart in a huge fireball will ever again forget for a moment that the uncertainties of space flight extend far beyond problems with a balky door handle or delays caused by a few days' bad weather.

The question now, of course, is what went wrong. Was the explosion, sudden and apparently without warning, caused by human error or by technological failure?

With no survivors from the shuttle crew, the question won't be easy to answer. But it must be answered, and given NASA's elaborate and highly sophisticated telemetric systems, it surely can be. And while NASA is asking questions, it should investigate whether maintenance and safety standards suffered as a result of the pressure the space agency has imposed upon itself to keep launching shuttles in rapid succession.

There is a poignantly ironic aspect to the Challenger disaster: It occurred as America was celebrating the successes of the unmanned Voyager 2 space probe.

That remarkable explorer of distant space, launched eight years ago, arrived in the vicinity of the planet Uranus last week after traveling 2 billion miles. It reached its destination just one minute and nine seconds ahead of schedule. Once there, it sent back astonishing new information about Uranus and its moons — especially the tiny Miranda, which one scientist described as a bizarre combination of all the geological features of the solar system.

As Voyager has demonstrated, unmanned space exploration is an enormously rewarding and scientifically profitable enterprise. The glib response to the juxtaposition of Voyager's triumph and Challenger's disaster would be to argue the advantages of unmanned space exploration over manned space travel. But this is not the time to make that judgment.

NASA must put its shuttle program on hold until it determines precisely what went wrong on Tuesday and, if possible, why. But decisions about future shuttle flights or other types of manned space exploration should not be made hastily, while the nation is numbed by the shock of witnessing the tragedy that engulfed Challenger and its seven crew members.

Meanwhile, let the names of these men and women be added to the roll of those throughout history who have risked their lives — and in many cases sacrificed them — in perilous voyages that sought to penetrate the unknown in the search for knowledge:

Christa McAuliffe, 37, teacher.
Francis R. Scobee, 46, commander.
Michael J. Smith, 40, pilot.
Judith Resnik, 36, crew member.
Ronald E. McNair, 35, crew member.
Ellison S. Onizuka, 39, crew member.
Gregory B. Jarvis, 41, crew member.

NASA's Titan and Delta Rockets Explode

A Titan 34-D launch rocket and its secret military payload exploded five seconds after launching April 18, 1986 from Vandenberg Air Force Base in California. The payload was reported to be a photographic reconaissance satellite expected to be to be put into polar orbit around the Earth. The failure and subsequent grounding of the vehicle until the cause of the accident was found and corrected left the United States bereft of a means to put big intelligence satellites or other heavy gear into space. In 1985, most of the Pentagon's critical payloads had been lofted into orbit aboard the Titan or the space shuttle, which was also grounded following the explosion and loss of the Challenger. It was the second consecutive Titan 34-D failure. Another of the rockets, also reportedly carrying a photo reconaissance satellite, exploded August 28, 1985. That mishap was traced back to two separate failures, either one of which would have doomed the mission—a massive leak of one of the rocket's liquid fuels and failure of a fuel pump. The KH-11 reconnaissance satellites reportedly lost in the Titan explosions normally operated aloft in pairs. Currently, the U.S. had only one of them in orbit. The "implications" of this latest explosion were "very serious," one security expert observed. "America's spy satellite fleet is basically grounded." The expert, James Bamford, pointed out, however, that the U.S. also possessed early-warning satellites and electronic intelligence satellites that could "pick up all kinds of signals" and were capable of providing warning of "any kind" of enemy attack. The accident also delivered another staggering blow to President Ronald Reagan's so-called "Star Wars" program, the Strategic Defense Initiative (SDI), the research program into space-based antimissile defenses. The SDI project was conducting several research projects from the shuttle. After the shuttle disaster, SDI officials altered some of the plans assuming that the Titans might be available. More than 70 persons were treated for various injuries, mostly eye irritations, after the blast April 18. One of them was hospitalized overnight.

An unmanned Delta rocket lost power about 70 seconds after liftoff at Cape Canaveral, Fla. May 3, 1986, veered out of control and was destroyed by remote control signal from an Air Force range safety officer. It was the third consecutive failure of a major space launch for the U.S. since the January Challenger explosion. The Delta rocket was carrying a $57 million weather satellite when it was destroyed. It was the first failure for the Delta in 44 launches. Over the years, the Delta, designed to lift medium-weight payloads of up to 2,800 pounds had achieved a 94% success rate in 177 previous launches, with 11 failures. However, the last Delta launch took place in 1984. NASA had been phasing out the rocket in view of the satellite-launching capability of the shuttle. Only three Delta rockets were left in NASA's inventory. They were grounded indefinitely, pending an investigation of the latest failure.

The Washington Post

Washington, DC, May 6, 1986

THE UNITED STATES has four major vehicles for launching objects into space. One is nearly obsolete, and now in succession the three others have failed. Is that a fluke, or does it mean there is something fundamentally wrong in the space program? NASA, the agency most involved, prefers the theory of coincidence. Its spokesmen suggest it would be wrong to aggregate the three events and try to read a pattern into them. Maybe so.

The most recent failure, which occurred on Saturday, involved a Delta rocket carrying a weather satellite. The Delta had always been reliable; only 11 of 177 previous flights had gone awry, and more than 40 consecutive launches had been successful. But this time the first-stage engine prematurely shut off about 70 seconds into flight, the rocket started to tumble, and a safety officer blew it up. That followed by two weeks the failure of an Air Force Titan rocket that blew up five seconds after launch, the second Titan to explode in two attempts. Before Titan came the Challenger accident.

The Titan and Delta accidents would have drawn scant notice absent what happened to Challenger and what has been unearthed about NASA since. (The previous Titan accident last August was only a five-paragraph story on the major wire services.) But if the tendency four months ago in Congress, press and public was to lead cheers for NASA, the pendulum now has swung the other way. The post-Challenger investigations have shown that for years federal auditors complained of poor management at NASA; that in the case of the shuttle, serious questions were raised within the agency about safety and design, and were brushed aside; and that decision-making at certain critical points had about it all the orderliness and decorum of a clamorous crap-shoot.

Last week it was reported that as early as 1978 NASA engineers had expressed concern about the design of the leaking O-rings that are thought to have brought down Challenger; that a memo in 1979 labeled the joints these rings were meant to seal "completely unacceptable"; that in 1985 officials imposed a "launch constraint" on the problem, meaning no launches were to take place until it was fixed; and that a Marshall Space Flight Center official, Lawrence Mulloy, then signed six successive waivers to let the flights proceed anyway. The laws of probability seem to have taken it from there.

At Cape Canaveral yesterday, NASA staged a ceremony marking the 25th anniversary of the first manned American space flight, by Alan Shepard in 1961. NASA is justly famous for its PR. This was to remind us all of the glory days of the past and take the edge off the present unpleasantness. A year ago it might have worked; it always did before. Now PR isn't good enough. That is the least of what you would hope the agency has come to understand.

DESERET NEWS

Salt Lake City, UT, May 6-7, 1986

A piece of folk wisdom known as Murphy's Law says: "If something can go wrong, it will."

The nation's space program is feeling the awful truth of that declaration this week as NASA officials try to figure out how to get objects into earth orbit after three consecutive disasters.

First the Challenger shuttle blew up Jan. 28, killing the seven astronauts on board, and grounding the entire shuttle program for an indefinite period — probably at least another year.

Then a Titan 34-D military booster exploded two weeks ago while trying to loft a spy satellite into orbit. The Titan 34-D has been grounded for at least six months while an investigation is conducted.

Last week, a Delta rocket, the Old Reliable of satellite launches with 43 straight successes, had to be destroyed when it tumbled out of control while trying to launch a weather satellite. The Delta system also will be grounded for a half-year for an investigation.

As a result, the U.S. is left without a reliable way to put satellites into orbit. Before the Challenger disaster, a decision had been made to have the shuttle handle virtually all satellite launches. The Delta rocket program had been all but abandoned. Only a few of the Deltas were left and it had been 18 months since the last launch.

Where does the U.S. go from here? Apparently nowhere for the time being. NASA has three old Atlas Centaur rockets, but they are all reserved for Navy payloads. The Air Force is modifying some old Titan ICBM's for use as booster rockets, but it will be 1988 before they are ready.

Obviously, the U.S. will have to revive the shuttle program and the Delta program as well. It's vital that the nation be able to put important communication, weather and intelligence satellites into orbit. Yet none of it can happen quickly.

For the next two or three years, the U.S. will have to sit on the sidelines and watch the Soviets, and the Europeans — with their successful Ariane satellite launcher — dominate the space show. How the mighty have fallen.

The Hartford Courant

Hartford, CT, May 3, 1986

The failures in the U.S. space effort have disheartened all Americans, but especially people in the Connecticut River Valley, birthplace of the Yankee ingenuity that helped make America the world leader in technology.

Many New England companies and their subsidiaries are involved in the work that's made the U.S. space program go. Over the past two decades, they've manufactured parts, equipment, electronic components, engines, spacesuits and rocket boosters — all of which have helped put human beings, and their artificial surrogates, into space.

Ingenuity is more than a tradition here — it's the source of our regional pride, and the stuff of legend.

"I am an American," the adopted New Englander Mark Twain wrote in "A Connecticut Yankee in King Arthur's Court," his great satiric comparison of feudalism with democracy. "I was born and reared in Hartford, in the state of Connecticut. . . . So I am a Yankee of the Yankees — and practical.

"I went over to the great arms factory and learned my real trade . . . learned to make everything; guns, revolvers, cannon, boilers, engines, all sorts of labor-saving machinery. If there wasn't any quick, new-fangled way to make a thing, I could invent one — and do it as easy as rolling off a log," Twain's narrator bragged.

A century later, such expressions of faith in American know-how have come to have a curiously hollow ring. Optimism about American technology, once invincible, has suffered a series of blows. American industry has seemed to lose its steam. Machine tooling, steel, automobiles, electronics — all appear to be either in trouble or relocating, and at a bewildering pace.

Still, Americans could point confidently to the country's pre-eminence in aeronautics and avionics — until this year.

The cruelest blow came Jan. 28 in Florida, with the explosion of the space shuttle Challenger. Then came another sickening explosion, this one at Vandenberg Air Force Base in southern California, when an unmanned Titan 34D rocket blew up.

The Air Force says it will focus its Titan investigation on the possible failure of rubber O-ring seals, which are suspect in the Challenger explosion. It will also look at the Titan's solid-fuel boosters, propellants and electronics.

These failures have cost lives, time and money. They have cast doubt on the future of America's space program itself. They have extracted a heavy price in the confidence and pride the space program once commanded.

But confidence is the component most crucial for Americans to retain. Crushing as the recent disasters are, their causes will almost certainly be found. And if causes can be discovered, so can remedies. Quick and new-fangled methods of making and fixing things may be scarcer than they were in Mr. Twain's day. Yet isn't it still America's way to summon the will to find them?

The Arizona Republic

Phoenix, AZ, April 25, 1986

IN the continuing vortex of controversy surrounding the U.S. air raid on Libya, a potential disaster has gone largely unnoticed.

Last week a U.S. Keyhole-11 high resolution photo-reconnaissance spy satellite was destroyed in the launch failure of an Air Force Titan missile in California, a double disaster.

The explosion leaves the United States with only one KH-11 series satellite in orbit, and the failure of the Titan booster is the second in a row. With the grounding of the space shuttle and the two Titan failures, the Air Force is left with no way to put a second satellite into orbit.

The KH-11 is a reliable piece of hardware and should keep functioning until either the shuttle or the Titans come back into service next year. Also, there are other spy satellites in orbit: Ferrets locate radar emissions; Rhyolites listen in on telecommunications; and Chalets monitor rocket telemetry while others look for rocket exhaust plumes.

The backbone of the system, however, has been coverage by two KH-11s which radio back detailed photos of the Soviet Union, monitoring compliance with arms-control agreements, watching troop and materiel movements, as well as providing secondary reconnaissance of other global trouble spots such as the Mideast and Central America. The Air Force has only had one working KH-11 in orbit since August.

In the nuclear age, both superpowers rely on satellites to gauge each other's strategic forces and intentions. The United States more than the Soviet Union relies on satellite reconnaissance, and it is troubling to have just one KH-11 circling over the Russian missile fields.

The Air Force's problems are compounded by the fact that there are no more of the $500 million KH-11s left even if they could be put in orbit. The one that was destoyed last week was a test model adapted for use when another was destroyed in a Titan explosion last August. The next-generation satellite, the KH-12, is too large for the Titan booster and must be carried aloft in the shuttle.

Having just one KH-11 is worrisome, but not disastrous so long as it keeps functioning. Should it fail or be taken out by the Russians' operational anti-satellite system, the United States could be in real jeopardy. Satellites are vital to the national security. They provide the margin of warning necessary to maintain the credibility of our nuclear deterrent.

Steps should be taken to expedite either the return of the shuttle to service or the development of the heavy-lift booster the Air Force asked Congress for last year.

The Dispatch

Columbus, OH, May 7, 1986

Those believing a critical breakdown has occurred in NASA's quality control and decision-making processes got dramatic and unwanted support recently when yet another satellite-carrying rocket failed barely a minute into its mission. The mission was aborted and the satellite destroyed.

This was the third major failure of the space program this year. First, the space shuttle *Challenger* exploded, killing its seven-member crew. Then a Titan 34D — the shuttle's major, unmanned backup — exploded as it tried to lift a military payload into space.

The National Aeronautics and Space Administration desperately wanted a successful launching of a Delta rocket last weekend — but it got more of the same. The rocket suddenly lost power 71 seconds after blastoff. Launch officials destroyed it moments later. The failure, only the 12th in 177 attempts, could not have come at a worse time for the country or NASA. It resulted in the destruction of the $30 million rocket and its $57.7 million weather satellite payload.

"You have to question whether something is fundamentally wrong with NASA's quality control and maintenance procedures," said Marc Vaucher, a program manager at the Center for Space Policy Inc., a consulting group. "Obviously, you can't have three widely different hardware systems go bad and blame it only on the hardware. It's got to be an issue of quality control and support."

That comment echoes what others have been saying ever since the investigation into the *Challenger* disaster revealed waste, inefficiency, and communications failures in the NASA chain of command. The subsequent failures only reinforce the belief that something is seriously wrong in an organization once viewed as a Cinderella agency.

The Delta rocket failure must encourage investigators to double their efforts to find out what is wrong and how it can be corrected. This nation's space goals are too important to be mistreated any longer.

THE INDIANAPOLIS STAR

Indianapolis, IN, May 11, 1986

There's an old song that asks, "Will you love me in December as you love me in May?"

Well, the American people loved the space program a lot better last December than they love it this month.

Last December, the glory days were still mounting. The series of astronaut probes, the man on the moon, the exploration of Mars and those fabulous rings around Saturn were splendid steps along a pioneer trail. Nothing seemed too tough for those modern-day wonders, the NASA scientists, engineers and technicians.

The shuttle program — with its promise of the big pay-off — was gung-ho. Sure, there were disappointments, a few lost satellites and in-flight goofs. But there were no disasters. Everything was coming up roses and big appropriations.

Then came the Challenger flight and an explosion that shocked even the most intrepid visionaries. The investigation that followed — conducted hour after dreadful hour in the glare of disappointment and mistrust — revealed that the space program's superstars had commonplace faults.

They presumed too much. Awestruck by their own track record, they forgot fundamental rules, ignored tedious but essential rote chores and neglected to talk to one another in the straight-forward manner that clears confusion and squares doubts.

The supermen fell from grace onto the harsh plains of disillusionment, their landing made all the bumpier by the blowup last month of a Titan rocket carrying a valuable spy satellite and the forced destruction last week of a Delta rocket carrying a $57 million weather satellite.

So it was that only a hundred or so people showed up the other day for ceremonies marking the 25th anniversary of the first manned American space flight. It was not a time for celebrating.

Nor is it a time for despair and recriminations. Confidence has been shaken but not lost. The challenge is to make the most of past mistakes and remember that successes far outweigh the failures.

The American lead in space must not be forfeited. There are competitors — friendly and unfriendly — eager to push ahead of us, particularly in the area of heavy satellite launching. An indefinite suspension of U.S. launches could curtail the flow of information from vital defense satellites and inhibit research into President Reagan's Strategic Defense Initiative.

The hard questions, the corrective action must go on. So must the space program. That is where the future lies. It still beckons to those willing to brave the unknown.

Wisconsin State Journal

Madison, WI, May 9, 1986

Twenty-five years ago this week, Alan Shepard became the first American in space when he was stuffed into the cramped quarters of Friendship 7 and rocketed aloft for a 15-minute, up-and-down flight.

In the years after Shepard's historic ride, the U.S. space program was an "up" that rarely came down. With the exception of a 1967 launch pad fire that claimed the lives of three Apollo astronauts, the National Aeronautics and Space Administration charted one breakthrough after another.

Nothing can diminish the significance of those milestones, but it is clear that 25 years after the dawn of the American space age, NASA has run short on credibility.

Three times this year, U.S. attempts to reach space have proved disastrous. First came the Jan. 28 Challenger explosion, which claimed the lives of seven crew members and grounded the space shuttle program for 12 to 18 months.

A presidential commission investigating the Challenger disaster appears headed toward the conclusion that NASA, beginning four years *before* the first shuttle launch in 1981, routinely ignored warnings from engineers about critical booster seals and proceeded with an unrealistic launch schedule to attract military and commercial customers.

The Challenger explosion was followed by the April 18 blast that destroyed a Titan 34D rocket carrying an Air Force reconnaissance satellite. Two other Titans, one of the nation's oldest operational rockets, were lost in August 1985.

Then came the May 3 loss of the unmanned Delta rocket, which was blown up by Air Force safety officers after its engine shut down and threw the rocket into a high-speed tumble off Cape Canaveral, Fla. A $57.5-million weather satellite aboard Delta No. 178 was destroyed.

A military Atlas rocket launched Feb. 9 stands as the nation's only launch success in months, although the pictures of Uranus delivered in mid-January by Voyager 2 (some 2 billion miles and eight years from home) were a dazzling technological display.

What has gone wrong with the nation's space program? For starters, it appears the people at NASA do not want to admit that *anything* has gone wrong; that recent failures do not illustrate a pattern of mismanagement and muddled goals.

The defensiveness of NASA in the face of adversity indicates that a shake-up is in order. Rather than throwing more money at the agency — or, worse yet, turning it over to the military so the Pentagon can throw more money at the agency — Congress and the administration should re-examine what made NASA a success in the first place.

A leaner, back-to-basics NASA run by civilians and scientists with a shared vision of space exploration will make the next 25 years as rewarding as the last.

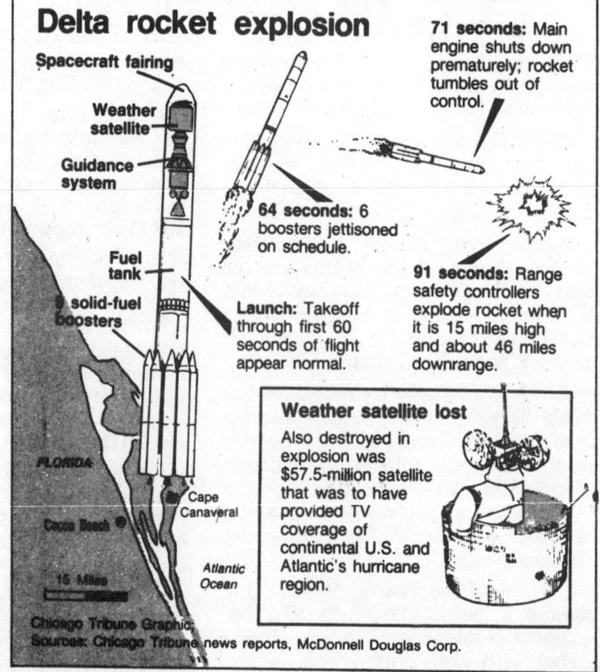

NASA SUCCESSFULLY LAUNCHES ANOTHER INVESTIGATION.

ARGUS-LEADER

Sioux Falls, SD, May 7, 1986

Space flight had become routine for Americans.

Since its exciting start in 1961, manned U.S. spacecraft explored the heavens with almost dull efficiency.

That changed Jan. 28, when the space shuttle Challenger exploded into a fireball after liftoff, killing all seven crew members.

Editorial

Now, instead of rebounding, the U.S. space program is encountering more setbacks.

On April 18, an Air Force rocket thought to be carrying a spy satellite exploded.

Saturday, an unmanned rocket carrying a weather satellite was destroyed by ground command after it went out of control.

Since the start of the year, NASA has had only one successful space flight: the Jan. 12 shuttle flight that carried Florida Congressman Bill Nelson.

Recent failures make it hard to focus on success. But Americans must not allow the space program to remain grounded.

The string of disaster and mishaps has inflicted serious psychological damage and eroded confidence in the U.S. space program.

As hearings and investigations have revealed, there are management and oversight flaws in NASA that must be corrected. But Americans must not allow setbacks to jeopardize the nation's future in space.

For our children's sake, we must remain committed.

We should not waver from the National Commission on Space's 50-year goal of establishing manned space outposts from the Earth to Mars.

We should not give up on plans to develop an aerospace plane that would carry people and cargo around the world in low orbit.

These are trying moments. But let's not forget the joy and hope that accompany success, like the excitement of that day in 1969 when Americans landed on the moon.

Let's not forget 25 years of mostly successful space exploration.

Before Jan. 28, there had been 24 successful shuttle flights.

Before Saturday, 43 Delta rockets had been launched successfully.

Nothing can remove all the risks that accompany manned space flight. But that's never stopped us before.

Setbacks didn't stop us 19 years ago, when three astronaunts died in a launching pad fire while preparing for an Apollo flight.

Problems must not stop us now either.

The Idaho STATESMAN

Boise, ID, May 6, 1986

Twenty-five years ago, Alan Shepard became the first American to fly in space, ushering in a proud era of U.S. space exploration and achievement.

Today, almost 25 years to the day after that historic flight, the U.S. space program is grappling with three major failures in less than four months, failures that force a reassessment of the U.S. space program.

NASA's problems began on Jan. 28, when the shuttle Challenger exploded, killing all seven astronauts. Virtually from the start, the investigation pointed to one cause: glaring safety defects – and a willingness at NASA to look the other way.

Then last month, failure struck again. An unmanned Air Force Titan rocket, the chief rocket used to launch spy satellites, exploded.

On Saturday, failure number three: An unmanned Delta rocket carrying a weather satellite lost main engine power, went out of control and had to be destroyed.

The immediate effect of these disasters is that NASA has suspended all three programs, crippling America's ability to launch commercial, scientific and military satellites.

The long-term effect is a continuing loss of confidence in NASA and growing doubts about its quality-control.

NASA officials predict that investigations ultimately will blame the Titan and Delta failures on improperly manufactured or installed components. That remains to be seen.

As one space policy consultant told *The New York Times*, "You have to question whether something is fundamentally wrong with NASA's quality-control and maintenance procedures. Obviously you can't have three widely different hardware systems go bad and blame it only on the hardware. It's got to be an issue of quality control and support."

Reagan Names Commission to Study Space Shuttle Disaster

President Ronald Reagan appointed an independent commission February 3, 1986 to investigate the January 28, 1986 explosion of the space shuttle Challenger, in which all seven crew members lost their lives. The commission, headed by William P. Rogers, former secretary of state and former attorney general, was named "to take a hard look at the accident, to make a calm and deliberate assessment of the facts and ways to avoid repetition." The panel would attempt to determine "the probable cause or causes" of the disaster and report back to the President in 120 days. The Challenger's crew knew the risks involved yet "chose to go forward, not reluctantly, but eagerly and with the thumbs up," the President said. "And we owe it to them to conduct this investigation so that future space travelers can approach the conquest of space with confidence and America can go forward with enthusiasm and optimism, which has sparked and marked all of our great undertakings," he said. Former astronaut Neil A. Armstrong, the first person on the moon, was named vice chairman of the panel. The President named 10 other members and reserved an option to appoint eight additional members including Sally K. Ride, the first U.S. woman in space and Charles "Chuck" Yaeger, the test pilot of "The Right Stuff" fame.

William Rogers, commission chairman, charged Feb. 27 agency officials had abandoned "good judgment and common sense" in handling safety problems prior to the fateful launch. On the third and final day of the hearings, the commission learned that officials of Rockwell International, manufacturer of the shuttle itself, as apart from the rockets that propelled it into orbit, had warned space agency officials that it was unsafe to launch the Challenger because of icing conditions at the launch site in Florida. Rockwell was the second major contractor to have recommended against launching. The presidential panel already knew about the no-launch position of engineers from Morton-Thiokol Inc., builder of the shuttle's booster rockets. Differing testimony from Thiokol engineers and NASA officials had been heard the first two days of the hearings, Feb. 25-26. Thiokol engineers testified in Washington before the panel Feb. 25 that they had been pressured by space agency officials to present a positive launch recommendation. The early recommendation to postpone the launch was eventually reversed by Thiokol managers. The next day, four key NASA managers denied having pressured the rocket builder for a launch. They told the panel that they had not considered the Thiokol engineers' recommendation of the situation valid. Rogers said it was his opinion, and the opinion of other members of the commission, that NASA's decision-making process was "flawed" because warning against launching by the two contractors were not fully comprehended or were not reported to the senior officials who gave final approval for the launch.

THE BLADE

Toledo, OH, May 26, 1986

THE presidential commission investigating the destruction of the space shuttle Challenger has uncovered many troubling factors contributing to the worst accident in the history of manned space flight.

There is evidence of faulty engineering and inadequate testing of the crucial rocket boosters. Fail-safe procedures were apparently short-circuited. Clear warnings from within the ranks of the aerospace bureaucracy were not heeded.

Unfortunately, a recent disclosure by the commission may make those errors appear even more intractable than they ought to be. Some individuals connected with the accident appear to be more interested in punishing those who bring problems to light than in correcting them. The commissioners, particularly Chairman William Rogers, are understandably disturbed by this possibility.

The source of new concern comes from reports that Allan McDonald, an aerospace engineer who provided some of the most important testimony to the commission, has been stripped of responsibilities by Morton Thiokol, Inc., manufacturer of the shuttle's solid-rocket boosters.

Mr. McDonald placed himself at odds with his corporate superiors when he told the commission that strenuous objections to the Challenger's final launch were overruled by managers at Thiokol and the National Aeronautics and Space Administration.

His early testimony strongly suggested that those responsible for the shuttle launch decision could be faulted for extremely weak judgment. However, the bureaucratic shunning of Mr. McDonald, who now finds himself without a staff and responsible for undefined special projects, raises the possibility that the space program may also be encumbered by inexcusable stupidity.

Rather than responding honestly to the troubling sequence of decision-making described by Mr. McDonald, executives at Morton Thiokol appear more intent upon punishing the bearer of a message they did not want to hear.

Detroit Free Press

Detroit, MI, May 12, 1986

NASA IS in worse trouble than most people would have predicted even after the Challenger disaster. Four months of official inquiries into the space agency's affairs and three other rocket launching failures later, public confidence in the U.S. space program is seriously eroded.

A big part of NASA's problem is technology and quality control, a basic nuts-and-bolts, dollars-and-cents affair. Some $3.5 billion dollars have been wasted in non-competitive procurement schemes identical to those devised and used by the Pentagon. Safety standards have lapsed. Sloppy designing and shoddy work have crept into the agency's spacecraft fleet. Bureaucratic and political wishful thinking of new management personnel gradually overtook the cautious pragmatism of the program's original technocrats who guided NASA to its glory in the 1960s.

Those problems, grave as they may appear at the moment, are not insurmountable. A solid management shake-up, internal reform and closer oversight of NASA's operation by Congress would help correct the agency's most obvious institutional failings. But NASA also suffers from an acute crisis of purpose, and this problem is much harder to address.

After the historic triumph of the Gemini and Apollo programs, the U.S. drive into space lost a lot of its momentum and imagination-stirring qualities. The dream faded and the consequences are disastrous. All trips into space are highly technical ventures, but NASA was successful as long as it was technologically lucky and managed to be the embodiment of American ingenuity and a vehicle for the human urge to push back the boundaries of knowledge. To bounce back, the agency must find a way to tap this mysterious source of inspiration. How?

We've heard of the proposal to build a space platform in Earth's orbit, and there was some talk of perhaps inviting the Soviets to join us on an expedition to Mars. So far those ideas have had few takers; they have not excited the public imagination. Hence the very real danger that by default and in the absence of better alternatives, President Reagan's Strategic Defense Initiative, or Star Wars, could become NASA's next raison d'etre. And we could not think of a worse answer to NASA's search for a purpose. The space program needs to be geared to civilian, peaceful uses, to scientific exploration and discovery. To capture our imaginations again, the space program should be seen as an attempt to open up new worlds, not divide and fence-in the old one.

The Washington Post

Washington, DC, May 13, 1986

THE NIGHT before the launch of the space shuttle Challenger on Jan. 28, engineers at Morton-Thiokol begged officials not to go ahead. They feared precisely what appears to have occurred: that cold weather at the launch site would stiffen troublesome O-rings used as seals in the spacecraft's boosters; that the stiff rings might not seal, letting hot gasses break through the boosters' thin walls; and that a leak of that kind could cause a catastrophe. But NASA officials insisted on launching anyway, and Morton-Thiokol management finally took NASA's side. Thiokol's chief engineer was told to "take off your engineer hat and put on your management hat." He did, the launch was approved, the shuttle blew up, and seven people died.

That is hardly a record of which the company can be proud. You can bet it won't make the annual report. It suggests that the desire to curry favor with a big-bucks customer overcame all other elements of judgment, and that the worst kind of organizational imperatives inside both NASA and Morton-Thiokol were allowed to crush the truth. Bad enough, but now Morton-Thiokol has one-upped itself. It appears to have put on the shelf two of the engineers who resisted that night and three weeks later went public to disclose what had happened and to make clear that the O-rings had long been recognized as a weak spot in the boosters and a source of potential disaster. The two are Allan McDonald, who was director of Thiokol's solid rocket motor project, and Roger Boisjoly, who had been put in charge of a task force on the O-ring problem.

In closed-door testimony May 2, the transcript of which was released last weekend, Mr. McDonald told the presidential commission investigating the Challenger accident that he has been given a new title, "director of special projects," and that "the people that all work for me work for somebody else." He was asked by chairman William P. Rogers if he felt he had been reassigned "because of the testimony you gave before the commission?" "Yes, I do," he replied. "I feel that I was set aside so that I would not have contact with the people from NASA again because they felt that I either couldn't work with them or it would be a situation that wouldn't be good for either party." "So you were in effect punished for being right?" Mr. Rogers asked. "I feel I was," Mr. McDonald responded. Mr. Boisjoly said in response to similar questions that while he remained "seal coordinator . . . I too have been put on the sideline in that loop with relationship to the customer.

Company officials quickly said that there had been no effort to punish the engineers for testifying. There had been a general reorganization after the accident, they said, and a lot of people had had their responsibilities changed, but "we haven't demoted anyone," and "well, everybody can't be in charge." Mr. Rogers found this unconvincing, and rightly so. He blistered Thiokol management, calling its behavior "shocking." In the same hearing, just as correctly, we believe, he admonished NASA for having "pretty well glossed over" and "almost covered up" the problem with the O-rings at various points along the way.

We tell our kids in this society not to lie. We encourage them to believe that if they tell the truth they'll be supported. We owe them the example of practicing what we preach. The most important thing at issue in this case is no longer how well the space program has been managed nor even whether Challenger should have gone up that day. It is the maintenance of and insistence upon integrity. Go get 'em, Mr. Rogers.

St. Petersburg Times

St. Petersburg, FL, May 13, 1986

The night before Jan. 28's ill-fated (and ill-advised) launch of the space shuttle *Challenger*, Morton Thiokol engineers Allen G. McDonald and Roger Boisjoly, among others, pleaded with their superiors to postpone the mission. The two engineers feared — correctly — that weather conditions increased the likelihood of the catastrophic failure of an already suspect rocket joint.

Their vehement objections were overruled by Thiokol and NASA superiors, and *Challenger* was launched the next morning. Investigators are blaming the failure of a joint between segments of the right booster rocket for the accident that killed all seven crewmembers.

Have Morton Thiokol and NASA executives reprimanded the officials who ordered the launch over the engineers' objections? Have McDonald and Boisjoly been commended for the courage and expertise they displayed in their futile attempts to postpone the mission?

The answer won't come as a surprise to anyone familiar with the way in which such closed bureaucracies often work in times of crisis: Incompetence, even criminal neglect, may be tolerated; public disclosure of that incompetence is strictly forbidden. McDonald and Boisjoly consequently have been demoted — "punished for being right," in the words of William P. Rogers, chairman of the commission investigating the *Challenger* disaster. And as if being right weren't reason enough for being punished, McDonald had compounded his crime by testifying candidly before the Rogers commission.

The corporate and governmental officials who spent years ignoring their experts' warnings about the safety of the shuttle are mostly still on the job. A big part of those officials' jobs now consists of attempting to rationalize away the increasingly obvious irresponsibility of those who decided that schedules were more important than safety in the shuttle program.

Lawrence B. Mulloy, the NASA official who routinely approved the shuttle for flight despite the growing evidence of its dangerous structural flaws, was transfered to new duties last Friday — not because of his performance with the shuttle program, but because his testimony before the commission proved embarrassing for his NASA superiors.

And what of the Morton Thiokol officials who demoted McDonald and Boisjoly? Does the company have a general policy of discouraging even the most important bad news — and then penalizing those who nevertheless have the courage to point out problems when they see them? If Morton Thiokol officials' behavior prior to the shuttle launch was similar to their coverup in its aftermath, their corporate myopia — along with NASA's bureaucratic blindness — must bear heavy responsibility for the *Challenger* tragedy.

WORCESTER TELEGRAM.

Worcester, MA, May 21, 1986

Morton Thiokol, the manufacturer of booster rockets pinpointed as a highly probable cause of the Challenger disaster, seems determined to pile one error atop another. The company appears to have learned precious little from the tragedy.

When two company engineers testified before the presidential commission investigating the accident, it appeared that the company was ready to disclose everything it knew and cooperate in determining its cause. The engineers' testimony was damaging. They said they argued against the ill-fated launch because unusual cold weather might further erode the rockets' O-ring seals. The seals had caused prior concern, but the manufacturer, instead of making improvements, petitioned the National Aeronautics and Space Administration to approve the O-rings. Only in view of a series of stories about agency mismanagement and complacency, disclosed during the inquiry, can one understand how NASA went along with the manufacturer.

Morton Thiokol's image as an honest broker changed when the company "reassigned" — translate that as "demoted" — the two engineers who spilled the beans. Trying to explain that the job change was part of a company reorganization, rather than punishment, only adds insult to injury.

The ensuing congressional uproar is justified. "We believe that these apparent disciplinary actions put your company in the worst possible light," members of the Senate space committee told top company officials. Massachusetts Rep. Edward Markey introduced legislation that would prohibit Morton Thiokol from receiving a NASA contract until the General Accounting Office finds that the two engineers are either restored to their former jobs or determines that the reassignments were not retaliatory in nature.

The rocket manufacturer's conduct indicates that at least one major NASA client that has enjoyed an inside track to lucrative contracts cares more about covering up its mistakes than about making a fresh start. If that attitude is allowed to prevail in the space program, it could easily endanger its future.

The Star-Ledger

Newark, NJ, May 19, 1986

Even in advance of its final report, the presidential commission investigating the causes of the Challenger tragedy has made clear it is thoroughly dissatisfied with the conduct of officials at the National Aeronautics and Space Administration. With each announcement, the criticism is more sharp and the cause for concern more pronounced.

The latest incident indicates that in addition to its inefficiency and incompetency prior to the disastrous accident, NASA is also guilty of the additional sin of arrogance after it happened. All this has caused commission chairman William Rogers to accuse NASA of "almost covering up" serious recurring problems with Challenger's booster rockets.

A transcript of one of the commission's hearings shows Mr. Rogers expressed anger when two engineers from Morton Thiokol Inc., manufacturer of the booster rockets, testified they were reassigned after they informed the commission they had argued for a delay of the flight because of cold weather. They suggested the reassignments were punishment for their testimony.

Mr. Rogers accurately commented that it appeared they were being "punished for being right." He went on to say that if their warnings had been heeded, the tragic accident might not have occurred.

The commission also criticized Lawrence Mulloy, formerly the manager of the booster rocket program. Documents showed he had imposed a special safety requirement that the rockets be repaired before the next mission, then waived the safety requirement to permit flights to proceed.

Conduct such as this makes it difficult, if not impossible, to restore confidence in the space program. The reassignment of the two engineers seems to indicate that even after the loss of life, NASA is still playing the old game of paying more attention to how things sound than how things work.

Final judgment as to who is to blame and what is to be done must obviously await the commission's final report. But even now, there is enough evidence to suggest that the defects at NASA go far beyond correction through some "quick fix."

It will take deep-rooted, fundamental changes in attitude, in policy, in organization and in personnel if the United States is once again to have a space program in which the public can have confidence.

St. Paul Pioneer Press & Dispatch

St. Paul, MN, May 13, 1986

Restoring public confidence in the National Aeronautics and Space Administration will require more time and effort than fixing the design flaw in the solid rocket booster joints that apparently triggered the Jan. 28 Challenger disaster. Engineers may need several months to design a new joint and sealing mechanism. But it may take years of flawless launches before the public puts a fallen and badly tarnished NASA back on its pedestal of credibility.

The most damaging information in the Challenger probe came Saturday when officials released a transcript of a closed hearing held May 2. During that hearing, members of the presidential panel studying the accident angrily accused officials from both NASA and Morton Thiokol Inc., the company that builds the rocket boosters, of repeatedly ignoring warnings of serious problems with joints in the months before Challenger's fatal launch.

In addition, panel chairman William Rogers claimed that NASA "almost covered up" evidence that would have grounded the shuttle program. Mr. Rogers also labeled as "shocking" Thiokol's treatment of the two company engineers who first led investigators to the rocket-joint problem. Shortly after testifying earlier that Challenger was launched over objections from engineers working on the rocket boosters, the two high-ranking Thiokol workers, Allan McDonald and Roger Boisjoly, were stripped of authority and reassigned to meaningless tasks.

NASA's coverup attempt, Thiokol's retaliation and the safety stonewalling that occurred on both sides are reprehensible.

Space exploration, by its very nature, is risky business. The smallest error or oversight can lead to instant catastrophe. NASA's primary mission was to minimize the risks.

During the space program's youth, the space agency was extremely successful. Launch buttons were never pushed until absolutely everything was "A-OK." Not every mission worked flawlessly, of course, but when disaster struck, as it did on the Apollo 13 moon mission, it came without a hint of warning.

Testimony before the Rogers commission indicates that some NASA officials had ample warning about leaking rocket booster seals and the catastrophic damage it would cause, but, for unexplained reasons, did not send the data to higher authorities. The surprise, it now turns out, is not that Challenger was destroyed, but rather that an earlier shuttle did not disintegrate in flames.

The horrendous behavior caused the needless deaths of seven brave astronauts, destroyed a $1.5 billion shuttle and crippled the nation's space program. Beyond that, the tragedy demonstrated that the rock-solid, safety conscious NASA of old has mushroomed into a giant self-centered bureaucracy capable of swallowing vital safety warnings rather than risk violating satellite delivery schedules and providing ammunition to opponents upset with space budgets.

However terribly disconcerting it may be, NASA's safety degeneration clearly is not an indictment of manned space exploration, nor of the shuttle program. America is a nation of explorers and visionaries. Space represents the crowning adventure and serves as the ultimate stimulus for technological innovation. If America is to remain a scientific leader, NASA must be fixed so that space exploration can resume with confidence.

THE DAILY HERALD

Biloxi, MS, May 20, 1986

NASA is struggling to recover from the disastrous explosion of *Challenger* last Jan. 28 and the confidence-shaking revelations of failures in the launch decision-making process.

New administrator James Fletcher has come on board with a clean broom, evidenced by the decision to investigate whether NASA officials had a hand in the job changes of two Morton Thiokol engineers who had opposed the launch decision.

The two, Allan McDonald and Roger Boisjoly, deserve commendations for their straightforward testimony to a presidential commission.

Both NASA and the solid rocket booster manufacturer must be held to account. Space program safety is a paramount consideration, not to be sacrificed to petty administrative politics, if such is the case.

The Houston Post

Houston, TX, May 13, 1986

The reassignment of two Morton-Thiokol engineers who opposed the Jan. 28 launch of the space shuttle Challenger is distressing. Both men say they believe the moves were retaliation for testimony to a presidential investigating commission.

Morton-Thiokol, of Brigham City, Utah, is the prime contractor for the shuttle's solid rocket boosters. A failure of a joint between booster segments is believed to have caused the explosion a little more than a minute after launch that destroyed Challenger and killed the seven people aboard.

The company, whose reusable boosters functioned nearly flawlessly on 26 previous launches, denies the reassignments were retaliatory. But one of the engineers notes that employees he supervises also report to another superior, and the other characterizes himself as having been put on the sidelines.

The aftermath of the Challenger disaster, and of subsequent failures of unmanned satellite launching rockets, is no time for recrimination. Let's hope this isn't. The most knowledgeable people have to be in positions in which they can have an impact on the redesign of elements of the boosters; the restoration of the space program demands no less.

There is every indication that those two engineers are in that category. Indeed if their objections had been honored, the Challenger disaster might have been averted.

DESERET NEWS

Salt Lake City, UT, May 15-16, 1986

With unerring accuracy, Utah's Morton Thiokol Inc. has managed to shoot itself in the foot just at the time when it can afford no more wounds from any source.

The firm has had no shortage of financial and morale problems since the tragedy involving the Challenger shuttle, which exposed defects with the booster rockets made by Thiokol as well as with the decision-making process of space shuttle flights.

Now Thiokol has made itself look petty and vindictive by reassigning the two engineers. These are the two who told investigators how the company's engineers unanimously opposed the January 28 launch after learning about cold temperatures at the Kennedy Space Center in Florida. But they all were over-ruled by company managers.

Engineers Allan J. McDonald and Roger M. Boisjoly also disclosed that the joints connecting the segments of the booster rockets had been a continuing source of trouble for several years but that NASA continued to launch the shuttles despite evidence that the seals became damaged in flight.

Thiokol, of course, insists it isn't punishing McDonald and Boisjoly. Instead, the firm calls their reassignment just part of a company reorganization.

But that isn't how their reassignments look to McDonald and Boisjoly, who complain they have been given smaller staffs and less authority and are not allowed to examine information about the Challenger accident. That certainly would strike most people as being a demotion.

Nor does the treatment ot McDonald and Boisjoly seem benign to William P. Rogers, chairman of the presidential commission investigating the Challenger accident. He has blistered Thiokol, calling its treatment of the two engineers "shocking." "It would seem to me," he continued, "they should be promoted, not demoted or pushed aside."

Likewise, the situation has so disturbed some senators that they are talking about calling Thiokol before a congressional committee to explain itself. At least one lawmaker also intends to push for a law to protect whistleblowers like McDonald and Boisjoly who testify before federal investigators.

Even if Thiokol's statement about its reasons for reassigning McDonald and Boisjoly could be taken at face value, the firm has made itself look incredibly naive about what constitutes good public relations. Thiokol already has enough image problems just now without needlessly adding to them.

What's needed is a vigorous investigation of how the two engineers have been dealt with by Thiokol. Then the firm would do itself a favor by adopting a company policy to the effect that honesty on the part of its employees is an important as loyalty.

THE SAGINAW NEWS

Saginaw, MI, May 13, 1986

Nearly four months after the explosion of the shuttle Challenger, media and official investigators still are asking what went wrong.

Part of the answer lies with some of the questioners.

Critical news reports during a six-week span of delays and aborted launches created "98 percent of the pressure" to send the shuttle into that frigid January sky, said Richard G. Smith, Kennedy Space Center director.

"The press would say, 'Look, there's another delay . . . Here's a bunch of idiots who can't even handle a launch schedule . . . You think that doesn't have an impact? If you think it doesn't, you're stupid," Smith said.

The "98 percent" overlooks the serious flaws disclosed in NASA's own procedures. But he has a point.

On the eve of the fatal launch, for example, Dan Rather opened his CBS newscast with this:

"Yet another costly, red-faces-all-around space shuttle launch delay. This time a bad bolt on a hatch and a bad-weather bolt from the blue are being blamed. What's more, a rescheduled launch for tomorrow doesn't look good either. Bruce Hall has the latest on today's high-tech low comedy."

Two weeks earlier, Rather said the launch "has been postponed so often . . . that it's now known as mission impossible."

The New York Times called the delays "a comedy of errors" and ABC's John Quinones reported, "Once again a flawless liftoff proved to be too much of a challenge for the Challenger."

Then, disaster — and suddenly the space agency was portrayed as launch-happy.

NASA has learned some hard lessons recently. One of them is to resist pressure, from the media or elsewhere, to override its own best judgment.

The lesson for journalists also is a hard one. A delay for safety's sake is not a failure. Never should caution again bear the label of comedy.

The Idaho STATESMAN

Boise, ID, May 31, 1986

The president, with NASA's urging, stands ready to commit the nation's resources to construct a $2.8 billion replacement space shuttle, as well as an $8 billion space station. But before those costly and precedent-setting decisions are made, the administration and Congress must set a new generation of space goals.

It has been 25 years since President Kennedy committed this nation to land a man on the moon by 1970. That goal reached, NASA talked Congress into placing all its space eggs into four space shuttles. To meet budgetary and economic constraints, NASA lobbied for a phase-out of expendable launch rockets and got Congress to promise that shuttles would carry all payloads, military, commercial and scientific.

The nation then turned to other interests, awaiting the promised day when shuttles would become as routine as flights to Chicago. Jan. 28 changed all that. Subsequent explosions of Delta and Titan rockets have left America stalled in its capability to place satellites, much less humans, in space.

Now NASA is pushing for a quick commitment to replace the Challenger. NASA officials say they accept the necessity of also relying on expendable rockets to deliver payloads. But for reasons of institutional inertia, NASA isn't about to settle for less than the four shuttles it once had.

Government, the scientific community and the aerospace industry agree that the United States is at a crucial turning point in its space program. Our scientific space research has taken a back seat. No major scientific missions have been launched since the 1977 Voyager probes. Other nations and private enterprise are competing to launch commercial payloads. And the 1990s timetable for a space station is fast approaching.

It's clearly time the administration and Congress stood back from the pressing and often competing space special interests and hammered out an achievable, affordable set of goals for space research that will carry this nation well into the 21st century.

A well-defined space program not only would set priorities for the types of space vehicles, shuttles or rockets in which America should invest, it would give the American people a clear sense of where we're going in space and how we'll get there.

Multi-billion dollar decisions must be made within the context of a comprehensive, long-term space program.

Shuttle Commission Blames NASA in Challenger Disaster

The presidential shuttle commission severely criticized NASA June 9, 1986 in reporting the results of its four-month investigation of the January 28, 1986 Challenger space shuttle disaster. The commission's report blamed the disaster directly on the system and bluntly stated that it was avoidable. The cause of the accident, as indicated from the outset, was a leak of superhot gases from a faulty seal in Challenger's right-side booster rocket, the panel concluded. The seal did not function, the panel said, because of cold weather. Later, after sealing temporarily, it gave way again because of strong winds and vibration. The escaping gases and flames burned through metal casings and struts and breached the shuttle's hydrogen and oxygen units, which then engulfed the craft in a giant fireball. But the problem of the booster rocket joints was a long-standing one, the commission pointed out. "The space shuttle's solid rocket booster problem began with the faulty design of its joint and increased as both NASA and contractor management (Morton-Thiokol) first failed to recognize it as a problem, then failed to fix it and finally treated it as an acceptable flight risk," the report said. The report was particularly critical of NASA middle managers, especially at Marshall Space Flight Center in Huntsville, Alabama, for ignoring mounting evidence that "a potentially disastrous situation" was building with the booster rockets.

The commission recommended a major overhaul of the shuttle's management to put control firmly at the top of the system. The commission advised NASA to take positive action to reduce "management isolation" at Marshall Space Center. Secondly, it urged creation of a special shuttle safety operation to detect and track hazards. The panel encouraged a greater role for astronauts in the safety operation. The commission concluded that "there was a serious flaw in the decision-making process" leading up to the launch. "A well-structured and managed system emphasizing safety would have flagged the rising doubts about the solid booster rocket joint seal," it said.

The News and Courier

Charleston, SC, June 3, 1986

Charles S. Locke, chairman of Morton Thiokol Inc., the company that developed the booster rockets for the space shuttle, is the kind of man who makes Marie Antoinette's recommendation that starving peasants eat cake sound compassionate.

For him the loss of the Challenger and seven of America's truly best and brightest people was merely "this shuttle thing" that he believes "will cost us this year 10 cents a share."

His display of crass insensitivity was not a slip of the tongue. It reflected the attitude of Morton Thiokol's management, confirming an already established pattern of corporate unconcern.

Mr. Locke is also on record as objecting to his employees traveling at company expense to appear before investigating commissions. That helps to shed a little more light on the reassignment of two Morton-Thiokol engineers who told the presidential commission investigating the Challenger disaster that they had opposed the fateful launch because they believed that rubber seals on the booster rockets might fail because of the prevailing low temperature at the time of the liftoff.

The company has not bothered to explain why the men should have been given other responsibilities and has merely admitted "a public relations mistake" in its handling of them. What is clearly lacking among the company's executives is a sense of accountablity. While it is true that Morton Thiokol is a private corporation and is under no obligation to explain why it has taken a particular course of action, it should not be forgotten that the company is also a virtual monopoly.

It has been standard practice for years to insist that private enterprises that are also monopolies, as is the case with utilities, should be made accountable to the public by making them subject to regulatory commissions.

Not all space companies are as arrogant as Morton Thiokol appears to be, however, and it is questionable whether a regulatory commission to check on companies that have a monopoly hold over NASA would do much more than add another layer of bureaucracy and further delay and dilute decision making. NASA, however, should try to break the kind of monopoly situation that has made Morton Thiokol so arrogant, so unaccountable and so impervious to public opinion. Mr. Locke himself should be made aware that the general public does not think of the shuttle disaster as "a thing" that can be written off at a cost of 10 cents a share.

THE LINCOLN STAR

Lincoln, NE, June 10, 1986

The immediate emptiness of failure seems inevitably to be filled with troubling contemplation of what might have been. And so it is with the presidential commission report on the tragedy of the space shuttle Challenger.

This analysis of the operations of the National Aeronautics and Space Administration looks at yesterday from the vantage point of today and finds a multitude of human weaknesses. The findings are aimed at the nation's space program but are valid considerations throughout the public and private sectors of our national life.

The details of the report can be seen in the news columns, as they catalog a litany of heavy personal pressure, of management too isolated from daily reality, of lack of clear direction, of growing complacency and absence of adequate communication. Individuals at fault were not named in the report, partly because it was more of an institutional kind of situation that contributed to the Challenger disaster in which seven people were killed.

"AN ACCIDENT that didn't have to happen," was the clear verdict of the commission after four months of investigation. Such a conclusion is both disconcerting and encouraging.

Peace of mind is hard to come by when you realize that the loss of seven lives need not have happened. But our spirits and our confidence are lifted by the realization that the nation has the clear potential to embark upon new and more successful efforts in space in the years ahead.

We should realize, too, that in seeking higher achievements, there will always be risks. In reaching new plateaus, there can be no absolute security and mistakes will not always be avoided.

No institution, no individual nor any single segment of our society has a corner on mistakes. Mankind's progress through the centuries has been the product of success through the disappointment of failure, and that is not going to change.

IN THE NATION'S space program, however, we have now come to better appreciate the reality of our limitations and the imperative of taking nothing for granted. Good management is not just something talked about in textbooks or seminars; it is the stuff of which satisfying results are achieved.

Nor is good management as individualistic as it is collective. The space program will rise from its current low point with the cooperation and dedication of NASA, the White House, Congress and its many private contractors or it will stumble along in ineptitude.

NASA itself will do the same in direct proportion to the ability of its top leadership to assemble and inspire a team with the potential and determination to get the job done. When all these things finally come together in a new and exciting program for the future, this country's space program will begin to achieve the rich promise of a still beckoning frontier.

THE ARIZONA REPUBLIC

Phoenix, AZ, June 11, 1986

THE United States has a habit of being at its best when things are at their worst.

The commission investigating the Challenger disaster is a good example. The panel, headed by former Secretary of State William Rogers and launched within days of the shuttle's destruction, has performed with rigor, thoroughness, fairness and with surprising dispatch.

The commission's findings are scathing. The technical failure of the solid rocket booster joint that led to the catastrophic end of mission 51-L was rooted in the history of a space agency which has lost its sense of direction, purpose and technological challenge, becoming complacent and victim to bureaucratic inertia.

Former astronaut Sen. John Glenn, D-Ohio, notes that after the Apollo moon landings and after the initial testing of the shuttle system, the NASA *esprit de corps* changed from "can do" to a *laissez faire* "can't fail." The shift was a prescription for disaster.

The Rogers commission's report is just one of the needed correctives for the floundering American space program. Yes, the rocket booster joint must be redesigned, the shuttle flight schedule adjusted for safety and NASA management restructured.

More than this, however, NASA needs a new set of goals like those established by John Kennedy when he committed the United States to putting men on the moon and returning them safely before the end of the 1960s.

That vision attracted the best minds in America to NASA, and the agency led a great adventure exploring the frontiers of space, the moon, technology and human endurance.

NASA is not designed to run a bus line into space, which is what the shuttle program is. It should be an advanced, state-of-the-art, limit-pushing, blue-sky program populated with the best, most innovative people in America.

In addition to implementing the Rogers commission's recommendations, the United States must rededicate itself to space exploration; to putting a permanent manned space station in orbit; to returning to the moon; to go to Mars and beyond.

Birmingham Post-Herald

Birmingham, AL, June 5, 1986

If the National Aeronautics and Space Administration does not feel chastened already, it surely ought to when its managers read the report of the presidential commission that investigated the shuttle Challenger disaster.

The report, which has gone to the printers and is expected to be presented to President Reagan this weekend, is said by commission sources to lay the blame for the loss of Challenger and its crew on the failure of NASA to correct known design flaws while letting the shuttles fly.

One source described the commission's conclusion on NASA's management system as one "that swept all the bad news under the table."

As expected, the report concludes that the Challenger blowup was caused by a leak in a joint connecting segments of a solid-fuel rocket booster. The report describes repeated warnings about the joints, which should have prompted NASA to redesign them and thereby have avoided the Challenger tragedy.

The commission recommends a reorganization of NASA management structure and instructs the agency to put safety requirements into every phase of flight operations.

It recommends a greater role for astronauts, engineers and contractors in approving launches. It was brought out in the investigation that engineers for the company that built the solid-fuel rocket boosters had recommended against launching Challenger in the cold weather prevailing on Jan. 28, the day the shuttle exploded.

The commission also recommends procedures to prevent NASA managers from signing waivers clearing shuttles for flight when there are documented problems affecting flight safety.

It suggests consideration of an escape system that would allow astronauts to bail out under certain conditions.

It also recommends a more centralized NASA management structure that would enable the Washington headquarters to keep closer tabs on all phases of the agency's operations.

NASA deserves the blistering the commission has given it. For the seven astronauts who lost their lives, it's a closing of the barn door after the horse was stolen. But the commission's work should lead to a safer program for those who will follow them into space.

THE BLADE

Toledo, OH, June 14, 1986

THE presidential commission headed by William Rogers has performed brilliantly in solving the mystery of the Challenger disaster and charting a course for the future safe operation of America's space program.

When the 13-member commission was appointed four months ago, many Americans believed that the cause of the accident would never be determined with certainty. They foresaw loose threads that would flutter over the years, inviting endless speculation about hidden causes, possible sabotage, and the like.

But the Rogers commission has reached a firm and compelling conclusion that should leave no room for debate. Challenger, which would cost $3 billion to replace, and seven human lives were lost because of the failure of a joint seal in the shuttle's right solid-rocket booster. The faulty joint allowed a jet of white-hot rocket exhaust to burn through the surface of Challenger's huge external fuel tank like a blow torch.

The panel said that the joint, or perhaps the entire solid-rocket booster, must be redesigned before the three remaining space shuttles can fly. The National Aeronautics and Space Administration is working on a new joint design. Thus there will be a relatively rapid technological fix to prevent recurrence of what the commission identified as the immediate cause of the accident.

But what of the contributing causes that the commission emphasized so heavily in its 256-page report? What of NASA itself, and the space agency's management failings and lack of proper regard for safety?

NASA engineers recognized as early as 1977 that the joint design for the solid-rocket boosters was deficient. Additional evidence came after the first shuttles began to fly in 1981, and NASA and Morton Thiokol, manufacturer of the boosters, saw joint damage in recovered boosters.

The commission uncovered a long "paper trail" of documents warning that continued flights invited a catastrophe. Yet NASA persisted, accepting the joint seals as just another unavoidable risk in an approach that one commission member likened to a high-technology version of Russian roulette.

NASA clearly has major management problems to solve. The Rogers commission has made recommendations that NASA should follow to the letter. High among these are a return to the "safety-first" philosophy that guided the agency during those pressure-filled years of the Apollo program.

Indeed, the commission found that if NASA had maintained such a vigilant safety program, it might have met the pressures of a heavy shuttle-launch schedule without encountering disaster.

Renewed attention to safety is especially important in this era of the space shuttle. Each launch involves — in addition to human lives — an enormously costly vehicle critical to national security. NASA and the country cannot afford another accident.

The Kansas City Times

Kansas City, MO, June 3, 1986

Retired Air Force Brig. Gen. Charles "Chuck" Yeager was among the high-profile members appointed to the presidential commission investigating the Jan. 28 accident of the Challenger space shuttle. His name, however, will not appear on the panel's report to be presented to President Reagan this week.

You remember Chuck, the first person to break the sound barrier and the one they called the fastest man alive. Chuck, Chuck Yeager. Rugged-looking actor Sam Shepard played him in the movie version of "The Right Stuff." He's the same ace pilot rejected by NASA when it launched its hunt for Mercury astronauts. The Chuck of the television commercials and the best-selling autobiography, "Yeager."

Apparently he was more absent than present for the Challenger proceedings. He appeared briefly at one hearing and observed the shuttle obviously should not be launched in cold weather. Wow, Chuck! We think you were on to something there, pal. Isn't that the warning Morton Thiokol engineers gave to their vice presidents and NASA executives?

We're sorry that Chuck, whose opinion was based more on deductive reasoning than on testimony during the hearings, wasn't around to provide more input to the Rogers Commission, whose task was a very serious one. A reporter asked him if his failure to participate in the hearings was due to previous commitments. "You got it," he replied. Asked whether he was bothered by the omission of his name from the report, he said, "Nope."

Even so, this episode makes you wonder whether this was Chuck Yeager's way of getting back at NASA, doesn't it? No! Not Chuck! We like his style.

ST. LOUIS POST-DISPATCH

St. Louis, MO, June 8, 1986

The 200-page report from the presidential commission on the Challenger space shuttle disaster will not be officially released until tomorrow. However, many of its findings have already been made public, and they confirm fears that the management at NASA has been held prisoner to bureaucratic momentum.

One commission member stated pointedly that NASA officials "were like lemmings rushing over a cliff." There appeared to be an unspoken ethos in NASA that nothing would get in the way of the space shuttle's overly ambitious launch schedule. Upper-level NASA managers never received warnings from lower-level engineering personnel on critical and potentially fatal technical and design problems. But there are indications that even if these warnings had made it to the top, they would have had little impact on NASA's planning and decision making.

"Nobody wanted to ground the [shuttle] fleet," one commission member noted. "Nobody wanted to say, 'Let's stop the music and fix it.'" It was the old Army Air Corps "can-do spirit" run amok.

Understandably, a major overhaul of NASA's management structure is recommended by the commission's report. Among the changes it suggests are better in-house communication on safety concerns, a new safety panel, more and tighter quality control and greater involvement by the astronauts in safety issues.

What is most disturbing in the commission's report is that the design flaws and problems with the fatal O-ring joints on the shuttle's solid-fuel rocket boosters are not unique to the program. In many ways, the commission's report clearly indicates that the space shuttle was a disaster waiting to happen. Besides a redesign of the solid-fuel rocket booster joints to make them "no less" strong than the rest of the rocket casing, the commission recommends that the shuttle's wheels, brakes and steering systems and the main engines be given special attention if another shuttle tragedy is to be prevented.

Moreover, the commission warns that there is no "quick fix" that will return the shuttle to operation soon. Acting on the report's recommendations could keep the shuttle grounded well into late 1987 or beyond. At the very least, NASA's plans to launch the next shuttle in mid-1987 appear optimistic.

The commission's report on the Challenger tragedy will provide grist for many mills. No doubt it will have a major impact on the future of both the shuttle program and NASA in general. It also raises many other questions, among them: Should the U.S. build a replacement shuttle or should the country jump ahead to a hypersonic aerospace plane? Should the U.S. return to using expendable, unmanned rockets for carrying satellites into space? Should the shuttle be limited to high-priority military missions?

In so many ways, the presidential commission's report on the Challenger disaster is only a prologue and not a conclusion.

The Augusta Chronicle

Augusta, GA, June 11, 1986

The technical reason for the tragic Challenger explosion that killed seven astronauts was the failure of a seal in the space shuttle booster rocket to do its job properly.

But the most shocking finding of the Rogers Commission's investigation of the tragedy was that the seal's failure was not a bizarre, freak accident. It was inevitable — in the report's words, "rooted in history."

The seven who died Jan. 28 had no way of knowing that they were fated to be the losers in a game of "Russian Roulette," again the report's words, that dated to 1977 when design problems with the booster rocket joint were first noted, but never corrected.

Due to several factors in the management structure — diffused responsibility, poor communications, pressures to launch — NASA (National Aeronautics and Space Administration) carelessly allowed itself to drift away from the "safety first" philosophy that governed the early years of U.S. space travel.

The commission makes many specific, constructive recommendations, but basically they boil down to only a few. On the technical side, the faulty solid rocket motor joint and seal must be completely redesigned, regardless of time or cost. On the management side, a chain-of-command must be established to improve communications and create accountability in decision-making.

The commission also advises that the "amateur astronaut" program — taking teachers, journalists, etc., along on shuttle flights — be put on hold indefinitely.

The commission should be commended for its work. Its criticisms and recommendations are balanced and sensible. It blamed the management system for what went wrong, eschewing the temptation to get into finger-pointing. Nothing is to be gained in blaming specific people unless there is evidence of corruption or criminal negligence, which there isn't.

The nation, too, must take a more balanced view of the space program. For years it seemed that NASA could do no wrong. Since the Challenger explosion, it seems like NASA did nothing right. The truth lies somewhere in between.

The U.S. space program, let us not forget, has a remarkable safety record considering that space travel is still in its infancy. It would be naive to believe all the risk can be taken out of it or that future loss of life can be averted.

An oversight committee, which NASA has already established at the commission's recommendation, should help greatly reduce those risks and tragedies. If such a committee had been in operation from the start of the program, the Challenger disaster likely never would have happened.

Newsday

Long Island, NY, June 10, 1986

After months of public hearings, the report by the presidential commission investigating the Jan. 28 space shuttle accident contains few surprises. But its meticulous compilation of mishaps and warnings going back many years shows with shocking clarity that Challenger was "an accident rooted in history."

As expected, the panel blamed the tragedy on the rupture of a joint in the right solid-fuel booster rocket. It sharply criticized the National Aeronautics and Space Administration and Morton-Thiokol Inc., the rocket's manufacturer, for glossing over a long history of problems and concerns about the O-ring seals in the joints.

"Even the most cursory examination of the failure rate should have indicated that a serious and potentially disastrous situation was developing on all solid rocket booster joints," the report said. "If the program had functioned properly, the Challenger accident might have been avoided."

But instead of heeding test and flight data that showed damage to the O-rings, NASA and Morton-Thiokol "accepted escalating risk apparently because they 'got away with it the last time,'" the commission concluded.

That's a damning indictment of an attitude that cost seven brave men and women their lives and seriously set back the nation's space program. Those responsible should be held to account. But the task that now confronts the government is how best to get the United States back in space and to ensure that the gross errors and misjudgments of the past are not repeated.

A series of sensible recommendations by the commission form a good basis for congressional hearings that begin today. They include:

- Totally redesigning the booster rockets' fuel segment joints and establishing an independent review board to oversee the operation.
- Completely overhauling NASA's management structure, emphasizing accountability and an increased astronaut role in the launch command chain.
- Establishing a safety organization within the agency to report directly to the administrator.
- Studying astronaut escape systems and improved landing margins. The boosters are not the only safety problems identified; others include the main engine, brakes and landing gear.
- Improving communications to make sure critical information reaches top managers. In the case of Challenger, serious reservations expressed by engineers were never conveyed to those who made the decision to launch.
- Setting a realistic flight schedule. The commission said NASA had "stretched to the limit" trying to meet its launch timetable.

The commission did not recommend building a new shuttle to replace Challenger. It did warn against reliance on a single launch capability, noting that it "should be avoided in the future."

Congress should pay particular attention to that advice. Only a handful of unmanned, expendable rockets remain. Building a fleet of new ones would be far wiser — and safer — than replacing the lost shuttle.

DAYTON DAILY NEWS

Dayton, OH, June 11, 1986

The prestigious panel headed by former Secretary of State William P. Rogers had to do some special engineering work to keep itself from flying apart, but it flew and landed a lot more skillfully than a lot of people expected it to at takeoff.

Thanks to its success in working at an objective arms length from the National Aeronautics and Space Administration, the Rogers Commission quickly tagged the technical problem — a faulty O ring — that caused the destruction of the space orbiter Challenger and honed in on the administrative problems — sloppy quality control in NASA.

The reform of NASA is likely to continue along the lines set down by the blue-ribbon panel. The reforms include a redesign of the booster rocket, streamlined program management and tighter safety reviews that give shuttle manufacturers and astronauts a say-so about launching.

The recommendations include consideration of a crew escape system useful for descents, though such a system would not have saved the Challenger crew. Perhaps such a system is possible, but it may be that safety on the shuttle, like safety on a passenger plane, mainly has to be built into the shuttle's basic flying ability, not in its ability to break into parachuting capsules or whatever.

The panel called for NASA to establish a flight schedule more "consistent with its resources," which means the shuttle program can no longer pretend to recover a lot of its expenses with space experiments and satellite missions for private enterprise.

For the decision to build a cheaper-than-ideal shuttle system, Congress can take some blame. For pressure to launch, the public, media and the political system can share some blame. NASA did not operate in a vacuum and it will not be reformed in a vacuum.

Reagan Orders New Shuttle; Ends Commercial Payloads

President Ronald Reagan ordered NASA to build a fourth space shuttle to replace the lost Challenger, it was announced from the White House August 15, 1986. The president told NASA to drop out of "the business of launching private satellites...NASA and our shuttle can't be committing their scant resources to things that can be done better and cheaper by the private sector," Reagan said in a statement. "Instead, NASA and the four shuttles should be dedicated to payloads important to national security and foreign policy, and even more, on exploration, pioneering and on developing new technologies and uses of space." The decision on replacing Challenger had been pending ever since the spacecraft was destroyed and its seven astronauts killed in an explosion shortly after launching on January 28, 1986. The three remaining shuttles were grounded. Details of the financing of the new orbiter, estimated to cost $2.8 billion over five years, remained to be worked out. White House spokesman Larry Speakes, at a briefing on the announcement, said that funds for the new shuttle could come from various sources, such as savings from the current moratorium on space launches, or other NASA programs and other federal agencies. Speakes said the space-station project, which Reagan strongly favored would not be affected. Speakes said the Administration would request Congress to authorize $272 million for fiscal 1987, which would begin October 1, to start construction of the new shuttle. Speakes said the agency would phase out its commercial launch activity. Currently it has 44 commercial or foreign launches scheduled. Of these, Speakes said, only 15 would be launched by 1992, with the scheduling to be worked out by an Administration task force.

Senators and analysts questioned the new policy, pointing out that the original space shuttle program had been initiated in part on the assumption that eventually it would be self-supporting as a result of the income derived from private, commercial or foreign satellites and other payloads. Speakes, at his Aug. 15 briefing, also addressed another concern arising from the decision to end commercial launches by NASA--that the Pentagon would fall heir to the payload-space left available by the changed schedule. "It was simply not true" that the new policy would contribute to the "militarization of space," Speakes assured reporters. For the "forseeable future," the Defense Department would continue to use approximately the same NASA launch capacity that had been planned for it originally, he said.

Roanoke Times & World-News

Roanoke, VA, August 26, 1986

PRESIDENT REAGAN has issued a replacement order for the destroyed space shuttle Challenger. At the same time, he took the National Aeronautics and Space Administration out of the business of launching commercial satellites. That move kindled a controversy, but the president's action seems appropriate.

Reagan has acknowledged that NASA, having brought the United States into the space age, should now devote itself to government missions (such as military launches) and to developing new technologies and uses of space. Commercial ventures should not be its concern. "NASA and our shuttles," said a presidential statement, "can't be committing their scarce resources to things which can be done better and cheaper by the private sector."

The president tends to believe that the private sector can do all things better and cheaper than government. That is a debatable topic. But whether the proposition is true isn't the central issue. On occasion the federal government sows the seed money and does the spadework in some field that's too chancy or costly for business to undertake. Some time after that, government should step back and let the private sector take over any commercial applications it wants to try and which are suitable for those companies to do. It may still be too costly for them to undertake, but that's an economics decision, not for Uncle Sam to make.

NASA got itself into the commercial area when the agency was trying to define its postlunar mission. To sell Congress, it needed to suggest a space vehicle that — unlike the manned rockets used for orbital flights and trips to the moon — could be reused, many times. It also needed to suggest a vehicle capable of multiple uses: military, scientific and business.

The shuttle was the result, and almost from the start, the project was in trouble. Construction ran very late and far over budget. In practice, these vehicles never delivered anything approximating the efficiency and utility envisioned for them. As performance lagged behind promise, pressure built for fewer delays, more launches and more interesting payloads — from members of Congress to private citizens such as schoolteacher Christa McAuliffe. This presssure led to corner-cutting and risk-taking, which culminated in the Challenger explosion.

There is grumbling in NASA and in Congress about the president's decision to shift commercial applications to the private sector. But it is one way of finding funds within the present space budget for replacing the Challenger, thus not adding to the deficits. And Philip Culbertson, NASA's general manager, sees it as a realistic move.

"It takes some of the pressure off of us," Culbertson said. "Until the accident happened, we maintained the goal that the system had to be cost-effective. This country is now going to have to get used to the thought that we are making a major investment for the future with the shuttle, but it won't have immediate returns to the Treasury."

Very few federal ventures do; very few are designed for that, except tax-collecting. It will take years for the private sector to get into this corner of space, and it will lose business to foreign countries meanwhile. But the federal space program should run better and at comparatively less cost if its mission has a smaller and more suitable focus.

The Des Moines Register

August 23, 1986

Des Moines, IA

If the Reagan administration learned nothing else from the space-shuttle disaster, it should have learned the folly of too-hasty decisions made under pressure. There are indications, in the administration's decision to remove the National Aeronautics and Space Administration from the commercial launching business, that the lesson was lost.

Pressure, it appears, dogged the program for 20 years before dooming it. The goal was to make the space-shuttle service pay for itself by carrying commercial satellites into orbit. In theory, the manned service would operate at less cost than expendable rockets.

But the shuttle was built for heavy duties, so the French entered the field to lure lighter payloads, thrust skyward by rockets built cheaply and easily altered to handle varying weights.

In order to compete, NASA had to send up more shuttle flights than it could safely handle. Pressure to hurry the schedule resulted in pressure to cut corners, pressure to ignore warnings — such as that from the engineers who said fuel-tank seals on the Challenger rocket might not hold under cold conditions. Finally, failure of those seals blew apart the Challenger, its crew of seven, and the long dream of economic justification for an exotic space-age venture.

In ordering an end to NASA's commercial role, President Reagan said private enterprise can do the job better at less cost. But NASA's long control of the field so discouraged potential competition that it may take the private sector five years to gear up. Some experts fear that private enterprise may never be able to catch up.

There's more bad news for those wanting payloads hoisted: The French service, Ariane, has been grounded for the rest of the year by an accident, and Ariane's prices are going up, thanks to lack of competition and a boost in liability-insurance costs.

But the worst news, many fear, is that the administration made a costly mistake in abruptly ending the space shuttle's mission rather than regrouping, learning from the Challenger tragedy and proceeding with caution. "I have very serious reservations that the White House did not think this through," said Senator Slade Gorton.

How ironic and tragic if the nation that sacrificed so much to achieve leadership in space should now, by default, let competitors harvest the fruits of that effort.

The Pittsburgh PRESS

Pittsburg, PA, August 21, 1986

President Reagan's decision to build a fourth space shuttle is intended to reinvigorate America's space program, which has fallen on hard times since Challenger blew up last January. Whether that goal would be achieved is debatable.

Significantly, the White House was vague about financing the $2.8 billion shuttle, and enormous budget deficits leave little room for new projects.

Will the money come from cuts in other National Aeronautics and Space Administration projects, such as a permanent space station and other scientific missions? If so, the space program could be set back, not advanced.

A major element of the president's plan is to get NASA out of the commercial satellite launching business. He said launching of commercial payloads "can be done better and cheaper by the private sector." This reverses a two-decade policy to make the shuttle program eventually self-supporting by launching satellites for private companies. Loss of this revenue will further squeeze NASA financially.

The White House thinks that taking commercial payloads off the shuttle will spur U.S. firms to develop private launching capability. This assumption is questionable.

Previously, American companies were discouraged from launching satellites because the shuttles were available. Now U.S. companies would have to catch up with foreign interests that already have launching facilities. It might take heavy U.S. subsidies to encourage them.

Mr. Reagan said NASA's shuttles would be reserved for "payloads important to national security and foreign policy, and even more, on exploration, pioneering and developing new technologies and uses of space."

Some quarters fear that the White House plan could lead to dominance of the U.S. space program by the Pentagon and a diminution of NASA's role. White House press spokesman Larry Speakes claimed, however, that "speculation that we are contributing to the militarization of space ... is simply not true."

Many experts believe it would make better sense to build more unmanned rockets to boost payloads into space, rather than to construct another shuttle.

In any event, Congress should take a hard look at Mr. Reagan's proposal. In recovering from the Challenger disaster, the nation needs assurance that it will get the maximum benefit from space dollars spent. Questions remain whether Mr. Reagan's plan is the way to go.

THE SACRAMENTO BEE

Sacramento, CA, August 24, 1986

If any careful thinking went into the Reagan administration's decision about the future of the shuttle program, it's undetectable. As a result of the fine work of the Rogers commission, appointed by the president to investigate the Challenger accident, and of the separate probes conducted by Congress, the New York Times, and NASA itself, the nation has been warned: We now know that a too-ambitious launch schedule can compromise safety and even efficiency; that it was a mistake to put all our space-program eggs in one shuttle basket; that it makes no sense to use risky and expensive manned shuttles for routine satellite launchings; that, in fact, the shuttle program never was the cheap, reliable launching service it was meant to be. The administration, however, has ignored all of that.

Its plan is to restart the three remaining shuttles by 1988, time only for minor fixes in the troubled booster rockets that failed on the Challenger, and to replace the Challenger itself with a new manned orbiter — at a cost of several billion dollars — by 1991, time only to introduce modest improvements in the same old technology. For the sake of those timetables, Morton Thiokol will remain the sole supplier of rocket boosters, despite all the talk about the need to line up alternative, competitive suppliers in order to improve quality and reduce costs.

That amounts to reconstituting the old shuttle program virtually intact; yet, at the same time, its original purpose is all but forsaken under the president's plan. Private industry is being told to build its own space service for launching commercial satellites, which the shuttle will no longer carry, and NASA's scientific research will no doubt be cut back severely in order to help pay for the Challenger replacement. Virtually the only client NASA will have left is the military, which meanwhile is planning to supply itself with unmanned launchers, so it won't be entirely dependent in the future on NASA's shuttle services. Why, under these circumstances, the existing shuttle fleet should be preserved — much less expanded — is not at all obvious.

NASA's clients now face a backlog of interrupted launchings, which had to be dealt with. The president did so by rescheduling some and abandoning others. But beyond that, some hard thinking was needed about what mission it makes sense for NASA to perform in the future; a plan was needed for creating, over the next decade, a diversified NASA capacity to serve that future purpose. And that the president didn't do.

Instead, the administration fell back on the most comfortable of its hobby horses — Star Wars and privatization. NASA's shuttles in the future will serve Star Wars, even if that's not what the program was designed for, while the commercial business it was supposed to handle will be privatized (how and at what cost to the nation's competitiveness in space are not addressed). And nothing else will be changed. In fact, a fortune will be spent not on forward-looking research and development, but on restoring the old shuttle fleet. The opportunity to raise a phoenix from the ashes of the Challenger was scuttled.

The State

August 22, 1986
Columbia, SC

AFTER ALMOST six months of delays, the President decided the National Aeronautics and Space Agency should build a new space shuttle to replace *Challenger*, which blew up last Jan. 28 with seven crew members aboard.

A new space orbiter is good news for the space program and for backers of the Strategic Defense Initiative, otherwise known as Star Wars, to develop a antiballistic missile defense system.

There was also good news for American space industries; the White House announced NASA would handle only defense, foreign policy or scientific payloads, leaving commercial payloads to the private sector. Six firms already have signaled their interest in commercial missions.

If this is a "go" from the President to revive NASA, it is like trying to launch a rocket that hasn't been fueled. The White House didn't say where the money for the new shuttle would come from — an estimated $2.8 billion by 1991, if Congress concurs. Initial funding for next year alone amounts to $272 million.

NASA and its backers on Capitol Hill say there isn't enough in its budget to pay the initial costs of a new shuttle. If the money were taken from NASA's programs, it would cripple space science and the space station.

The fourth orbiter will assure our space program's continuance; it has been at a standstill since Jan. 28. We can resume launches of our present three when the problems with the solid fuel booster rockets are solved, possibly next year. But the fourth space shuttle is needed to keep pace with an increased launch schedule of missions NASA is committed to.

We also endorse participation of private enterprise in launching commercial satellites and scientific missions. The economic impact will be quite large in the future, and so will spinoff technology.

But the Administration will have to provide specifics on where the additional money for the new orbiter will come from. Agreeing to build one has not removed uncertainties which plague NASA. Once again Mr. Reagan hasn't got his "want to" and "can do" together.

THE DENVER POST

Denver, CO, August 19, 1986

THE ORDER by President Reagan that NASA get out of the commercial-satellite launching business and proceed with building a new space shuttle — at a cost of $2.8 billion — adds a new round of questions to those that came after the tragic loss of the space shuttle Challenger. For openers, the White House announcement fails to satisfactorily explain where the funding for the new space craft will come from. Will NASA's other programs be squeezed to cover appropriations over five years? Will the Pentagon be asked to chip in a share large enough to fairly reflect the shuttle's military uses?

More importantly, with the three space shuttles still left out of service, why is a fourth one needed — especially if the president is ordering a reduction of the shuttle mission by prohibiting use of the giant orbiters for hauling commercial satellites in space?

There's an even more basic question. Why build a new shuttle — which will be nearly a duplicate of the design of existing space craft — at a time when the debate remains unresolved over the relative roles of manned and unmanned space flight?

Some leading space scientists argue persuasively that the glamor of manned space travel has seriously set back our scientific exploration of the heavens. In their view, manned probes are more expensive, and less versatile, than unmanned probes.

And besides, it has been about 10 years since any significant probe of deep space has been launched by the United States. A renewed push to build another shuttle will mean that no funds for vehicles into deep space will be available for another decade.

Consequently, the U.S. will suffer a 20-year gap in scientific knowledge to be gained by sending unmanned scientific explorations through the solar system and beyond. The price of such a lapse is high, considering that the generation of scientists who designed such probes as Viking (to Mars) and Voyager (to Saturn and Jupiter) will have aged into retirement before the next deep-space missions are begun.

The president spoke for all the nation last January, expressing our grief and determination to continue mankind's flights into space, after the Challenger's explosion on liftoff and the deaths of the seven crew members. But the emotions of that time don't require that a new shuttle of vague purpose be built merely to memorialize the lost astronauts.

In the decision to end the commercial launch service — a decision that surprised many of those involved in the space business — President Reagan said such work can be done "better and cheaper by the private sector." That may prove to be so sometime in the future but, for the moment, the president seems to have ignored a couple of space-business facts.

The shuttle orbiters were designed and built with the profitable commercial work as a basic use, and the private sector, at least in this country, is unprepared to assume a vigorous role in the launching of commercial satellites. Such business, very likely, would go to France or Japan, where ambitious private-satellite launching efforts exist.

It can't be said that the president is flatly wrong in what he's announced for the space orbiter program. Our problem is that he has raised more questions than he's answered, and we're waiting for those answers.

The Star-Ledger

Newark, NJ, August 24, 1986

President Reagan has made a decision to replace the lost Challenger with another space shuttle. In so doing, he has charted the course of the nation's space policy toward continued manned space flights.

The decision ends a debate that has been conducted with some heat since the explosion of Challenger last January. This does not seem like sufficient time for resolving so complex a question.

Exactly how great a commitment we are to have to a new space shuttle program remains to be seen. At the moment, the planning seems to be uncertain. The final price tag for the vehicle would be $2 billion and, in a time of budgetary cutbacks, it is difficult to see where the money would come from.

Cost is important, but it is not the only question to be resolved. More to the point is the question of whether manned space travel should be the central focus of our space program. An enormous financial commitment to a new shuttle would virtually require that it dominate our space research.

Mr. Reagan is said to have based his decision on the need for a four-orbiter fleet to counter Soviet advances and to support the space station, international commitments and advanced military flights. We are unquestionably in a space race with the Soviets but whether the replacement of the Challenger is the best strategy to adopt is very much open to question.

During the lengthy investigatory process of the Challenger tragedy, many prominent scientists said that manned space research frequently was not the best policy to pursue. Basic research goals are apt to be compatible with unmanned space research, according to a considerable body of expert testimony.

In connection with this, it is worth noting that Soviets have been far less dependent on manned space travel than we have. There is considerable reason to believe that our reliance on manned space flights has often stemmed from its high visibility, which was translatable to quick gains on the public relations front.

The most unsettling aspect about the President's decision is that an arguable, highly complex question has been settled with a minimum of public debate. The question of replacing Challenger was deserving of far more study and argument than it received from the Reagan Administration.

THE SPOKESMAN-REVIEW

Spokane, WA, August 16, 1986

The past week's announcement of an improved design for the space shuttle's solid-rocket boosters represents good news indeed for the nation's stalled space program.

The United States must restore its launch capability and if tests show that engineers have indeed produced a design that will prevent burn-throughs at the rocket's joints (the cause of the Challenger mishap), then perhaps the shuttle's hardware can be pronounced ready for use by 1988.

However, safe equipment is only half of the problem that must be resolved before another manned U.S. spacecraft is allowed to fly. The other half involves human oversight.

It probably is inevitable that at some point in the future, some engineer will discover another weak point in the shuttle and will argue that a scheduled launch should be scrubbed. When that happened in January, political and fiscal pressures prompted decision makers to reject the warnings and proceed with the fateful launch.

Those pressures resulted in part from Reagan administration insistence that the shuttle should pay its own way; that policy had produced a rigorous launch schedule and corner-cutting on safety. A new management philosophy is as essential to NASA's future success as is a new rocket design.

The Salt Lake Tribune

Salt Lake City, UT, August 19, 1986

For some reason President Reagan's decision to build a space orbiter to replace the destroyed Challenger brings to mind the phrase Pyrrhic victory.

To refresh memories, Webster's New World Dictionary definition is:

"**Pyrrhic victory:** *a too costly victory: in reference to either of two victories of Pyrrhus, king of Epirus, over the Romans in 280 and 279 B.C., in which his losses were extremely heavy.*"

This sort of fits the situation at the National Aeronautics and Space Administration. Officials at NASA and outside experts agree that the victory, a decision to rebuild the space shuttle fleet to four orbiters, came at a very high price.

NASA has to give up commercial ventures, which were a source of revenue for the space agency, along with paying for the new orbiter out of existing funding; the White House isn't going to ask Congress for any new money.

The result? Possibly another robbing of Peter to pay Paul because there is a real risk that some NASA projects will be curtailed or junked completely.

Another concern centers on the so-called "militarization" of space.

With commericial space ventures, such as the launching of private communication satellites, being shunted from the manned space vehicles to less expensive unmanned, solid fuel rockets it will leave the Defense Department as NASA's customer for at least half the shuttles' major missions.

Also the Pentagon has announced plans for a medium-size unmanned rocket to handle many of the payloads originally intended for the shuttles.

Conceivably, the Pentagon could become the driving force of the nation's space program, rather than just another shuttle user. Such an eventuality would be a far distance from what has been the vision and practice of NASA since its beginning; an open and above board, civilian-run effort to push back the frontiers of space.

On this score the White House advises Americans not to worry. Said Larry Speakes, the president's chief spokesman: "There has been some speculation that we are contributing to the militarization of space. That is simply not true."

Still, that "cost," as well as determining whether the price of restoring the shuttle fleet to four orbiters will be too high, may take years in assessing. The answers certainly won't be available before 1991 when the new shuttle will be ready to fly, a year or two later than most people hoped that Challenger could have been replaced. With a Chief Executive like President Reagan, NASA doesn't need a monarch like old King Pyrrhus.

Los Angeles, CA, August 18, 1986

President Reagan announced last week that the U.S. will build a fourth space shuttle. But the country still lacks something it needs far more: a coherent civilian space policy.

The president's decision continues the militarization of NASA. A few scientific experiments like the Hubble Space Telescope are designed specifically to be carried aboard the shuttle, but most of the shuttle's future launches will transport military payloads. These will include Star Wars experiments and modules for the space station, a laboratory for which no one has developed a mission.

Although the Pentagon will be the chief beneficiary of a fourth shuttle, it won't pick up the tab. Instead, the $2.8 billion in estimated construction costs will come from "savings" in NASA's budget due to delays in shuttle launches, from space science projects and from squeezing other federal agencies. Meanwhile, the Air Force is buying cheap, expendable Titan 4 rockets to launch spy and navigation satellites. NASA should lower costs by buying Titans, too, rather than building another shuttle.

Besides squandering precious dollars on a shuttle that uses 1960s technology, but won't be ready to fly until 1991 at the earliest, the White House also has decided to get NASA out of the commercial end of the space business. This is a victory for the aerospace industry and the influential Heritage Foundation, which have long argued that the space agency's monopoly over rocket launching should be eliminated.

Some privatization makes sense, but the administration's plan is to continue offering government-owned launch facilities at reduced rates so that industry gets the profits of orbiting satellites without the overhead costs. This is exactly the kind of thinking that got the shuttle into financial trouble.

The U.S. cannot ignore the military potential of space. And a carefully regulated private space-launch industry can provide many long-term benefits. But the country also needs a government-led civilian space program that focuses on scientific research and international planetary exploration. Building one more obsolete space shuttle takes us in another direction altogether.

THE PLAIN DEALER

Cleveland, OH, August 19, 1986

President Reagan last Friday made known his decisions concerning a major portion of America's space program and, in so doing, further clouded that program's future. NASA is to build another shuttle, he said, while at the same time it is to get out of the commercial launching business in deference to private enterprise. Conspicuously absent from the presidential pronouncement was any indication as to just where the $2.8 billion for a new spaceship is to be found, or how the fledgling American private rocket industry is to pick up the satellite-launching business.

That's not unusual, considering Reagan's history of avoiding anything that smacks of details. But it is unsettling to those who trouble themselves with budgetary responsibility. Among those individuals are NASA Administrator James C. Fletcher, who said yesterday he has no idea where the money will come from, and Sen. Slade Gorton, the Washington-state Republican who chairs the subcommittee on science, technology and space that oversees NASA's budget. One suggestion seeping from the White House was based on a dubious understanding of NASA financing: Since shuttle flights have been suspended, NASA thus must be saving money that could then be used to pay for part of a new shuttle. Gorton, well aware that NASA already has been hit by the Gramm-Rudman-Hollings budgetary cutbacks, pointed out that the space agency is losing money by staying on the ground. "The money doesn't exist, and NASA is being precluded from raising any revenue," Gorton is quoted as saying. "I don't see how we can approve this."

Equally baffling is how the president expects the private sector to jump into a commercial launching market from which it has been systematically excluded by federal subsidization of NASA's launch costs. Industry executives say it would take five years for privately owned companies to get on the launch pad—an interval during which many of the 44 backlogged commercial payloads will have sought, and probably acquired, a position on other nations' rockets. While privatization is a desirable long-term goal, it is not the immediate answer.

Further, if that is to be the course of America's space enterprises, what is the rationale for building another orbiter? As Gorton noted, launching commercial satellites and providing a platform for commercial experiments has been a key reason for the shuttle to exist. "I have very serious reservations that the White House did not think this through," Gorton said.

Then again, maybe it has. But that scenario is even more troublesome. More and more, it appears that the shuttle program is destined to become a Pentagon program in mufti with a major, if undefined and top-secret, role in the development of the Strategic Defense Initiative. The White House says it wants Congress to authorize $272 million in "new money" in fiscal 1987 for the new shuttle, and Gorton staff members say the senator will try to comply, despite his misgivings about the logic behind the request. If that is to be the case, then the new orbiter's funding should not come at the expense of what remains of NASA's basic science and research programs, but should be paid for by some combination of SDI and general military funding. That appears to be the only reasonable approach to a not very reasoned request.

IT'S NOT CHEAP.
IT'S NOT TESTABLE.
IT'S NOT PRACTICAL
AND IF IT'S NOT A BARGAINING CHIP... WHAT IS IT?
STAR WARS
HEY! I THINK I KNOW!
Mike Keefe
THE DENVER POST

Part III: "Star Wars"

The military potential of space attracted the attention of strategists in both the United States and Soviet Union almost as soon as the space race began. Spying from space was a commonplace feature of international intelligence operations by the early 1970s. The spy satellites have grown in sophistication until today's high-resolution cameras are able to record minute details of even the most secret military operations. The most advanced cameras can pick out a license plate from an altitude of 200 miles. The development of space weapons became a major priority in the United States with the emergence of the hunter-killer satellite. While the earlier conceptions of space weapons involved a simple rendezvous and destroy technique, more recent developments employ the laser as well as electromagnetic and atomic technology in their design. The laser was never seriously considered as a weapon in conventional warfare even though it has a wide range of applications in everyday life. But in space with no atmosphere to filter its beam, a laser could be quite destructive. The use of laser in attacks by anti-satellite spacecraft (ASAT) is only a small step from the spectacular space battles of science fiction.

Indeed, it was the name of the 1976 film, "Star Wars," which the Reagan Administration first adopted to push its space weapons program in 1983. The merits of the President's plan, later dubbed the Strategic Defense Initiative (SDI), has become a matter of intense worldwide debate. It is seen by its advocates as a means of ending the threat of nuclear devastation while opponents charge that the program is an exorbitant boondoggle whose stated objective is ruled out by the limitations of technology. Even more frightening, these critics contend, is that "Star Wars" defenses might so upset the fragile balance of forces between East and West that war might become more rather than less likely.

As the debate approaches fever pitch, "Star Wars" research has moved ahead quickly, consuming more than $3 billion in the last year alone, and giving unprecedented momentum to a broad range of advanced scientific programs. "Star Wars" research has led to the production and development of exotic new materials and technologies which promise to have particular importance for conventional warfare, fostering changes in land combat as radical as those wrought by the introduction of gunpowder in the Middle Ages. Spinoffs are, however, also finding their way into a myriad of civilian fields, including energy production, transportation, communication and medicine. As a result, science itself has gained new research tools from SDI projects. SDI's critics are quick to point out that the technological side benefits of "Star Wars" research could be had more cheaply and efficiently if they were pursued directly rather than as the unintended off-shoots of an extravagant military spending program.

The extent of Reagan's vision of space-based strategic defenses remains an open question. But that vision is being threatened. Congress, under the shadow of the increasing Federal budget deficits, has already voted significant cuts in SDI funds and scientists from hundreds of universities throughout the country have refused, on moral grounds, to become involved with "Star Wars" related research. Even the program's staunchest supporters admit that enormous technological obstacles still loom ahead. However, even if the a continental defense is never actually deployed, the long-term impact of SDI research programs promises to be enormous. In laboratories from Silicon Valley to Route 128, "Star Wars" is no longer a cute label or subject of debate as SDI is already yielding new technologies that seem destined to change the world.

Reagan Urges Antimissile Technology, Cites Soviet Buildup

President Ronald Reagan, in nationally televised speech March 23, 1983, made an appeal for public support of the Administration's proposed 10% increase in defense outlays. The bulk of the address was dedicated to a detailed examination of Soviet military strength. The President used charts, graphs and previously classified intelligence photos to illustrate Soviet efforts to build what he called "an offensive military force" around the world. "I wish I could show you more without compromising our most sensitive intelligence sources and methods," Reagan said. The photographs were of purported Soviet arms or installations in Cuba, Nicaragua and Grenada. Quoting a string of statistics comparing U.S. and Soviet production levels for missiles, bombers, submarines and tanks, the President stated that the Soviets were building "far more intercontinental ballistic missiles than they could possibly need simply to deter an attack."

Saying he wanted to share with Americans "a vision of hope," Reagan called for the development of an antiballistic missile (ABM) system capable of destroying Soviet missiles before they could reach their targets. The President described the development of an ABM system—which presumably would employ lasers, microwave devices, particle beams and projectile beams, and be based at least partly in space—as "a formidable task, one that may not be accomplished before the end of the century." Deployment of an ABM system would represent a fundamental revision in nuclear strategy. For the last 35 years, the U.S. has relied on the threat of massive nuclear retaliation against the Soviet Union to deter an attack and maintain a so-called "balance of terror" between the two powers. That policy had effectively prevented a nuclear war for three decades, Reagan conceded, but said that the Joint Chiefs of Staff and other advisers had convinced him of the "necessity" of developing an antimissile system that could prevent a Soviet attack from ever occurring. This new policy, the President said, "could pave the way for arms control measures to eliminate the weapons themselves."

THE MILWAUKEE JOURNAL

Milwaukee, WI, March 25, 1983

In his address to the nation Wednesday night, President Reagan revealed his support for a drastically new nuclear strategy aimed at "changing the course of human history." As broadly sketched, the bold initiative is attractive, yet it is also fraught with uncertainties and risks. The details need to be fully spelled out and exhaustively debated.

In essence, Reagan proposed to abandon the strategy of nuclear deterrence that has prevailed since the onset of the nuclear era. That strategy is based on the promise of retaliation: Each side knows that any nuclear attack would invariably provoke devastating reprisal.

Indeed, the topsy-turvy logic of stability in the nuclear age *requires* each side to leave itself exposed to the retaliatory power of the other. And, as Reagan noted Wednesday night, "this approach to stability through offensive threat has worked."

Now, Reagan has proposed a futuristic program — remindful of "Star Wars" and evolving over 20 years — to counter the Soviet missile threat with measures that are defensive rather than retaliatory. In other words, US and Soviet cities would be defended, not offered as hostages. The appeal of such a change, of course, is that US and Soviet military efforts would be devoted to the quest for more effective ways to defend lives, not destroy them.

Under the Anti-Ballistic Missile Treaty of 1972, both Washington and Moscow agreed to abandon almost all measures to shoot down attacking missiles. Research on the kind of anti-ballistic-missile defense Reagan seems to have in mind is not prohibited by the ABM treaty. But if the two sides were to seek to deploy an ambitious ABM system, some renegotiation of the treaty would be required. That would be a fateful step. The ABM treaty has been a key ingredient in a deterrence policy that has prevented nuclear war.

Furthermore, the ABM treaty was negotiated largely because both sides had concluded that ABMs won't work. That mutual recognition raises doubts about the Reagan initiative. The scientific problems associated with building an effective ABM system are immense, perhaps insurmountable. Scientists tend to believe that, when it comes to nuclear war, the offense can always overcome the defense.

The most obvious pitfall in the Reagan plan was recognized by the president himself. He acknowledged that if defensive systems are "paired with offensive systems, they can be viewed as fostering an aggressive policy, and no one wants that." What Reagan meant is that US attempts to defend itself could lead the ever-fearful Soviets to conclude that the US might be planning a first-strike and attempting to defend itself against the counterattack. The Soviets might be tempted to pull the trigger pre-emptively.

Hence, the task would be to build defensive systems while simultaneously dismantling offensive weapons, *and* to have the superpowers build and dismantle in concert, lest one side dangerously rattle the other and cause it to act rashly.

Unfortunately, Reagan has shown too little willingness to dismantle offensive weapons. In fact, on Wednesday night he vigorously championed an arms buildup that includes offensive weapons like the MX missile.

Nevertheless, the Reagan plan deserves serious discussion. However elusive, its goal is one that al' Americans can support: ridding the planet of offensive weapons that Winston Churchill once called the odious apparatus of modern war.

ST. LOUIS POST-DISPATCH

St. Louis, MO, March 25, 1983

In his speech Wednesday night on national security, President Reagan said that his concept of an absolute defense against nuclear missiles "holds the promise of changing the course of human history." Yet behind the noble rhetoric, his address was an intensely political one, intended to sell to the American people his extravagant military buildup. Moreover, Mr. Reagan offered a plan that can only be described as one of the most dangerous, ill-considered defense propositions that any president has ever put before the nation.

Make no mistake about it: Mr. Reagan is proposing to embark on a major escalation of the arms race in the only environment that is now relatively free of military activity. We can fight wars on land, in the air, on the sea and under the sea. Space remains the only arena in which the superpowers cannot now do combat. Mr. Reagan would change that by developing and deploying an antimissile system that, through exotic technologies, would ostensibly render the nation immune from atomic attack.

It would, of course, do no such thing. The Reagan plan would cost untold billions, involve a research and development effort that would rival the scope of the Manhattan Project (and provide less assurance of success) and, in the end, leave America no more secure than it is now. Indeed, if there were any prospect that Mr. Reagan's plan might actually work, the chances of nuclear war would be dramatically increased. The plan would certainly lead the Soviets to fear that, under the cover of such a system, the U.S. might launch a first-strike, a concern that would wreck the stability of the strategic balance.

The president's proposal is predicated upon the naive assumption that nuclear missiles represent the only threat at the disposal of the Soviet Union. Yet even if a foolproof antimissile system could be deployed in space, the nation would still be susceptible to attack from enemy bombers. A star wars defense system would do nothing to deter a cruise missile strike, for these weapons fly at tree-top level with extreme accuracy. Moreover, even a workable antimissile system would be vulnerable to decoys and other countermeasures. And, at bottom, no one could be certain that it would be effective against a massive ballistic missile attack.

In any case, as Mr. Reagan conceded, such a system would not be available for many years. In the meantime, he is saying that not a penny can be cut from his plan to increase defense spending by 10 percent. He is holding out the pie-in-the-sky prospect of nuclear immunity as an excuse to spend trillions in the next few years on the Pentagon. To that end, he reiterated the familiar litany of the Soviet buildup. His presentation included photos of Soviet military activity in the Western hemisphere — including one of an airstrip in Grenada that Canada (!) is helping to build.

Neither the people nor Congress should be fooled. The Reagan military program is excessive and must be cut back. His proposal to militarize space must be seen for the reckless new step in the arms race that it is.

The Oregonian

Portland, OR, March 25, 1983

When sophistication is added to nonsense, the result still is sophisticated nonsense. That is precisely what President Reagan offered the nation Wednesday in his proposed shift of strategic military doctrine away from massive retaliation toward a vague, ambiguous, undeveloped, Buck Rogers-style, anti-ballistic-missile defensive system.

The president's call for a national effort to develop means to defend the country against ballistic missiles was prompted by "great technological progress" and by "the bleakness of the future before us." Ironically, Reagan proposes to abandon a system — "stability through offensive threat" — that he conceded has worked for three decades.

The fatal flaw (a macabre but distinctly apt cliche) in the Reaganesque fantasy is the misperception that the defensive strategy would curtail the nuclear arms race. In fact, it would encourage it. The reason is that technology is not static; it evolves, occasionally takes quantum leaps; every measure produces countermeasures. Security, thus, is never assured, never foolproof, and never more than transitory. The perfect, impenetrable defense is fictional fluff, so offensive capacity must serve as a deterrent.

Yet, the mere prospect of one side gaining a decisive defensive edge, however temporarily, is so frightening, so militarily destabilizing and so ruinously costly in terms of weapons development that the United States and the Soviet Union 11 years ago limited their strategic competition to offensive nuclear weaponry. Neither adversary could afford then — and neither can afford now — to seek competitive advantage in both offensive and defensive technology.

The Reagan shift would ratchet uncertainty, and therefore fear, about an opponent's abilities and intentions to new highs. Chances for catastrophic miscalculation would soar.

How self-flagellating all this would be. Arms reduction agreements, avowedly the U.S. goal, will not occur if the United States simultaneously is working to neutralize the Soviet Union's remaining weapons. Thus, the Reagan shift, if continued, is a message to the Russians to stop bargaining in Geneva. They cannot afford to let the United States develop a system they cannot breach. Nor could the United States, if the tables were turned.

The fiscal enormity of the Reagan proposal must be noted. Jerome Wiesner, former president of the Massachusetts Institute of Technology, points out that much of the nation's economic difficulty stems from the fact that "about one-third of our best scientists and engineers work on military technology." The Reagan plan necessarily would increase this investment of treasure and brainpower enormously. A new arms race in defensive weaponry would so distort our financial and technological capacity that the nation could not and should not bear it.

Reagan's best service to this nation and to mankind would be to stick to the negotiating table and to shun the tail-chasing futility of a second front to the arms competition.

THE ATLANTA CONSTITUTION

Atlanta, GA, March 25, 1983

A book title of recent years ranked the ways of dissembling: "Lies, Damn Lies and Statistics." In his Wednesday night television show to rally support for a record peacetime defense budget, President Reagan went all the way — to statistics. The talk was among Reagan's most skillful — and, sadly, one of his most misleading.

The issues Reagan addressed are far too serious to be given such highly charged treatment. His talk was more a stump speech than the careful, sober weighing which the nation expects from a president on a matter as fundamental as defense.

Reagan played loose with the public, trying to frighten Americans with distorted force comparisons rather than involve us in the complex calculations which are necessary in order to fashion a sound defense policy.

In the process, the president attacked the historic bipartisan base for defense, even as he claimed to be working to re-establish it. And for mere tactical advantage in his troubled dealings with Congress, he risked destabilizing the U.S.-Soviet military balance — already dangerously tenuous — by raising the remote possibility of a sci-fi defense against Soviet missiles.

Make no mistake: The Soviet military buildup of recent years has been strong and, if unanswered, would pose potential threats to U.S. interests. A steady, purposeful U.S. response is necessary.

The president was huckstering misimpressions, however — as when he put forth the idea that the deployment of 1,200 Soviet SS-20 missiles has gone unanswered by U.S. forces. U.S. Pershing missiles will be introduced into Western Europe later this year, and, in the meantime, the United States has counters in its submarine-based missiles — far superior to the Soviets' — and in the developing cruise missiles. The president mentioned none of the U.S. systems.

Reagan was unwilling to represent even the politics of the situation accurately. He condemned Wednesday's House vote — by "liberals," he said — as limiting defense increases to 2 or 3 percent. Actually, it would cut the increase from the 10 percent Reagan wants to 4 percent.

And although the House vote was indeed partisan, the president is only slightly less trouble in the Senate, where his own party is in the majority. The Senate's Republican-controlled budget committee voted last week to hold defense to a 5-percent increase.

That figure is likely to be boosted, but neither the House figure nor whatever one the Senate adopts will be final. Both are bargaining positions for further negotiations with the White House.

None of the moves under way in Congress to contain defense spending is meant to, or would, bare the breast of America to the Soviet dagger.

Instead, if in the imprecise way which is the only one open to Congress in such complex matters, both Republicans and Democrats are asking the president to proportion defense to budget realities — in effect, to conduct the tough-minded strategic review his administration has forgone in favor of buying the Pentagon wish-list and hoping it will add up to a coherent defense.

These issues are, as the president said, as basic and serious as any the nation faces. Reagan had an opportunity Wednesday night to illuminate them. It is a pity that he chose to muddy them up instead

Houston Chronicle

Houston, TX, March 25, 1983

President Reagan spoke to the American people last Wednesday night on the subject of national security. He sought support in getting his national defense program through Congress, declaring more funds are needed to match the Soviet buildup. He also proposed an expanded effort to develop a high-technology defense against ballistic missiles.

The Soviet reaction was swift and angry. Soviet leader Yuri Andropov called Reagan a "liar" who is on an "insane" path of casting aside the principle of deterrence. He accused Reagan of derailing the second strategic arms limitation treaty, which the Carter administration dropped after the Soviets invaded Afghanistan.

Why such a bellicose response? What's going on here?

For one thing, the Soviets are not very accustomed to straight talk from the West. In the past, the communists have screamed about "capitalistic warmongering" and "imperialist aggression" to their heart's content. It was the role of the United States to talk softly and bring new compromises to the conference tables. The Soviets could put 300 SS-20 missiles in place aimed at European capitals, but when NATO agreed to match those weapons, the Soviets encouraged a "peace campaign" to halt NATO's efforts. So one thing that is going on is a battle for world opinion.

For another thing, Reagan's proposal to seek a defense system based on high-technology challenges the Soviets where they are the weakest. The Russians trail badly in computer and microchip development, and such expertise is the main target of the Soviet import and espionage efforts. What they can't buy, they would like to steal. In addition, the Soviets have for years been working on laser and energy beam research, so their protestations about U.S. efforts endangering peace sound very hollow.

President Reagan has said repeatedly that he wants a disarmament agreement and that his defense budget increases are aimed at achieving that goal. His "zero option" for intermediate range missiles in Europe was rejected by the Soviets. The various U.S.-Soviet arms negotiations have gotten nowhere lately.

Thursday, President Reagan is to make a speech in which, according to advance reports, he will set forth his views on the intermediate range missile talks in Europe, and perhaps on other arms negotiations. The president is still working on this speech, but it is safe to assume that he will stick to his stated desire to negotiate arms reductions, but from a position of renewed strength. From the way Andropov reacted, he may be getting Reagan's message.

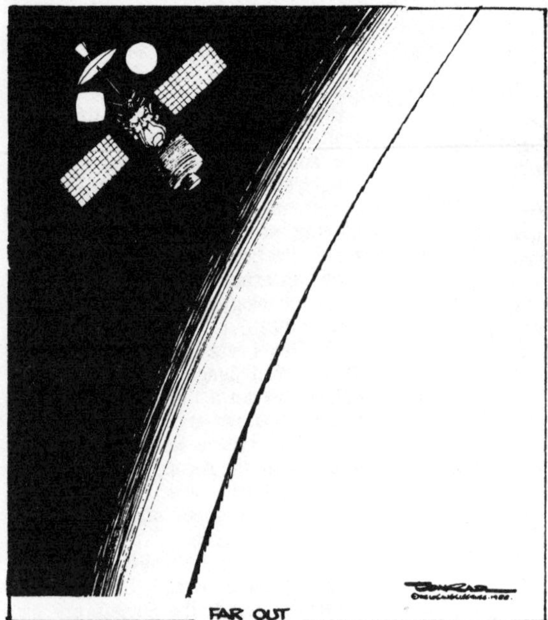

The Boston Herald

Boston, MA, March 26, 1983

Guess who said this about President Reagan's plan to develop "Star Wars" weapons to destroy hostile missiles before they can penetrate our air space?

"Thus, what is being talked about is a new attempt by the U.S. to achieve superiority in strategic arms over the Soviet Union and to upset the existing rough balance of power."

Tass, the Soviet news agency, said it. Strangely enough, though, it said nothing about the fact that Russia began research and development of this kind of weaponry long before the president announced publicly his intention to marshal our scientific and engineering genius to catch up.

In short, if the Soviets hadn't started on it first, there'd be no necessity for our having to make up ground on them now.

Somehow we don't expect Tass will mention that — but then, they're still telling the world the Afghans love having the Red Army "visit" them.

THE BLADE

Toledo, OH, March 31, 1983

PRESIDENT Reagan's call for a strong American effort to develop highly sophisticated anti-missile weapons is not as wild-eyed an idea as some of his critics are making it out to be. In typical Reagan fashion the President has advanced a thought-provoking proposal which merits serious consideration.

It is, to be sure, a radical departure from the MAD concept (mutual assured destruction) under which both superpowers face the certainty of a nuclear response to any first-strike action one might consider. Under that deterrence idea, defensive weapons against nuclear missiles have had no meaningful role.

That does not mean that plans should not be considered now for possible development some time in the future. Under the President's proposal, defensive systems might eventually become a significant part of the national defense picture.

It is true that deployment of anti-ballistic missile systems is outlawed by the SALT I treaty of 1972 with the Soviet Union, except for one permissible installation on each side. But that does not prevent research and development of such systems which, in fact, has been going on in the scientific communities of both countries for quite some time.

Scientists in the United States are divided on the practicality of the Reagan plan, with some referring to it as a pipe dream of "death rays" designed to knock out Soviet missiles before they can reach their targets. Others, however, see it as well within the realm of possibility. The Pentagon already has contracts out for the design of space-based laser systems, for example, some of which are well advanced.

Anyone who thinks these kinds of weapons are beyond the bounds of believability might recall that many scientists — while accepting the theory — were uncertain that the atomic bomb would work, either, at the time it was being developed. And Americans listened with incredulity when President John Kennedy pledged the United States in 1961 to put a man on the moon and return him safely within a decade. But the technical and scientific genius of this country was brought to bear on the problems involved with unprecedented intensity, culminating in the moon walk of Ohioan Neil Armstrong and his companion, Edwin Aldrin, in 1969.

It is known that Soviet scientists are proceeding with their own advanced research on anti-missile devices. That alone suggests that the United States cannot afford to ignore that facet of defense. On the other hand, it does not mean that both powers cannot proceed with efforts to achieve reductions in offensive nuclear missiles, such as have been going on in Geneva.

If there is one thing that arms races of the past have demonstrated, it is that today's breakthrough is tomorrow's passe system. President Reagan is not talking about pie-in-the-sky "Star Wars" weaponry in encouraging this kind of advanced research. He is outlining attainable scientific achievements which might one day be essential to this nation's defense.

RAPID CITY JOURNAL

Rapid City, SD, March 27, 1983

Depending on who is voicing the opinion, President Reagan's call for a national effort to develop means to defend against a missile attack was either a visionary view of the future or dangerous talk that could lead to new dimensions in the arms race.

A proponent of the switch in emphasis from offense to defense is Dr. Edward Teller, the physicist who created the hydrogen bomb. He contends that if the U.S. invests enough in anti-ballistic defensive measures, the arms race could be directed from offensive to defensive weapons and the threat of nuclear holocaust would be lessened.

Teller points out that because the machinegun and trenches of World War I made mass offensive attacks impossible, defensive strategy prevailed in the post-war period. However, in 1940 the German army skirted the Maginot Line, considered to be the ultimate in defense. Thereafter, offensive weaponry began to dominate, culminating with nuclear weapons.

In the U.S. and Soviet arms buildup, both sides have concentrated on developing offensive nuclear weapons. The theory of Mutual Assured Destruction (MAD) evolved as a formula which might prevent each country from attacking the other.

Teller feels if the technology of defense can be developed to the point that it cannot be defeated except by a much greater effort on the part of the offense, then war will not be winnable for the side that started it. The only course left would be for both sides to turn to a defensive strategy.

However, plans to deploy a new generation of weapons in space is seen by some scientists as posing a greater threat than the security they seem to offer on the surface.

Basis of that view is that the development of offensive weapons is much cheaper than defensive weapons and that for every defensive innovation, there is a much cheaper way for offensive weapons to circumvent it. If that is true, there's the possibility political leaders might decide they could launch an attack and defeat attempted retaliation with defensive weapons. That would pull the nuclear "trip wire" tighter because both sides would have to seriously consider launching the first strike.

Whether there will be a switch in emphasis from offensive to defensive weapons will depend on technological developments. Recent advancements in radar, long wave infrared sensing, and computer technologies have encouraged researchers to suggest a number of very accurate anti-ballistic missiles could be developed.

Whatever course this country follows, the search for defensive technologies will take years or even decades. But until other ways are found to lessen the nuclear threat, pursuing that search would be a prudent course.

DESERET NEWS

Salt Lake City, UT, March 25, 1983

Can the United States build a foolproof defense system that would make Russia's missiles obsolete?

That is the difficult task assigned to U.S. scientists this week by President Reagan in his national TV talk. His call for an effective anti-missile defense that would offer "a new hope for our children in the 21st century" was a surprise in a talk devoted mainly to statistics on Russian-U.S. military strength that Americans already largely know.

The president himself is well aware of the technical and diplomatic difficulties ahead in pursuing an anti-missile defense. And he acknowledged as much. But he said that U.S. technology had reached a level of sophistication "where it is reasonable for us to begin this effort."

Ideally, it would be much better to develop an iron-clad defense system that would render Russia's missiles obsolete than to hold the Soviet population hostage to an atomic attack — the principle behind mutual deterrence. Certainly that helps to project an image of President Reagan as devoted to peace and ultimate demilitarization at a time when he is insisting on a major beefing up of U.S. military forces.

But developing an effective anti-missile defense is going to be both difficult and expensive. The most promising weapons to accomplish that task are not other missiles, but high-energy lasers stationed in space and particle-beam generators within the earth's atmosphere.

Laser weapons operate by transferring an intense light beam at a target to destroy or disable it, or to knock it off course. Laser beams, however, are diffused and refracted by the earth's atmosphere. Hence, laser weapons work better in space.

Particle-beam generators, on the other hand, utilize a beam of concentrated radiation sub-atomic particles, chiefly neutrons and gamma rays. These particle beams tend to degrade across space unless they interact physically with the surrounding atmosphere to "replenish" particles within the beam. Obviously they work better within the earth's atmosphere.

In theory, orbiting laser stations would attack rising missiles during their boost phase, when they are most vulnerable. And particle-beam generators would destroy any incoming missiles that escaped the laser attack.

To be effective, such defense systems need ultra-sophisticated "eyes" to tell them when an attack is launched and to ferret out incoming missiles. But these eyes are inherently the most vulnerable elements of the system.

Donald M. Snow, writing in the Christian Science Monitor on anti-missile technology, commented: "It is not far-fetched to envision space cluttered with laser-armed satellites, dummy satellites, satellite killers, and satellites designed to destroy the killers."

Such a system certainly won't come cheap, and in fact might be far more expensive than even the projected U.S. arms buildup. One of the first casualties of such a technology almost certainly would be the Antiballistic Missile Treaty (ABM) of 1972, which expressly forbids basing missile defense in space.

Nor can the Russians be expected to sit idly by while the U.S. develops such a system. Every advance in weaponry brings a dramatic effort to develop counter-mea-sures. That has been the rule since soldiers fought with swords and protective breastplates.

Anti-missile systems are not the answer, any more than bigger and better weapons are. The only way to effectively insure world survival is still at the disarmament table.

THE SUN

Baltimore, MD, March 25, 1983

The president's razzle-dazzle defense of his military buildup needs to be played back in slow motion. To strategize space-age nuclear missile defenses, beat up again on the Russians, unveil pictures of Communist installations in Central America, present a slick slide-show on the Pentagon budget and deflect public attention from an admininistration drubbing in the House — to do all these things in a half-hour of prime-time television is being Ronald Reagan.

Let's start with the Buck Rogers part of his Wednesday speech. Mr. Reagan announced a change in U.S. nuclear strategy from retaliation against an enemy strike to one of intercepting and destroying incoming missiles with laser beams or other futuristic devices. Although such an active defense system is a current favorite among conservative arms specialists, it would be wrong to attach an ideological label to the concept. It is, basically, a reversion to the ballistic missile defense proposals of the Johnson administration, proposals that were all but abandoned under the SALT I treaty because they were basically unworkable. Since then, technology has made such vast strides that Mr. Reagan could envisage an effective missile defense by the end of the century.

Although the president couched his strategy shift in words of peace, critics were quick to revive 1960s warnings that effective missile defenses might tempt a superpower to launch an offensive first strike.

Why did the president speak up now about a system two decades in the future? His more immediate objective may have been to increase the current billion-dollar-a-year program for Star War research. But we suspect he wanted to break out of a sterile military budget debate, where he has been losing ground, in order to identify with a program likely to excite the public's imagination.

The presidential slide-show purporting to prove that the Soviet Union is pulling ahead of the United States in a chilling array of weapons sectors was vintage Reagan — the kind of stuff that swung public support behind the president's impressive 1981 and 1982 boosts in defense spending. With Congress threatening to halve his 10 percent hike for fiscal 1984, Mr. Reagan put pressure on legislators by trotting out classified pictures of Soviet and Cuban military installations in Cuba itself and in Grenada and Nicaragua. This close-in look at the Soviet threat held an added bonus: It put more bite in administration requests for military assistance to El Salvador and other Latin friendlies.

The president's purpose in all this is to get the country behind a military buildup he considers crucial to the nation's security. His goals may be laudable but his methods are something else. If his new missile defense strategy is as epochal as he says, it should have been the subject of a separate speech. If the Soviet threat in the Caribbean is as pressing as he suggested, it hardly warranted being paired with a partisan attack on Hill Democrats. Mr. Reagan, in short, may be overdoing it. Too much razzle-dazzle can ruin the best of shows.

Andropov Calls for Space Weapons Ban, Reduction in Nuclear Warheads

Yuri Andropov, the Soviet Communist Party general secretary, April 27, 1983 called an international agreement banning weapons in outer space. "Recent developments," Andropov charged "have demonstrated that the use of space-based military technology is being assigned an even greater role in United States strategic plans, including those announced by the top United States leadership." The Soviet leader was apparently referring to U.S. President Ronald Reagan's recent call for development of new antimissile technology. The U.S. denied it was contributing to an arms race in space. "The Soviets possess the world's only operational satellite interceptor, which they continue to test," a U.S. State Department spokesman said. "Other than a commitment to develop an antisatellite capability to match the Soviets, we are not planning any space weapons system."

A groups of prominent Soviet scientists April 9 had criticized President Reagan's proposals on antimissile defense systems as an "illusion" that could only lead to a more intensified arms race. "Proceeding from the understanding of the basic nature of nuclear weapons, we declare in all responsibility that there is no effective defense means in nuclear war and their creation is not practicably possible," the 244 Soviet scientists wrote. "The attempt at creating so-called defense weapons against the strategic nuclear force of the other side will inevitably grow into the emergence of a new element strengthening the American first-strike potential," the scientists continued. "By his statement the President is creating a most dangerous illusion that may turn into an even more threatening spiral of the arms race."

Los Angeles Times

Los Angeles, CA, May 5, 1983

If Soviet leader Yuri V. Andropov meant what he appeared to be saying in a Kremlin speech this week, we may be seeing the beginning of a breakthrough in the Euromissile negotiations and possibly in the strategic arms reductions talks as well. President Reagan was right to welcome the Andropov speech as a positive development that deserves serious consideration.

Unfortunately, Andropov's remarks were ambiguous; they may turn out to mark nothing more than another attempt to block the planned deployment of U.S.-made missiles in Western Europe without abandoning the Soviet advantage in medium-range missiles that now exists.

Andropov appeared to offer a deal whereby the Soviet Union and the Western allies would count nuclear warheads in calculating an equal balance, instead of counting just the missiles themselves.

If taken at face value, this is extraordinarily good news. The Soviet offer comes at a time when there is a developing American consensus, now supported by the President, that calls for basing nuclear arms limitations on the number of warheads rather than the number of bombers or missiles. However, the Reagan Administration and allied governments are being cautious in their interpretation of the Andropov initiative—and appropriately so—until the ambiguities in the new Soviet negotiating position are satisfactorily explained.

The Soviets have close to 1,000 warheads deployed aboard 247 triple-warhead SS-20 missiles that are targeted on Western Europe, plus approximately an equal number of older single-warhead missiles.

The West has nothing comparable to the SS-20s in range, accuracy and destructive power, but plans to begin deploying an offsetting force of cruise and ballistic missiles late this year in the absence of an agreement that would make the new missiles unnecessary.

In December Andropov offered to reduce the Soviet Euromissile force to 162, the number of single-warhead missiles deployed independent of NATO by the British and French. But, since the SS-20s carry three warheads each, that still would have left the Russians with an enormous advantage.

Depending on the interpretation, Andropov's latest offer could mean that he is now willing to remove all but 97 Soviet SS-20s in order to balance off the number of warheads on each side. In rough terms that should be satisfactory to the West.

The Soviet leader's speech, however, did not deal adequately with a number of troubling questions.

To begin with, the French and British strongly object to having their missiles, which are not controlled by NATO, being counted as part of a U.S.-Soviet agreement. The Soviets cannot be expected to accept the distinction. If they are seriously interested in an agreement, however, they must be willing to fuzz up the issue. Instead, they seem determined to exploit it in order to cause trouble within the Western alliance.

More important, Andropov gave no hint of willingness to destroy the surplus SS-20s. Merely moving them beyond the Urals—where they could be brought back on short notice or used to threaten Japan, South Korea and China—would be unacceptable.

Andropov was not clear on whether he would reduce the number of Soviet warheads by removing the necessary number of triple-warhead SS-20s or by removing two warheads from a larger number of SS-20s. If the latter were the case, serious problems of verification would arise.

There are indications that Andropov wants to count bombers as well as missiles in the balance. Although reasonable in theory, experience suggests that disagreement over what planes should be included would make the negotiations hopelessly complex.

Finally, one passage of Andropov's speech indicates that he might require an actual reduction of French and British nuclear forces as part of a deal. If so, that is hard to square with his supposed willingness to accept a genuine balance.

It is easy to speculate that the Soviet leader's ambiguity-filled offer is really designed not to produce agreement but to increase public pressures on Western governments to make one-sided concessions.

But it is possible that the Soviets have reluctantly concluded that the NATO countries are determined to redress the nuclear balance in Europe one way or the other—and that, this being so, Moscow's interest lies in striking a balance at lower rather than higher levels of missiles and warheads.

The best place to find out is at the negotiating table.

WORCESTER TELEGRAM.

Worcester, MA, May 7, 1983

The question is: Did Yuri Andropov make a new move on the nuclear chess board, or didn't he? Did he lift his hand clear, or does he still keep a tentative finger on the knight, or bishop or pawn?

Whatever else he meant by his dinner speech in Moscow the other day, he certainly captured the attention of the Western powers. If he really is offering to put a mutual cap on nuclear warheads as well as launchers, he is opening up a channel that may lead somewhere.

Warheads, not rockets, are what must be counted these days. When the Soviets decided to install hundreds of new SS-20 rockets in Eastern Europe, they threatened the balance of nuclear power. Each SS-20 carries three warheads. The NATO countries responded by announcing plans to install new Pershing missiles and cruise missiles in Western Europe. The Soviets have promoted an intense campaign in Western Europe to block any such move. They see the Pershing as a particular threat, because it has enough range to reach the western sector of the Soviet Union, possibly even Moscow, from West Germany.

Some dismiss the Andropov speech as nothing but a propaganda ploy. Maybe so. But most arms control proposals start out as propaganda ploys which become the basis for future negotiations. That certainly was true of President Reagan's "zero option" initiative.

The official responses in Washington and the European capitals have been cautious, as they should be. As it stands, Andropov's offer is too vague and has too many loose ends. But there is no doubt that the Soviets want to keep the Pershing missiles out of Western Europe, and they may be willing to make a deal to do so. An agreed limit on nuclear warheads would be a good way to start.

The Miami Herald

Miami, FL, May 7, 1983

SOVIET leader Yuri Andropov offers to hold all Soviet intermediate-range nuclear weapons in Europe — missiles, warheads, bombs deliverable by airplane — to numbers possessed by NATO nations. In so doing, he is asserting the principle of nuclear equality. On its face, his proposal seems entirely reasonable.

To be sure, it presents difficult practical problems. First, certain ambiguities demand clarifying detail. The Soviet weapons most involved are mobile SS20s. It is not clear from Mr. Andropov's statement whether he would achieve balance by destroying extra SS20s or perhaps by simply moving them outside the European zone. Merely moving them will not suffice. Because of the SS20's range and easy mobility, it would remain a threat to Europe even if pulled back.

Next, there is some confusion in Washington as to just how Mr. Andropov proposes to count. Would each side's equal ceiling apply separately to each category, such as missiles, warheads, aircraft? Would instead there be allowance for, say, more aircraft on one side balanced by more missiles on the other? Would equal numbers of warheads be counted per weapons category, or overall? These are the kinds of details that negotiations, which resume May 17 in Geneva, can and must resolve.

Most complicating is the political conundrum posed to the Western allies by Mr. Andropov's formula. He insists that Soviet nuclear deployments in Europe must balance all European nuclear deployments aimed at the Soviets. Viewed from Moscow's perspective, that is so fair as to be inarguable. A warhead is equally as threatening, after all, whether it is made in the U.S.A., Britain, or France.

Britain and France insist, however, that their nuclear forces must remain outside the Geneva negotiations, and the United States agrees. Those talks are between the Soviets and NATO, as led by the United States. The British and French nuclear forces are not under NATO command. Nor, in the view of London and Paris, are they "intermediate" weapons. Rather, they are national deterrents lined up against not only the Soviet medium-range missiles, but the Kremlin's entire arsenal.

The political complexities posed here are obvious, yet they exist entirely within the West's camp. The Soviets are in the position of facing three loaded guns: they obviously must take each into account. It will not suffice for the West's response to be, "Balance the U.S. forces and ignore the rest." That's simply not credible.

A prime Soviet goal is to prevent deployment of any U.S. intermediate-range nuclear missiles in Europe, which is to begin in December unless an accord is reached first. A prime NATO goal is to include some U.S. missiles within Europe's forces, whatever nuclear ceiling is reached, to avoid "decoupling" Europe from the United States. To achieve NATO's goal and an agreement limiting warheads within the Soviets' fair principle of equal forces will not be easy.

The Soviets are making important concessions under the threat of the U.S. deployment. That in turn will build pressure in the West for concessions in response. This just might result in a new treaty limiting nuclear forces in Europe. Pray that it does

The Times-Picayune
The States-Item

New Orleans, LA, May 5, 1983

Despite the warmonger tag many are attempting to pin on the Reagan administration's military policies, one of the goals of those policies is to draw the Soviet Union to the arms-control negotiations table with serious, realistic proposals. The negotiations game — now being played primarily at Geneva with intermediate-range missiles in Europe as counters — has proceeded through several proposals and counterproposals.

The Reagan proposals have been pertinent; the Soviet proposals, dilatory and propagandistic. Now General Secretary Yuri Andropov has made a new proposal that is not necessarily reasonable — indeed, it seems to be based on an idea that has already been rejected — but is worth exploration when negotiations reopen on May 17.

Mr. Andropov offers to count warheads rather than missiles and says "We stand for the U.S.S.R. to have no more missiles and warheads mounted on them than on the side of NATO in each mutually agreed-upon period." The initial State Department response was properly cautious, not dismissing out of hand a statement it could noncommittally call ambiguous.

Counting warheads, the department spokesman said, was "a sign of progress" since the Soviet SS-20 missiles that created the new problem of missile balance in Europe carry three warheads each. But counting in the missile forces of NATO nations — France and Britain are the two in Europe with such forces — has been rejected by both nations and the United States. The argument is that those missiles are part of national forces that are not under NATO command and are a different class of missiles that should be discussed instead in the strategic arms control talks.

The Soviets have already offered to reduce the number of missiles targeted on Western Europe to the number of French and British missiles targeted on Russia. The timing of that proposal — during the height of the European anti-nuclear movement preceding the March West German elections — stamped it as more propaganda than serious offer.

The new offer, though perhaps little more than the first in slightly different words, comes after the conservative Christian Democrats won in Germany, NATO chiefs of state confirmed support of NATO missile policy and as time continues to tick away toward the deployment of Pershing 2 and cruise missiles in Western Europe later this year. We can at least assume it is serious enough to justify pursuing to see if the negotiations can finally get off dead center.

The Wichita Eagle-Beacon

Wichita, KS, May 5, 1983

The latest offer by Soviet leader Yuri Andropov on the deployment of intermediate-range nuclear weapons in Europe is a promising development for several reasons. His agreement to count warheads instead of just missiles signals a basic change in the Soviet stance on the matter, making progress in the crucial resumption of arms negotiations in Geneva much more likely. Finally, it appears both sides may be able to talk in terms of comparing apples and apples, instead of apples and oranges.

In the past, the Soviets have insisted on basing strength calculations on launch vehicles only. This would work to their advantage, since their SS-20 missiles, unlike NATO weapons, carry multiple warheads capable of striking widely distant targets. Counting warheads is a more accurate and meaningful way of assessing either side's striking power.

Another encouraging aspect of Mr. Andropov's offer is that he is talking about significant reductions in that firepower, rather than trying to establish a status quo arrangement. While the process of proposals and counter-proposals is vital, it's equally important to note the intermediate weapons talks now seem aimed in the right direction: namely, toward fewer weapons on line.

Numerous points remain to be worked out, — for example, the disposition of Soviet missiles removed; whether to include British and French weapons in the NATO count; and the possible modernization of the NATO armory with new U.S.-supplied Pershing and cruise missiles. No one should expect a clean-cut breakthrough that answers all the questions at one time. Even given that, this latest proposal must be regarded as a positive development, on which additional accord may be built.

THE DENVER POST

Denver, CO, May 12, 1983

SOVIET LEADER Yuri Andropov seems to have made a genuine concession toward arms control in the West. But at the same time, he's racing ahead with a threatening buildup of Soviet nuclear weapons in Asia.

Even ignoring the Asian offset, Andropov's May 3 offer to cut the number of nuclear warheads in Europe has many flaws. But it does show movement toward a compromise. That's a sign the policy of the Reagan administration and its stauncher European allies may be beginning to bear fruit.

While the U.S. has continued to seek a "zero option" that would remove all medium-range theater nuclear weapons from Europe, it recently agreed at the urging of West German Chancellor Helmut Kohl and other allies to accept an interim agreement with equal numbers of NATO and Soviet missiles.

Initially, the Soviets seemed to be hoping that the anti-nuclear movement in NATO nations would allow them to stop deployment of U.S.-built cruise and Pershing II missiles without any offsetting Soviet concession. They've often pretended to support the "nuclear freeze" movement in the West.

The catch was that the Kremlin began deploying its modern SS-20s in Europe in 1976, while NATO has yet to deploy its first Pershing II or cruise. Thus, the Soviet Union's idea of a "freeze" was to leave its estimated 243 modern SS-20 missiles in Europe in place while NATO would also be "frozen" at its present level: zero.

But candidates leaning toward Andropov's position were defeated in West German elections. That prompted Andropov to offer to reduce the SS-20s in Europe to the same number of missiles as those that are deployed by France and Great Britain if the U.S. would spike the Pershing-cruise plans.

That offer wasn't acceptable for many reasons. The British and French say their relatively small arsenals are national deterrent forces only — not under NATO command. They are mostly sea-based and not accurate enough to be counterforce weapons able to offset the SS-20's first-strike capability. Each SS-20 carries three warheads to the single warhead on its Franco-British counterpart. Finally, it wasn't clear whether the Soviets would scrap the SS-20s removed from Europe or merely redeploy them in Asia.

But on May 3, Andropov refined his offer to match the British and French on a warhead-for-warhead basis. And he seemed to concede for the first time that the U.S. could at least station a limited number of missiles in Europe. That principle is more important to our allies than the specific number of missiles stationed there. They see the presence of a U.S. "tripwire" force as lending overall credibility to the U.S. "nuclear umbrella" over NATO. If that is accomplished, it would seem possible to reach some acceptable agreement on the Anglo-French missiles.

Still, the biggest flaw in Andropov's latest move is the continuing threat that any SS-20s removed from Europe will merely be trundled across Siberia to glower at Japan and China. U.S. intelligence experts say the Soviets now have 108 of the triple-warhead missiles facing Asia and are building enough new sites to double that total.

The U.S. position has been that any SS-20s removed from Europe should be scrapped. We should stay firm on that point. Otherwise, whatever progress we make in removing the nuclear cloud over Europe will be offset by darkening Asia's own hope of survival.

FORT WORTH STAR-TELEGRAM

Fort Worth, TX, May 6, 1983

The word is *patience.*

Soviet boss Yuri Andropov bolstered the world's hopes for some progress in European arms reductions when he offered this week to reduce his nation's westward-aimed nuclear missile forces not just to the number of missiles possessed by the British and French but to the number of warheads in those nations' relatively small nuclear forces — in exchange for a halt to NATO deployment of U.S. missiles in Europe later this year.

Counting warheads rather than delivery rockets is a step in the right direction, and something the Soviets had refused to do earlier. This change means, in effect, that the Soviets are admitting that their multi-headed missiles give them a huge present edge in European missile power, something else they had denied earlier. U.S. figures count almost 700 Soviet missiles in the European theater as opposed to none for the United States.

It also indicates how badly the Soviets wish to prevent deployment of those NATO Pershing and cruise missiles, which would somewhat balance the scales.

Thus patience, combined with an attitude of accommodation to further Soviet yielding on this issue, would seem to be the best plan for the West.

The flaws in Andropov's still-hazy offer involve verification and the details of how the Soviets would reduce their missile force.

The West should remain unwilling to accept a mere pulling back of SS-20s, since those missiles could quickly be returned to Western Russia and retargeted on Europe. Nor can it look favorably on a plan of removing two of the three warheads on the 243 SS-20s targeted on Europe, since constant on-site verification would be required to be sure that the excess warheads had not been refitted by the Soviets.

Aside from that, Andropov still ignores the fact that the British and French strategic missiles pose no threat to the Soviet Union similar to the one presented to Western Europe by Soviet missiles. Obviously, neither the British nor the French would conceivably contemplate a first strike against the Soviet Union.

The United States and its NATO allies also point out that France is not even a NATO member and that the British missiles (64 of them, submarine-based) are not under direct NATO control, while the Pershing and cruise deployment would be under NATO control.

Linking Soviet missile cuts to the British and French forces, on a strictly numerical basis, may make sense to Andropov and Co., but not to Western military thinkers.

Still, Andropov's latest offer is a sign of progress, and the West's response should be tailored to encourage further Soviet concessions, while not backing away from the NATO deployment plan until the best possible reduction plan can be reached.

The ultimate objective remains not counting missiles, but eliminating missiles.

The Oregonian

Portland, OR, May 15, 1983

Yuri Andropov has offered to negotiate the reduction of the Russian midrange nuclear missiles pointed at Europe, an encouraging sign that progress can be made at the upcoming Geneva talks.

There is a catch, of course. The Soviet leader wants to include the British and French nuclear weapons in the U.S. Pershing II count. This won't do, because the European nuclear weapons are strategic, not midrange like the Pershings or the Soviet SS-20s zeroed at Europe, and are to be compared to large numbers of Soviet long-range nuclear weapons not now on the table.

That is the long and the short of the Andropov proposal, and it is unacceptable to Britain, France and the United States. Both the long- and the short-range missiles ultimately have to be on the table, as it matters little which one destroys your hometown.

THE CHRISTIAN SCIENCE MONITOR

Boston, MA, May 5, 1983

Serious arms negotiation cannot take place publicly. So it may be assumed that Soviet leader Yuri Andropov's offer on nuclear Euromissiles is designed more as a political maneuver than as a negotiating position. Mr. Andropov made his offer at a dinner in honor of East German leader Erich Honecker, and there is little doubt it is designed above all to try to sway West German opinion. His hope is that enough public pressure can be brought to bear on the Kohl government to influence US policy and prevent deployment of the new NATO medium-range missiles, scheduled to begin in December of this year.

Political considerations notwithstanding, Mr. Andropov's new bid does represent an encouraging step forward in the US-Soviet effort to work out an agreement on Euromissiles. For the first time the Soviet Union has stated its willingness to negotiate equal ceilings on nuclear warheads in Europe; previously Moscow had called for equality only in missile launchers – which would give it the edge because the Soviet SS-20 missiles have three warheads each. Inasmuch as it is warheads that carry the destructive power, making them central in calculating the East-West balance makes sense. That is therefore a significant Soviet concession and the Reagan administration has duly welcomed it.

As with any public offer, however, the Soviet proposal leaves many questions unanswered. It says nothing, for example, about whether Moscow would include the missiles in Asia or whether it would scrap the mobile SS-20 European launchers or simply move them to the Asian theater. There are other ambiguities as well. Yet there is enough give in the Soviet position to warrant detailed discussion – and a counterproposal from the United States – when the superpowers resume their negotiations in Geneva on May 17.

It does not seem reasonable, for instance, for the US to continue refusing to negotiate limits on the non-NATO British and French missile or bomber forces or to refuse to allow the Soviet Union to keep some missiles targeted on the Far East. Mr. Andropov insists on both, and from the Soviet standpoint that is not surprising. The independent British and French forces may pose little threat to the USSR at present, but Britain and France are embarked on a massive modernization program that will add hundreds of warheads to their arsenals. The Russians can hardly ignore those weapons. By similar logic, the Soviet concern about a threat from the Far East also will have to be taken into account.

This is not to absolve the Soviet leadership of having created the current arms problem in Europe to begin with. If the Russians had not begun installing their controversial SS-20s – after they had already stated that a nuclear balance existed in Europe – they would not now be faced with the prospective deployment of new NATO missiles capable of reaching Soviet targets in a matter of minutes. Thus the onus is also on them to negotiate, and it can be said that, without NATO resolve to proceed with the deployment of 572 Pershing II and cruise missiles, there would be no incentive for Mr. Andropov to talk.

But it is clear that, to reach an arms control agreement, both sides will have to compromise – the United States no less than the Soviet Union. Certainly this is the direction in which most Europeans and Americans want the superpowers to move. In the United States the pastoral letter just approved by the Roman Catholic bishops denouncing nuclear war as well as the strong support in the US House of Representatives for a bilateral nuclear freeze point to the depth of concern about the present arms buildup – and a desire to do something about it. In West Germany the Catholic bishops recently took a more cautious stand on nuclear arms, but they too came out in favor of arms control.

With his latest bid, Mr. Andropov not only courts the Europeans but opens the next chapter in the US-Soviet negotiations. The public on both sides of the Atlantic can hope it will be a chapter marked by bona fide bargaining and successful signing of an agreement.

THE SACRAMENTO BEE

Sacramento, CA, May 6, 1983

Soviet leader Yuri Andropov proposes counting warheads on missiles in Europe rather than launchers as a new approach to an arms-control agreement at Geneva. President Reagan correctly calls this a positive step and says it will be given "serious consideration." While the proposal at this stage only represents a talking point, and shifting the count from missiles to warheads would still entail all sorts of difficulties, it's no small matter that the two leaders have found common ground at last on anything relating to arms control.

The idea behind warhead counting, as opposed to counting launchers that carry up to 10 warheads each, is that it offers a path toward nuclear balance that moves away from such monster, multiple-warhead weapons as the American MX and the Russian SS-18, both of which represent deadly counterforce — or first-strike — threats and a hair-trigger balance of terror.

That Andropov sees the value in shifting to the warhead count for missiles in Europe may mean readiness to consider that procedure in the more important U.S.-Soviet negotiations on limiting, and eventually reducing, each side's arsenal of powerful intercontinental strategic missiles. Reagan would be smart to pursue the idea vigorously at Geneva, even though the Soviet leader has attached conditions — among them the inclusion of British and French nuclear weapons in the count — that Washington, which has no control over British and French forces, can hardly accept.

The president in fact might go Andropov one better, and at the same time bend toward the nuclear-freeze resolution voted by the House on Wednesday, by proposing to explore at Geneva the so-called "build-down" approach to arms control, in which each side would junk two older missiles for each new one deployed.

Both tracks — counting warheads while reducing the number of launchers — would result in shifting to fewer single-warhead missiles. It would be harder for either side to knock all the other's out in a single pre-emptive strike if they're deployed, as assumed, on mobile carriers. Moving in stages to strategic forces of that kind would reduce the chance that either side would be tempted to risk nuclear war.

Whether Andropov's proposal can lead to a breakthrough on European missiles will require further clarification of what he intends to do with the 250 or so Soviet SS-20 medium-range, triple-warhead missiles now aimed at Europe. Since they are mobile, if they are not destroyed, they could easily be redeployed to threaten Europe, while moving them to the eastern part of the Soviet Union would present a threat to Western allies in Asia.

Those problems, however, shouldn't discourage pursuit of the promising warhead-count approach as a general principle for arms-control negotiations. If that idea can be applied successfully in an agreement on Euromissiles, it could pave the way for a still more important breakthrough on strategic arms. There are few enough points at which minds might meet on arms issues in Washington and Moscow. This one should be nurtured with the greatest care.

Reagan Dismisses ASAT Treaty; Scientists Assail Antimissile Plan

United States President Ronald Reagan April 2, 1984 informed Congress that he saw little use in attempting to negotiate with the Soviet Union a comprehensive ban on antisatellite (ASAT) weapons. Reagan's position was detailed in a report mandated by Congress in the defense appropriations bill for fiscal 1984. Under that legislation, the President had to submit a plan for a negotiated agreement on antisatellite weapons before Congress would release $19.4 million in funds for testing air-launched antisatellite missiles. The President told Congress that the small size of ASATs would make verification difficult. He also contended that the Soviets would be unwilling to negotiate a limit until the U.S. developed its own ASATs. Reagan argued that for reasons of "U.S. and allied security," the U.S. had to continue its efforts to "protect against threatening measures." The report detailed Soviet progress in ASAT technology, which it said far outstripped that of the U.S. The U.S.S.R. was said to have deployed a system consisting of landbased rocket projectiles. In addition, the report stated that the Soviets had "ground-based test lasers with probable antisatellite capabilities" and nuclear-armed antiballistic missiles that could be used to attack U.S. satellites.

A group of prominent scientists March 21 had warned that President Reagan's dream of an ultra-sophisticated antimissile defense system would be prohibitively expensive and technologically "unattainable." The study was issued by the Union of Concerned Scientists. It had been prepared by a nine-member panel that included Hans Bethe, a Nobel laureate in physics and one of the builders of the first atomic bomb. The panel, after a detailed analysis of Administration proposals, came to the conclusion that an advanced ballistic missile defense system would cost hundreds of billions of dollars to implement, could be easily frustrated by the Soviets and would probably escalate the arms race.

The U.S. Air Force had conducted its first test of an antiballistic missile (ASAT) near Vandenberg Air Force Base in California on January 21. The missile, launched from an F-15 jet fighter, flew towards a point in space rather than a specified target and then dropped into the Pacific Ocean. The weapons, 17 feet long, consisted of a two-stage booster and a miniature "homing vehicle" designed to destroy an enemy satellite by smashing into it at high-speed. Pentagon officials, saying that the test was classified, would not disclose details of the weapon's performance.

THE ATLANTA CONSTITUTION

Atlanta, GA, March 26, 1984

Exactly a year ago this past Friday President Reagan made his astonishing "Star Wars" speech, urging the military establishment to develop an impenetrable death-ray defense against Soviet nuclear missiles. To mark that inauspicious occasion, so to speak, a panel of eminent scientists has just released a report stating flatly that such a system simply can't work.

The study, undertaken by the Union of Concerned Scientists, tries to be generous. Even assuming that it would be possible to overcome the awesome technical and engineering feats inherent in creating laser or particle-beam weapons that would destroy just-launched Soviet ICBMs, the scientists say that the Soviets could employ relatively cheap and simple countermeasures that would permit a high proportion of their missiles to evade our defenses.

Even if only 5 percent of their warheads filtered through our anti-ballistic missile shield, the scientists calculated that the toll would be catastrophic — about 60 million American deaths.

The sense of the UCS report is that the project is doomed to failure even before the Pentagon begins its announced research and development program, estimated to cost about $26 billion before the end of the decade.

Let's say for the sake of argument that the UCS report may be way off base, that its signatories — whose numbers include numerous Nobel Prize-winners and former presidential science advisers — are biased or shortsightedly unwilling to make the enormously expensive gamble that our best minds could somehow nuke-proof America.

Suppose such a system were feasible; what in fact would we have?

First of all, it would be in direct violation of the Anti-Ballistic Missile Treaty, a not-insurmountable problem for Reagan's super-hawk supporters, who would be willing to abrogate the pact in order, they would say, to protect the nation from attack. But then that would surely rule out *any* U.S.-Soviet weapons-control negotiations for the foreseeable future and ensure a no-holds-barred superpower arms race.

Further, the mere possibility that America might be within reach of a foolproof defensive system would be destabilizing in the extreme during the period before it became operational, raising the risk of a pre-emptive Soviet strike before our guard was up.

In short, attainable or not, President Reagan's ABM fantasy is a recklessly expensive no-win gamble, one that he should shelve as quickly and quietly as possible.

WORCESTER TELEGRAM

Worcester, MA, April 3, 1984

Should the United States try to protect itself from hostile intercontinental ballistic missiles by developing a way to blow them up before they reach their targets?

That question is becoming a big issue. It may surface in the presidential campaign. It certainly will be passionately debated, pro and con, in the years ahead.

Those opposed scornfully dub it a "Star Wars" idea that can't work and shouldn't be tried if it could. Those in favor say it will free humanity from the ominous menace of nuclear immolation.

As expected, the "experts" are arrayed on both sides. There are those who say it is technologally feasible to protect our missile sites from a first strike by ground-based devices. Others say we can protect our whole country by mounting defensive weapons in space. But still others doubt that the idea can be made to work either from the earth or from space.

The second part of the proposition is whether such a defense should be mounted if it is found feasible. Although that seems a separate issue, those experts who say the thing can't be done tend to be the same people who say it shouldn't be done, even if feasible. The argument here is that defense against nuclear weapons is "destabilizing." This school of thought argues that any major attempt by the United States to protect its nuclear missiles would be seen by the Soviet Union as a move toward a first-strike capability by the United States. Those who hold that view fear that a space-based defense might tempt the Soviets to launch a first strike before the United States had the thing in place.

In the coming debate, it may be useful to keep these distinctions in mind. The Reagan administration wants to commit $25 billion toward research on defenses, including laser beams. Would that be a huge waste? Possibly.

But if the research did uncover some promising possibilities, what then? Should the United States cease and desist for fear the Soviets would feel threatened?

For our part, we are more impressed with the first argument than with the second. The idea of mounting a defense — either on earth or in space — against missiles traveling 12,000 miles an hour, stretches credulity. It just doesn't seem likely.

But if there is some breakthrough, via lasers or some other development hardly guessed at, should this country move ahead? Our reaction to that is that we should keep open as many options as possible. Perhaps treaties and arms control agreements will make this sort of defense unnecessary. But until such treaties are signed and sealed, we see more reasons than not to try to make our land inviolable.

The Star-Ledger

Newark, NJ, April 6, 1984

President Reagan is on solid ground in proposing a worldwide ban on the production, possession and use of chemical weapons. Of all the unpleasant contents in the Pandora's box of weaponry, chemicals are considered by most people to be the most vicious and reprehensible. Much of the world must join the President in wanting to see these foul tools of warfare done away with forever.

It is also true, as the President charged at his most recent news conference, that the Soviet Union has a "massive arsenal" of these noxious weapons. The Soviets are particularly vulnerable to the accusation that they have been using them in their long and costly war in Afghanistan.

Mr. Reagan did not disclose details of his Administration's proposal other than to say that it would entail an effective procedure for monitoring and verification. This is, quite obviously, an essential feature of any arms control proposal.

While the President deserves high marks for his suggestion, there is little reason for optimism about a favorable Soviet response. While the Soviets have previously spoken favorably in principle of some kind of an agreement in this area, dealings with them have taught us that there is apt to be a wide gulf between their words and their deeds.

Unlike the Soviets, the United States does not have an arsenal of chemical weapons. In each of his three years in office, President Reagan has asked Congress to appropriate money for the production of these weapons, but Congress has refused. It seems unlikely, given the emphasis that the Soviets have been placing on weapons production, that they would be willing to give up the wide lead they now hold in this area.

That could present us with a difficult choice. The President stated that he still wants production funds for such weapons, because "you've got to have something to bargain with." The United States has not manufactured any chemical weapons since 1969.

To return to the manufacture of chemical weapons would be a highly regrettable reality, but one the nation may have to face. If the Soviets continue not only to manufacture and store such weapons but to actually use them, it is difficult to see how we can refrain.

Mr. Reagan asserted that the Soviets and their allies had used chemicals against "defenseless peoples" not only in Afghanistan but in Laos and Cambodia as well. He also referred to the documented assertion that Iraq has used chemicals in its long war against Iran.

The President's assertion that the increased stockpiles abroad "have serious implications for our own security" cannot be taken lightly. Congress may well have to reconsider its position and, despite serious moral reservations, permit at least a measure of production of our own stockpile of chemical weapons.

THE SACRAMENTO BEE

Sacramento, CA, April 5, 1984

President Reagan's refusal to enter negotiations with Moscow on banning space weapons and his decision to push ahead instead with development of a sophisticated anti-satellite weapon are both ill-advised and dangerous. Despite his rhetoric about pursuing arms control, they indicate that he still believes not in negotiations to reduce nuclear dangers but in striving to build a war machine superior to that of the Russians.

Not only is the U.S. anti-satellite weapon now being tested superior to the crude one the Soviets have developed, and therefore not merely a matching of strength. It is also an integral first step in Reagan's far more ambitious "Star Wars" scheme for deploying laser guns in space to knock out Russian missiles. Thus, on two levels Reagan would abandon the longstanding doctrine of deterrence — which relies on the realization that there is no effective defense against mutual nuclear destruction — for a two-pronged offensive and defensive capability that can only force the Soviets to undertake the same course.

In themselves, anti-satellite weapons are destabilizing because they threaten the eyes and ears in space on which both sides rely for early warning against attack and for key military command and control communications.

The Soviets recognize this. The new Russian leader, Konstantin Chernenko, has in fact urged a moratorium on such weapons development and invited negotiations for a total ban. Reagan's refusal on the ground that absolute verification of any accord would be almost impossible is specious. As with any arms agreement, small loopholes may be unavoidable, but none are likely to be large enough to permit either side undetected to do the extensive testing required for such an advantage.

Since a successful nuclear first-strike would become more possible with the ability to eliminate an enemy's satellites, anti-satellite weapons pose a destabilizing offensive threat.

Reagan's Star Wars scheme, though theoretically a defensive policy, raises equally difficult problems. He would deploy in space a system of Buck Rogers-type laser and X-ray weapons to knock out Soviet missiles before they reach the United States. But many reputable scientists and defense strategists consider such a space canopy technologically impossible for the foreseeable future even at the staggering, administration- estimated cost of more than $200 billion by the year 2000.

They point out that such a system would impel the Soviets to match it and to develop a new generation of offensive missiles to defeat it. If both sides proceed on that course, little incentive would remain for trying to halt the nuclear arms race through negotiations.

Reagan's proposals for the anti-satellite program and a Star Wars defense now face new examination and debate in Congress. The lawmakers can and should take the first step to halt this madness. Some months ago, they refused further to fund the anti-satellite program unless the president makes a genuine effort to negotiate with the Soviets a total anti-satellite weapons ban. They should stick to that position.

THE LOUISVILLE TIMES

Louisville, KY, April 9, 1984

With the recent appointment of an Air Force general to plan a space-age defensive shield against nuclear attack — the "Star Wars" project — President Reagan has begun a potentially reckless initiative.

The project, unveiled a year ago, contemplates the use of space-based lasers and particle beams to shoot down incoming missiles. It has been ballyhooed as a way to defend the United States against attack and as a deterrent to a Soviet first strike. The plan has special appeal to an administration that shows little enthusiasm for diplomacy.

This "high frontier" defense is plausible, however, only if it is totally effective in preventing enemy missiles from reaching this country. Aside from a handful of true believers, almost no one seriously claims that this is possible. Even some Pentagon officials have doubts, and with good reason. Armies throughout history have discovered that the most impressive walls and fortifications can be breached by a resourceful enemy.

The Russians, through innovations of their own, would surely find ways to punch holes in this one. If just 5 per cent of the Soviet warheads got through, 60 million Americans would perish, according to an estimate by the Union of Concerned Scientists.

The myth of a technological solution to the problems of war and peace is one of many defects in the Star Wars plan.

By attempting to construct futuristic missile-killing devices, Mr. Reagan would also undermine the SALT I treaty, which limits missile defense systems on both sides. Other arms control efforts, most of them dormant, would be in even greater jeopardy.

The theory behind the SALT I limitations is that anti-missile weapons are "destabilizing," since superior defenses could allow one nation to attack the other without fear of retaliation. The nuclear deterrent works, in other words, only if both sides are vulnerable. While that doctrine need not stand unchallenged forever, Mr. Reagan must be cautious, as his own Scowcroft commission advised, about upsetting the few arms control mechanisms that restrain the superpowers.

Another danger Mr. Reagan glosses over is that the Russians will have an incentive to strike first if they think the United States can erect a successful laser defense. They would have reason to fear, after all, that once the shield were in place, they could be at the mercy of our offensive missiles.

The project, which will cost $26 billion at first and billions more later, will be a drain on the Treasury. Better uses for these defense dollars are plentiful.

The key point, though, is that Mr. Reagan and his advisers have fallen into the trap of believing high technology is a better answer to human conflict than the difficult process of hard-nosed, step-by-step diplomacy.

Columnist Ellen Goodman, who is no weapons expert but wise in the ways of human nature, identified the problem when she wrote recently:

"There is something fundamentally perverse about pinning our hopes for the future on hardware. It prevents us from resolving conflicts, discourages us from thinking about the real reasons for the arms race."

Rather than squander billions on a high-risk project that would add new strains to superpower relations, Mr. Reagan should look harder for a way back to the negotiating table.

The Wichita
Eagle-Beacon

Wichita, KS, March 31, 1984

It's becoming increasingly clear that President Reagan's dream of a space-based missile defense, which he articulated in a speech a year ago, is not a realistic alternative to conventional defense systems. A group of distinguished scientists — including Hans Bethe, a Nobel prize winner in physics, and astronomer Carl Sagan — warned recently that Mr. Reagan's proposed high-tech defense not only is unrealistic, but could fuel the nuclear arms race. It is a warning the administration should heed.

Mr. Reagan's space defense sounds too good to be true: Orbiting satellites, equipped with laser or particle-beam weaponry, would track and destroy incoming Soviet missiles. This defense, the president argues, could make offensive nuclear weapons obsolete.

But the idea of a high-tech nuclear umbrella is full of holes. The scientists point to a fundamental flaw: Present missile technology gives a clear edge to offensive missiles; defensive measures are always at a disadvantage. Therefore, no matter what expensive defense the United States devised, it would be relatively easy for the Soviets to devise a countermeasure. This could lead to a new arms race in offensive weapons.

The president is asking Congress for a whopping $26 billion over the next five years — equivalent to the $30 billion Apollo man-on-the-moon program — just to study whether the idea is feasible. Most scientists agree it is not. Enormous technical problems must be overcome, and even if they were, such a defense still would be many years and upwards of a trillion dollars away.

President Reagan should abandon his space defense dream or research it at a more realistic level of funding. Meanwhile, the immediate dilemma of nuclear arms better would be addressed by reviving the stalled U.S.-Soviet arms talks.

The Tennessean

Nashville, KY, April 9, 1984

It is not surprising that the NATO defense ministers have expressed skepticism and nervousness about U.S. plans to prepare for war in space, for much of the same feelings exists in this country.

In Turkey last week Defense Secretary Caspar Weinberger explained the administration's plans to develop a comprehensive space-based missile defense and defended its decision not to seek a treaty with the Soviet Union banning anti-satellite weapons in space.

The defense ministers of the alliance were not reassured by Mr. Weinberger. Their general feeling was that such a defense would further destabilize the arms race, increase the threat of nuclear conflict, and remove from Europe the existing shield of defense.

It was a year ago that President Reagan outlined his plan, urging the effort for a defense system so effective as to render offensive missiles obsolete. He said at the time that if paired with offensive weapons, such a plan could be viewed as "fostering an aggressive policy and no one wants that."

Not many people really believe that a space-based system is now possible, or that it could do what is envisioned. A recent report by the Union of Concerned Scientists warned that a "thoroughly reliable and total defense" of American cities is impossible. The report noted that if only 5% of Soviet warheads were able to penetrate such a shield, at least 60 million Americans would die.

The blue ribbon panels named by the White House to study the idea both said that a perfect system cannot be assured and concluded that offensive weapons would still be necessary. The Pentagon's top research official said he couldn't envision a time when defense weapons would not be paired with offensive weapons.

So the issue is already being moved away from what Mr. Reagan originally said about pairing the systems. Nevertheless, he is going forward with his "star wars" scheme. He has appointed an Air Force general to head the project and is seeking about $2 billion for fiscal 1985 for research and development.

The administration points to what the Soviet Union has been doing in the field of anti-satellite work. It is true enough that Russia has been working on ground-launched anti-satellite weapons for some time. According to U.S. experts, their testing has not evolved into a very satisfactory system and it can only threaten objects in low orbit.

For once, Mr. Reagan is not asking for a new weapons system as a "bargaining chip"for negotiations with the Soviet Union. He is arguing against any such treaty and contending that compliance would be impossible to verify.

The Congress, however, looks more kindly on the idea of such a treaty. It last year banned any advanced testing of an anti-satellite system until the administration asserted it was ready to negotiate with the Russians on banning such weapons in outer space. That was a sensible position then, and it still is. Congress should stick to it.

DAYTON DAILY NEWS

Dayton, OH, April 1, 1984

With the appointment of a former Wright-Patterson Air Force Base officer as its head, the Star Wars program is off the ground, at least administratively. It probably was inevitable that laser-type defensive weapons would be explored. The debate about them now is almost entirely technical; the decision to do more research makes sense.

The research will not be cheap, because the time is near for some pretty expensive experiments. The administration is asking for $2 billion for next year and about $25 billion for the next five. If the payoff is — or even might be — permanent security, it's worth the price and more.

If. There has been a lot of knee-jerking on this one, both sides making the assumptions that please them. What's called for is open-mindedness.

The newly-appointed Star Wars chief, Air Force Lt. Gen. James Abrahamson, just may have been signaling the other day that his mind still is open, even if his boss's isn't. The boss, Defense Secretary Caspar Weinberger said, "I don't have doubts about our ultimate ability (to achieve) a thoroughly reliable and effective defense (if) our commitment is strong enough." So, when Mr. Weinberger in future years testifies in behalf of the program, it should be remembered that his confidence comes not from its findings, but from an ideological commitment that preceded the "in-earnest" (his phrase) part of the research.

All Gen. Abrahamson said was, "What I've seen in this country is that we have a nation that can indeed produce miracles."

Gen. Abrahamson has landed a choice assignment, no doubt about that. Technical people want to be on the leading edge of technology, and today that certainly means Star Wars.

People in the defense community talk about the nuclear era entering its third phase. The first was the age of the bomb; the second, now nearing its end, is the age of the missile. Both sides have more second-phase weapons than they would know what to do with under any circumstances, neither side has a defense against the other's weapons or any way of attacking most of them in place.

Phase three involves the transition from hardware to electron beams. Whether it will really make nuclear war less likely is entirely problematical. We don't know for sure that it won't, and that's good enough reason for some further inquiry. But we need not make a commitment now to see the Star Wars program through to the end; research and further thought may suggest that the most likely uses of beams would be offensive and destabilizing, not defensive, as the dream holds.

The Morning News

Wilmington, DE, April 4, 1984

PRESIDENT REAGAN told the Congress Monday that negotiations with the Soviet Union on curbing space defense weapons are not practical. Pravda will seize on the report to bolster its April Fool's Day claim that the United States does nothing to ease superpower tensions and that the Reagan administration is destroying "detente."

This is more "You're another," a game for children in a schoolyard, not for the world's most powerful nations.

Congress mandated Mr. Reagan's report when it approved development of an anti-satellite weapon system only after a "good faith" administration effort to reach "a mutual and verifiable ban on anti-satellite weapons" with the Soviets.

Such verification, the president said in effect, is pie high in the sky. He insisted that real deterrence can be achieved only if the new U.S. weapon offsets the "destabilizing advantage" given the Soviets by the systems they have now.

Critics of anti-satellite systems, chiefly the Union of Concerned Scientists, assert that reliable verification indeed is possible. It may be more relevant that the same group insists that such weapons won't work in the first place.

But you don't have to go far into space to see the problems. For example, the United States has had the Patriot system under development for 20 years as an anti-aircraft weapon. Plans described recently by Pentagon officials to turn this computer-radar-missile complex into an anti-*missile* weapon stirred fear that this would violate a 1972 Soviet-U.S. pact limiting anti-missile systems. On the other hand, the Soviets have in place the mobile, SA-12 air defense system, which the Pentagon contends already violates that treaty.

It was into this heated atmosphere that Pravda on Sunday fired its blast at Mr. Reagan. The official Communist Party newspaper declared that his administration's record is one of "actions aimed at destroying the system of international treaties . . . aimed at stabilizing the world situation and preventing the slide to nuclear war."

All this is a lot of aiming of weapons systems and firing of verbal blasts. Verbal blasts are much better than nuclear blasts and there is no more important topic for discussion than how to avert nuclear annihilation.

It isn't necessary to call the Soviet Union an evil empire to see the main difficulty in reaching viable agreements.

The Soviet Union is, more to the point, a closed society.

This one is, comparatively, wide open. It has no official paper to level *official* blasts at the Kremlin. In its turn, the Kremlin has no such critics as a Union of Concerned Scientists or free press to oversee any actions of its own that impair "detente" or hasten the "slide to nuclear war."

The Des Moines Register

Des Moines, IA, April 9, 1984

President Reagan issued an invitation the other day that we want to accept: He asked for suggestions on how to improve relations with the Soviet Union.

He didn't put it quite that way: "The ones who you hear yelling the loudest these days are the ones who put us behind the eight ball in the first place. . . . It's about time to get serious and ask those would-be leaders what they expect to use as incentives with the Soviet Union."

Since then, of course, he has called for a moratorium on debate after he has initiated a policy line, on the contention that it undermines American efforts abroad, but we'll take his first request seriously.

Not that we are the first to do so; many a suggestion has been offered. Lately, Reagan has even taken a couple. For example, he has stopped talking about winning nuclear wars, and he has stopped talking of the "evil empire." Restraint contributes to what George Kennan calls "civility of communication," an essential for stable relations.

Reagan took another suggestion on Wednesday — most recently and prominently made by Soviet leader Konstantin Chernenko. He will, the president said, present "a bold American initiative for a comprehensive worldwide ban on chemical weapons." Regrettably, he reverted after the announcement to the more familiar stance of dwelling on how far ahead the Soviets are and the need for a U.S. buildup first as deterrent and bargaining chip. Still, the proposal is progress.

•

One list of suggestions was offered by former Secretary of Defense Robert McNamara: No first use (already agreed to by the Soviets); no launch on warning; further reductions in tactical nuclear warheads in Europe and withdrawal of such weapons from "forward areas of Germany."

Reagan's political opponents have made many others, including a nuclear-arms freeze, various forms of "build-down" proposals, a comprehensive test-ban treaty and a ban on weapons in space.

Little daring or imagination would be required for U.S. ratification of existing treaties on underground nuclear testing and testing for peaceful purposes. Then there is always the still-unratified SALT II treaty. Or there could be an offer to merge the two sets of talks that broke down last year in Geneva, thus broadening possible areas of agreement.

•

Still, arms-control agreements may not be the best way to begin. Cultural and economic exchanges have usually come more easily. Here again, the Reagan administration has shown some movement of late: Foreign-policy officials on each side are discussing consulates in Kiev and New York, and other cooperative moves.

Far more could be done to cooperate on areas of mutual concern. There could be joint research on public health, nuclear fusion, oceanography and numerous environmental issues. More cultural, scientific and trade exchanges could be worked out.

One group has stressed the importance of Reagan's meeting with Chernenko simply to become acquainted and to clarify their perceptions and intentions.

It does not take great imagination to think of improvements. It does take a commitment, patience and a belief that improvement is possible. Too often, both sides seem to put of their energy into proving otherwise.

Roanoke Times & World-News

Roanoke, VA, April 1, 1984

INSIDE AND outside government, generals and scientists are raising doubts about the "Star Wars" missile defense proposed by President Reagan a year ago. Technically, it may not work; politically, it may deal another setback to arms-control talks. Meantime, however, White House and Pentagon are gearing up to pour tens of billions of dollars into this questionable concept.

As posed by the president in a televised speech March 23, 1983, the idea was very attractive: saving lives (by knocking down enemy missiles) instead of avenging them (by counterattack with our own missiles). Appealing to Americans' sense of wonder as well as to the hopes they place in science, Reagan called for a laser-type missile defense that would render nuclear weapons "impotent and obsolete."

The president noted then that development of such a system was only a possibility and would take years. Some others, though, grabbed the ball and ran headlong. "What we want to try to get," said Defense Secretary Caspar Weinberger, 'is a system which will develop a defense that is thoroughly reliable and total. I don't see any reason why it can't be done."

What seems thoroughly reliable and total is Weinberger's enthusiasm for anything that would increase Pentagon spending. The fiscal 1985 budget includes nearly $2 billion for research in this area, and another $24 billion is projected over the next five years.

The Defense Department should keep its options open in any category where national security might be affected. But an all-out push, trumpets blaring, to develop this kind of system could be both wasteful and dangerous.

Indications are that an impenetrable defense against missile attack cannot be achieved. There are cheap and relatively easy ways to defeat or confuse a defense that relies on lasers or other electronic means to detect and destroy missiles in flight.

For examples, the enemy could reduce the length of time his missile rockets burned; he could send out hundreds of decoy missiles; he could spread clouds of chaff to diffuse radar and other signals. He wouldn't have to confound the entire defense: It's estimated that if only 5 percent of the Soviet Union's warheads got through, 60 million Americans would die.

Points like that were made soon after the president's proposal. Now comes a 106-page report by a study panel of the Union of Concerned Scientists that documents the arguments against the "Star Wars" defense. The nine-man panel includes a number of weapons scientists. Among them are Hans Bethe, who directed the theoretical physics team on the Manhattan Project, and Richard L. Garwin, a member of the team that built the first thermonuclear bombs. The report was signed by 36 others, several of them Nobel Prize winners and former presidential science advisers.

The report concluded that, for some of the reasons cited above, a "total" defense system would not work. But it would cost hundreds of billions of dollars. If the United States chose to put chemical-laser "battle stations" in low orbit around the Earth, at least 2,400 would be needed to blanket Soviet missile fields. Only to lift the fuel for these stations into space would cost $70 billion.

Another alternative, favored by White House science adviser George A. Keyworth, is to link ground-based lasers with orbiting mirrors. Also impractical, said the panel; besides, the electricity for these lasers would draw from 20 to 60 percent of the nation's generating capacity. The cost of building special generators is put in the $40-300 billion range.

It may be significant that the Union of Concerned Scientists has opposed some of President Reagan's other military policies. But another key objection of the panel to this project — its potential effect on arms-control talks — is echoed by one of Reagan's advisers, retired Lt. Gen. Brent Scowcroft.

Head of the 11-member commission set up last year to study the MX missile, Gen. Scowcroft marked the end of his work with a nine-page report to the president this month. Therein, he endorsed continuing research on new anti-ballistic missile systems, but urged "extreme caution" in allowing any engineering development. This might breach the 1972 ABM treaty with the Soviet Union and imperil further negotiations.

That's no idle warning. Scary as it is, the peace has been kept between the two nuclear superpowers by mutual assured destruction: each side's knowledge that a first strike would lead to its own annihilation. An attempt by either country to achieve a comprehensive defense would upset that balance. It would wipe out any chance for strategic arms reductions and lift the arms race into another, even more frightening phase. The temptation to attack before the defense was in place could become irresistible.

The better idea is an old one: to defuse tensions and get the two sides back to bargaining that will steadily reduce the number of nuclear weapons. The fewer there are, the less need for exotic defenses.

AKRON BEACON JOURNAL

Akron, OH, April 9, 1984

DESPITE strong opposition from Congress and the scientific community, President Reagan appears determined to go ahead with his proposed anti-satellite weapons system before he begins to talk with the Soviets about controlling such weapons.

Mr. Reagan fears that an anti-satellite missile gap exists. National security, he argues, compels the United States to develop its own system before it negotiates with the Soviets. In addition, his report to Congress claims it would be difficult to verify an anti-satellite weapons treaty.

But the Union of Concerned Scientists argues that a treaty would be verifiable, and it is skeptical about the quality of the Soviet satellite program. The scientists agree with Congress that development of the system — Congress appropriated $19.4 million last year — should be accompanied by a good-faith effort to ban anti-satellite weapons.

The administration's report is part of its effort to add $143 million to the multibillion-dollar project this year. Last year Congress demanded the report as a way to get the White House to look at the problems before more money was spent.

Obviously, the administration has learned little over the past year. Negotiations have not been pursued. Mr. Reagan argues the Soviets are only interested in talking about controlling anti-satellite weapons in order to slow down development in the United States.

He may be right, but talks are still crucial. An unfettered arms race is a more dangerous prospect. Both sides have to realize that.

Congress would be wise to hold Mr. Reagan's anti-satellite system hostage to real progress toward negotiation. Anything less speeds up the arms race and threatens international security.

THE CHRISTIAN SCIENCE MONITOR

Boston, MA, April 4, 1984

IT might come as a surprise to many Americans, but President Reagan's futuristic 'star wars' missile defense system has a powerful lot of today's budget resources and development thrust in it already.

President Reagan was in effect acknowledging the momentum behind his administration's comprehensive space defense program when he told Congress this week that seeking a treaty ban on satellite weapons would be fruitless.

The White House says (a) the Soviets already are out in front on an antisatellite system, and (b) verification of compliance with any treaty would be nearly impossible. It likely well believes this.

But the larger point is that the administration's policy on space, missiles, and nuclear weapons is heavily driven by the military's views; arms control is secondary. Fundamentally, it's star wars that the administration is driving toward — a system that can destroy offensive weapons in flight. Many Reaganites think the Soviets are already off and running, so the United States must get on with its program too.

The President's Commission on Strategic Forces issued a report this week that tried to cut a reasonable line between the buildup and arms-control factions in the United States. It cautioned not to expect too much, too soon from talks. It criticized the nuclear freeze movement for potentially locking the US into destabilizing, outmoded systems. But it also warned against tampering with the 1972 antiballistic-missile treaty by pushing antimissile defense systems. It disputed the cynical view of some Reagan aides, that Washington is being duped by Moscow by arms negotiations, and it recommended a cautious step-by-step approach to arms control.

How odd and disappointing that the Democratic presidential race has so far mostly drawn a blank on this issue of US space/missile defense programs, or has tried to oversimplify it into which candidate was first to favor a nuclear freeze.

The 1972 antiballistic-missile treaty rightly allowed for research and some development on defensive systems: The critical line it set was between testing and deployment.

The defense budget already shows terrestrial and extraterrestrial sections. The 1985 and 1986 department of defense budgets contain several billion dollars for "strategic defense initiatives" — almost a defense budget in itself. Antisatellite weapons are discussed as part of the overall strategic requirement. The mandate for a package of measures — including F-15 launched weapons and antisatellite and space weapons — is broad and ambitious. Administratively, the program is institutionalized in a Pentagon space command.

So much momentum has gathered that it may soon be more a question of *which* defensive system to build, rather than *whether* to build one at all, say experts concerned about the Washington trend. Trying to control so much activity by treaty can only become more difficult.

One can't be naive about Soviet intentions. What they have been doing all these years in space with their docking maneuvers bears suspicion.

But Washington has a duty to examine its own steadily rising commitment to space weaponry — a commitment larger than has been acknowledged.

Newsday

Long Island, NY, April 6, 1984

President Reagan's latest arms-control initiative is a plan to end production and possession of chemical weapons, which were used as long ago as World War I and banned by international agreement in 1925. Vice President George Bush will submit a broad American proposal to a 40-nation disarmament conference in Geneva later this month — even though the new prohibition would be extremely difficult to police.

At the same time, Reagan is unwilling to resume negotiations with the Soviet Union to ban antisatellite weapons, which exist only in relatively primitive form and have never been used in war. He says compliance with an agreement would be too hard to verify.

The President sees no contradiction in these two positions, and perhaps that's not surprising for someone who has urged passage of a balanced-budget amendment while budgeting the biggest deficits in the nation's history. In fact, Reagan takes a similar approach to chemical weapons: Even though his plan would require their destruction, he wants Congress to fund production of new ones. To get a treaty banning them, he said at Wednesday night's news conference, "you've got to have something to bargain with."

How can this President speak with authority on what it takes to obtain any kind of arms-control agreement? None have been negotiated in the three years since he took office. Even though Congress has gone along with him on funding a new intercontinental missile and a new strategic bomber, Moscow has suspended talks on both strategic arms reduction and the control of intermediate-range nuclear weapons in Europe.

Iraq's use of poison gas against Iran has reminded the world how barbarous chemical weapons can be. But at least they don't threaten to upset the balance of deterrent forces on which the world's two superpowers rely to prevent a nuclear holocaust, as antisatellite weapons do. No matter how difficult it may be to verify an antisatellite weapons treaty, it's incredibly shortsighted of the Reagan administration not to make the effort.

The Philadelphia Inquirer

Philadelphia, PA, April 13, 1984

The world stands at a crossroads in its uses of outer space. Decisions the United States and the Soviet Union make now will dictate whether this frontier can be protected or turned into earthlings' next battleground.

One such decision came early this month when President Reagan told Congress he would press ahead with development of weapons, called ASATs, designed to destroy satellites.

Previous U.S.-Soviet talks on ASATs halted in 1979 after the Soviet invasion of Afghanistan. In 1983, the Soviets, despite the breakdown on nuclear arms talks, proposed a treaty dismantling existing ASATs and banning new ones. Congress had required the administration to show "good faith" efforts to negotiate a ban on ASATs, which the Soviets have and the United States is testing. But Mr. Reagan said the United States would not negotiate on an ASAT ban because it would be virtually impossible to verify.

This decision must be reconsidered. For one thing, the impossibility of verification is sharply disputed: Many leading American scientists say a workable ASAT treaty could provide adequate verification safeguards. More important, the very real difficulties of verification should be addressed at the bargaining table, not used as an excuse to avoid it.

Attention has focused on development of satellite-killer weapons because it is at a critical stage. Space satellites have become the nerve center of superpower military operations. But so far neither the Soviets nor the Americans can threaten satellites in high earth orbits where the most important early warning and communications systems are based.

The Soviet anti-satellite weapon at present is slow, ground-based and rather crude; most U.S. satellites are above its range. The United States is developing a swifter, much more sophisticated weapon, launched from an F-15 fighter plane. At its present stage of development the U.S. ASAT would not threaten high orbit satellites but it has the potential to do so. Many experts fear that perfection of the U.S. weapon will spark an unchecked race with the Soviets for better satellite killers.

Why is the administration so determined to press ahead with ASAT development? First, because it believes the United States must have a fully tested ASAT to deter the Soviets from using theirs. Second, because it says a ban cannot be verified.

Some experts contend an unmentioned third reason for the administration's push on ASATs is that a treaty banning them would hinder work on President Reagan's equally controversial "star wars" space defense plan.

The merits of tit-for-tat countering an inferior Soviet ASAT (as opposed to blocking its potential use through other forms of military threat), must be measured against the danger of a space weapons race. The United States has the most to lose in such a race. America depends on its more technologically advanced satellites more heavily than does the Soviet Union, both for peaceful scientific and commercial uses, and to provide communications, navigation, intelligence and targeting information for the military.

Moreover, while the United States has a lead in space technology, it could evaporate under a determined Soviet effort, as has happened before.

As for verifying a ban, no one would deny the difficulties. Nor can verification ever be perfect. But claims that reasonably effective verification is impossible are challenged by experts like the Union of Concerned Scientists, which includes former top CIA, military and intelligence figures. They say that any secret attempts by the Soviets to upgrade or add to its present system would require testing that would be detectable. They also believe that use of non-ASAT weapons or space vehicles to threaten satellites would be detectable and that a Soviet "breakout" (breaking the treaty and quickly assembling an illegal ASAT system) would be much harder than the administration asserts.

When calling verification "impossible" the administration often refers to a "comprehensive" ASAT ban. Even the most ardent ASAT foes don't demand a total ban because they know it can't be achieved. But there are ways to address the problem short of that, as even Mr. Reagan noted. Congress must continue to press him to negotiate on ASATs before the race for satellite killers moves beyond control.

The Houston Post

Houston, TX, April 5, 1984

President Reagan's reason for deciding this week to proceed with the development of a satellite killer has provoked concern among arms control advocates. The president said research on the anti-satellite device will continue because it would be difficult to verify the Soviet Union's compliance with a ban on such weapons.

The president's decision was seen as a rebuff of a Soviet initiative to negotiate a ban on satellite killers. The Kremlin has suggested that work on such an accord could lead to revival of the suspended nuclear arms reduction talks. Arms-control supporters worry that Reagan's action could keep superpower relations in the deep-freeze. They also point out that the president could apply the rationale behind his decision on the satellite killer — that we can't detect Soviet cheating — to nuclear or other arms control negotiations.

The verification issue was an increasingly important factor in the Strategic Arms Reduction Talks in Geneva before they were recessed last year and the Soviets refused to agree on a date to resume them. It is fair to ask how seriously we can take the Kremlin's offer to negotiate a satellite killer ban.

We have been working for several years on an anti-satellite weapon, but the Soviets are well ahead of us in the field. The administration contends that they have had an operational satellite killer system for the past decade.

The president's report this week was required by Congress to free $19.4 million it had voted to develop the U.S. weapon. But the administration is still barred from testing the weapon until it reports to Congress that it is making progress toward negotiations to ban satellite killers. Yet they are the vanguard of a new generation of space weaponry that could ultimately include Reagan's proposed Star Wars shield of satellite-mounted laser or particle-beam weapons. The feasibility of the Star Wars defense concept has been vigorously challenged scientifically. Even if it worked, it could cost half a trillion dollars.

How much better it would be to negotiate an arms control agreement that would allow us to save the vast amounts of money and resources that such a high-tech defense system would cost, and put them to more beneficial use. But until the Soviet Union agrees to tighter anti-fudging safeguards — including on-site inspection if necessary — we have little choice but to continue our development of sophisticated new weaponry.

Chernenko Renews ASAT Arms Ban Offer

Soviet President Konstantin Chernenko June 11, 1984 urged the U.S. to negotiate "without delay" a treaty banning the use and development of antisatellite (ASAT) weapons. Soon after taking office, Chernenko had appealed to the United States to accept the ban proposed in 1983 by his predecessor, Yuri Andropov. Chernenko's remarks came in response to questions by U.S. journalist Joseph Kingsbury-Smith. The interview was carried by the official Soviet news agency, Tass. "I would wish to underscore this," Chernenko said. "Agreement on these questions must be sought without delay while space weapons have not yet been deployed and while a breakthrough in the face of space weapons, unpredictable for its consequences, has not yet been made. Tomorrow may be too late." Chernenko said U.S. officials who stressed the difficulty of verifying an antisatellite accord were "consciously bent on having their hands free for pursuing the course of space militarization." Such a course, he added, could lead "only toward a sharp increase in the threat of war" and "cannot be allowed to happen."

White House spokesman Larry Speakes told reporters that "the door is not closed" to an "effective" antisatellite weapons accord. The Reagan Adminstration would "study" Chernenko's remarks, Speakes said, but "would like to see some movement" first in suspended negotiations on reducing nuclear weapons. The Soviets "talk about banning" space weapons, Speakes said, "because they have the only effective system in the world" and "they have a desire to preserve that monopoly."

The Philadelphia Inquirer

Philadelphia, PA, June 18, 1984

President Reagan and the Soviets are playing games with arms control that could play havoc with the world. The name of the riskiest game: Make me an offer you know I'll refuse.

Under pressure from Republican leadership to meet on arms issues with Soviet leader Konstantin U. Chernenko — "we've got to figure out some way not to blow each other up" pleaded majority leader Howard H. Baker Jr. (R., Tenn.) — the White House insisted any summit should be carefully prepared ahead of time to guarantee results.

But as soon as the Kremlin announced it favored the idea of a well-prepared summit President Reagan changed his tune. No tight advance planning was necessary; just a get-together was OK. But, the President said at his news conference Thursday, the Soviets had rebuffed him because *they* now insisted on "a very carefully prepared agenda" of items.

The fact is that a Reagan-Chernenko summit before the U.S. elections probably is a pipe dream, for several reasons. The atmosphere has been poisoned by Mr. Reagan's anti-Soviet rhetoric and hostile Soviet actions and strident anti-Reagan rhetoric. The Soviets are preoccupied with their leadership shakeout and fiercely unwilling to help re-elect Mr. Reagan by handing him a peacemaker image.

But both sides could make an end run around stalled talks on nuclear arms if they wanted to get serious. The Soviets have just renewed their offer to negotiate a ban on anti-satellite weapons, known as ASATs. If deployed, these weapons could knock out communications and military satellite eyes in the sky and move the arms race into a new phase in space.

The Soviets want to talk because U.S. ASAT technology far surpasses theirs, a fact which puts America in a good bargaining position. But they probably expected the President to turn them down because the administration had firmly vetoed talks on ASATs. Yet suddenly, with Congress pressing the administration, and Reagan aides anxious to erase a warmaker image, the White House has announced that it may put forward a draft ASAT treaty.

It should do so and the proposals should be serious. President Reagan was hardly reassuring when, in downplaying his challenge to the Soviets, he told Sen. Baker "If they want to keep their Mickey Mouse system, that's OK." (Would that the superpower arms race were no more serious than a Walt Disney space movie.) This is as good a time as any to test Soviet intentions, and to use an issue on which the Soviets want to negotiate toward ending the superpower impasse. It is not the time to play election politics or arms control games.

The Tennessean

Nashville, TN, June 14, 1984

THE Senate has put a restriction on the testing of a U.S. weapon designed to shoot down satellites until President Reagan certifies he has made a good-faith effort to negotiate a treaty with the Soviet Union on strictly limiting such weapons. That is not an overwhelming obstacle but it is a prudent one.

It seems apparent that the administration is not seriously interested in negotiations that would ban weapons in outer space, either anti-satellite missiles or the so-called "Star Wars" shield against Soviet missiles.

On Monday, Soviet President Konstantin Chernenko proposed an immediate resumption of negotiations to ban tests of anti-satellite weapons before there are "drastic developments" in the arms race in space. The Soviet leader said his government "consistently stands for keeping outer space peaceful" and that a resumption of talks should be made without delay.

Mr. Chernenko said a test ban could be verified "very effectively" through national technical means and any problems could be solved in negotiations. The Soviet leader referred to photo reconnaissance and other detection measures.

President Carter suspended these negotiations in 1980 after the Russians invaded Afghanistan. President Reagan has ruled out any resumption, saying that an agreement to ban weapons the Soviet Union already has and that the U.S. is trying to develop would not be in the best interests of this country.

So the White House has dismissed the latest Chernenko offer, saying that it contained nothing new. A spokesman said it is looking into "possible initiatives" in space arms control that could be verifiable, but not a comprehensive ban.

The Senate restriction would ban the next round of testing until Mr. Reagan certifies he is trying to negotiate the strictest possible limitations on anti-satellite weapons.

According to the Pentagon, the Soviets have deployed a workable system which could knock out low-flying U.S. satellites. But it is generally conceded that the Soviet system is cumbersome and inaccurate as it now exists and does not pose much of a threat.

There is grave concern among lawmakers and others that weapons systems moved to outer space increase the possibility of accidents and misjudgments that could set off a nuclear conflict that could rage both on earth and in space.

It would be to the advantage of both superpowers to reach agreements that would ban weaponry in space. Outer space may represent the "high ground" but it also represents an arena that ought to be kept free of weapons, nuclear and otherwise.

The effort to pursue an anti-satellite treaty ought to be one of the highest priorities of the administration. Instead, it continues to offer excuses and vague reasons why such an agreement couldn't be verified. At least the Senate has given it a warning that it ought to try for negotiations.

THE SACRAMENTO BEE

Sacramento, CA, June 1, 1984

The debate in Congress over the Reagan administration's anti-satellite weapons (ASAT) program soon moves to the Senate floor, where a number of influential Republicans have joined a large group of Democrats determined to press the White House for a reversal of its present shortsighted policy. That policy is to push ahead with a new arms race in space because, the administration argues, an agreement with the Russians to halt development of anti-satellite weapons can't be effectively verified.

By an astonishing 57-vote margin, the House recently rejected that position. The majority sensibly concluded that, although it might be possible to cheat on marginal elements in the development and testing of such space weaponry, no reliable ASAT system could be perfected and deployed without the other side detecting it well in advance. That's also the position of a great majority of American scientists and technological arms experts, including many now or formerly in various high positions in the Defense Department.

They have made an overwhelming case that the safety and security of this country demand a halt to the development of space weapons before that movement goes any further. The Soviets, too, have sought a moratorium and invited the administration to reopen negotiations on an ASAT treaty.

It would be folly for the United States to fail to grasp that opportunity, not only because it might produce an important agreement at a time when relations are dangerously chilled, but also because of the gravity of the consequences for this nation's security of going ahead with an ASAT contest.

A halt in such development now, when the Soviets have only a slow, crude rocket system that is launched to follow a low-orbiting target satellite and then explode in its vicinity, would leave the most important high-orbiting U.S. satellites invulnerable. Those are the ones that provide early warning against Soviet missile attack and link U.S. communications for military command and control operations.

The Soviet ASAT has failed in 11 out of 18 tests monitored by U.S. intelligence. It is so ineffective, and so easily detectable, that before it could be improved it would be possible for the United States to provide evasive or defensive measures for the less crucial, low-orbiting U.S. satellites that would be its only targets.

The American ASAT, likewise designed to target only low-orbiting Soviet satellites, is far more sophisticated. It consists of a computerized, satellite-homing rocket launched from high altitude by an F-15 fighter. Its development has reached the point that the Air Force wants to test it against actual objects in space later this year.

Yet to proceed with this weapon would compel the Soviets not only to improve their own model, but very likely to begin development of a new generation of ASATs — including, almost surely, weapons able to reach and destroy the vital, high-orbiting American satellites. Ironically, the United States is far more dependent on satellites than the Russians are. Our more open society makes it easier for the Russians to monitor military activities and weapons developments here through other forms of espionage. The Soviets' closed system, on the other hand, requires U.S. intelligence to rely heavily on its eyes in far outer space to monitor the testing and deployment of nuclear missiles and other forces.

Halting an arms race in space now, before it goes to the point of no return, is clearly in this nation's best interests. That concern becomes all the more urgent in an increasingly destabilizing arms race, because if satellites become vulnerable, the temptation becomes all the greater to rely on first-strike nuclear strategies. Those dangers argue powerfully for a moratorium and for taking the minimal verification risks an ASAT treaty might entail.

THE MILWAUKEE JOURNAL

Milwaukee, WI, May 26, 1986

Hurrah for the House — at least for those 238 members who outvoted 181 of their colleagues to block the Defense Department from space-testing an anti-satellite weapon against a target. The ban would prevail so long as the Soviets held off conducting similar tests of their ASAT weapon.

Having a weapon that could destroy orbiting Soviet satellites is an appealing idea — until you start to think about it. In our view, the prospect offers much more peril than security.

Remember when the Russians shot down that South Korean 747 passenger jet last year? They showed that they tend to panic easily, to overreact. If, in a crisis, the United States were to blind the Soviets by destroying their spy-in-the-sky satellites with ASAT weapons, or even threatened such action, the trigger-happy Russians could panic again, with truly catastrophic consequences.

It is sometimes said that the Soviets are "ahead of us" in ASAT weapons technology. Even if that were the case, it would not prove the need for this country to have an ASAT. In fact, the Soviet ASAT seems to be as slow, clumsy and inefficient as just about everything else that the Soviets make. If anything were likely to encourage the touchy, sometimes irrational Russians to improve their space klunker, it would be the challenge of an effective ASAT program in the Pentagon.

The satellites of both countries should be protected, not menaced, and the best way to do that would be through an international treaty. The Soviets have submitted a draft ASAT ban to the United Nations. The Soviet proposal, like the Soviet ASAT, is no doubt a lemon. But maybe — like the ASAT — it can be improved.

Why not try? If nothing else, the negotiating effort would be less expensive and less dangerous than shifting part of the arms race to outer space. It's hard to imagine how that could make the globe a safer place to live.

The Des Moines Register

Des Moines, IA, June 16, 1984

The United States relies very heavily on satellites for both military and civilian communication, and the Reagan administration is worried about Soviet progress in devices to "blind" satellites, jam their radars or blow them out of the sky.

Its response is to work on developing anti-satellite weapons of its own, fending off Soviet appeals for an agreement banning tests of such weapons and claiming that the Soviet moratorium on launching them (which has been going on for 10 months) is only a device to stop our progress while they are ahead.

Unfortunately, these weapons cannot be merely wished away, although Congress has been pressuring the president to try.

Last year, Congress approved an amendment banning tests of space weapons unless the president promised to try to negotiate a comprehensive ban on space weapons. The president did not like the idea. (There already is a treaty ban on placing weapons of mass destruction in outer space.)

This year, the House of Representatives wrote into its military-budget bill a ban on testing anti-satellite weapons in space as long as the Soviet Union abides by its moratorium.

Then the Senate, acting this month, wrote into its bill a ban on testing anti-satellite weapons unless the president certifies that he is "endeavoring in good faith to negotiate the strictest possible limitations on anti-satellite weapons."

Congressional restrictions like this cannot really bind the president, but they do indicate a frame of mind. Only the executive branch can negotiate international agreements; all the lawmakers can do is advise and consent. But their advice is worth listening to.

The administration gave some advice, too. The Senate debate was interrupted by a secret briefing requested by administration supporters. Then the Senate toned down the original amendment calling for "immediate" negotiations, but passed a compromise.

Arms-limitation and disarmament negotiations are enormously difficult. They have been going on regularly since 1899 (yes!) and almost continuously since 1945, often in several different conferences. Results have been skimpy and not universally admired.

George Will in Newsweek magazine suggested (tongue in cheek, we hope) that we give up trying: Anything the Soviets agree to will be against our interests, and even successful pacts simply turn the arms race into new channels.

Let's keep on trying.

WORCESTER TELEGRAM.

Worcester, MA, June 13, 1984

Last year, when President Reagan made his famous "Star Wars" pitch for building a defense against intercontinental ballistic missiles, the critics savaged him on two counts.

1. They said that a defense against incoming missiles was technologically impossible and would be a prodigious waste of money. "Like hitting a fly in space" was the way they put it.

2. They said that, even if it could be made to work, it would be "destabilizing" and a bad idea.

The Army may have knocked the first argument out of the sky. On Sunday it hit the fly in space. An experimental antiballistic missile fired from a Pacific island intercepted and destroyed a Minuteman 1 missile with a dummy warhead that had been fired from Vandenburg Air Force Base in California. The interception was accomplished more than 100 miles above the earth.

One successful trial does not make an effective antiballistic missile system, but it does take the idea out of the realm of science fiction. A technological defense against ICBMs can no longer be dismissed as absurd. It may be possible.

What about the second argument — that it would be destabilizing and dangerous? Those making this argument fear that the Soviets would see a U.S. ABM buildup as a long-range plan for a nuclear first strike against the Soviet Union. Faced with this dread possibility, Moscow would strike first, before the system was in place.

There is something to this argument, but perhaps not as much as some think. At any rate, it should not be accepted as gospel. It should be examined closely. At this point, the country should be neither committed for or against an ABM defense system.

On the same day that the Army made its successful test, Soviet leader Konstantin Chernenko called for immediate negotiations to ban anti-satellite and space weapons. He insisted that such a treaty could be verified, a point U.S. defense experts doubt.

If Chernenko is worried about the nuclear buildup, he should order his negotiators back to Geneva to the START and INF nuclear arms talks. A reduction of nuclear weapons in Europe is a lot more easy to achieve and verify than theoretical weapons systems in space.

There are plenty of ways for the Soviet Union to reduce world tensions. If tensions were reduced, there might be no need for any ABM systems.

THE ARIZONA REPUBLIC

Phoenix, AZ, June 16, 1984

SCORE one for the Senate, but not the United States.

By a vote of 61-28, the Senate has beaten, or, at least, delayed, the Reagan administration's efforts to test a U.S. weapon designed to shoot down satellites.

The senators did give President Reagan an option.

All he has to do is certify that he is "endeavoring, in good faith . . . to negotiate [with the Soviets] the strictest possible limitations on anti-satellite weapons."

Then, presumably, the testing of a satellite killer could go on as scheduled this fall.

Providing, of course, that the Senate or the House or a combination of the two didn't throw another strike at the administration.

The only thing good that can be said about the Senate's action is that it represents a compromise between Reagan's supporters and critics who would have imposed stiffer arms-control requirements.

Even so, it is a serious rebuff of the president's policy on space weapons and a threat to the development of needed technology.

Reagan has made his position clear.

He told Congress earlier that there is little point in seeking a treaty banning the testing or deployment of anti-satellite arms because of the virtual impossibility in verifying compliance.

There's nothing in the Soviets' book to indicate a change despite Soviet President Konstantin Chernenko's latest call on the United States to negotiate "without delay" a pact banning anti-satellite weapons.

The White House responded to the Chernenko overture by pointing out that the United States has made "concrete proposals" and is "ready to talk" any time the Soviets show "more evidence of their sincerity" on arms control, especially the stalemate in talks to curb nuclear weapons.

There is convincing evidence that what the Soviets really fear is that the United States will catch up with Russia in the field of anti-satellite weapons — ASAT — development.

When Moscow announced a voluntary moratorium against testing satellite killers, it did so with the knowledge that it was comfortably ahead of the United States.

According to *Air Force* magazine, Soviet anti-satellite capability has been amply demonstrated. President Reagan spelled out in an April report to Congress that the Soviets have an operational system that could be used any time, quick off the mark, against low-orbiting U.S. satellites.

Moreover, the Soviets are testing land-based lasers of ASAT capability and could also use as ASATs the nuclear-armed Galosh ABM interceptors emplaced around Moscow.

Faced with this clear and present danger, it's foolhardy for the United States not to proceed with its own ASAT program.

Further delays are in the best interests of the Soviet Union, not the United States.

The Boston Globe

Boston, MA, July 16, 1984

Space has long since been militarized, with surveillance and communication and weather satellites, but it has not yet been *weaponized*. Whether the weaponizing of space can be stopped is one of the great war-and-peace issues of 1984.

In agreeing to talk with the Soviets about a ban on antisatellite (ASAT) weapons, the Reagan Administration is trying to make a distinction between weaponizing space at low levels, relatively close to the earth, and at higher altitudes where the most critical US surveillance satellites ride. The Administration proposes a ban on high-altitude ASAT systems, but wants to let low-altitude systems run free.

This is a false distinction. If effective low-orbit ASAT systems are deployed they will inevitably spill into higher orbit, and space will become crowded with devices that are not only vital to the superpowers for intelligence, but also are capable of blowing each other up.

At this moment neither side has a meaningful satellite-killing system. For years, the Soviets have had a clunker ASAT capability; the system is slow and cumbersome and has failed many of its tests. The US system under development, however, will be fast, flexible and efficient. With this forthcoming system – homing devices on rockets fired from F15 aircraft – the United States could disable all lower-orbit Soviet satellites quickly. That is why the Soviet system, although already "operational," is less destabilizing than the yet-untested US system.

Eleven months ago, in August 1983, the Soviets proposed a ban on all ASAT weaponry. The Administration delayed answering, and now, since the proposal has been repeated, wants to confine the talks to high-altitude systems.

In hindsight, after the 1963 Limited Test Ban Treaty, arms controllers deeply regretted that an extra effort was not made to ban all nuclear test explosions through a "comprehensive test ban." The partial ban allowed testing to continue underground, and hence the warhead race continued unabated.

Americans will look back with similar regrets if the Administration manages to keep a toe in the door for the low-orbit weaponization of space. The technology needed to hit low satellites differs very little from the technology needed to hit targets in higher orbit. Rockets need to be bigger and guidance systems somewhat sharper, but it is a difference in degree.

The Rubicon in space is not, as the Reagan Administration pretends, between low- and high-altitude ASAT systems, but between the present state of affairs, in which the Russians have an ASAT "monopoly" so crude as to be meaningless, and the next step, in which satellite killers will spread, unbounded. That is why it is vital that US negotiators participate in the Vienna talks with the aim of freezing all ASAT development, high and low.

The Oregonian

Portland, OR,
May 27, 1984

The House of Representatives is to be congratulated for having voted to extend the moratorium on testing anti-satellite space weapons, a positive step predicated on the Soviets sticking to a test moratorium they have announced.

The Senate also will have to approve the action, attached to the 1985 defense authorization bill. It was a rare military defeat for President Reagan, who is hostile to the moratorium and who presides over an administration that has failed to respond to a Soviet proposal to ban all anti-satellite weapons. The Soviets have tested a crude device that, if made effective, would work only against low-flying satellites of no strategic value.

The president, in his so-called Star Wars speech, advocated efforts to develop strategic space weapons. He failed to recognize that these efforts shake the very foundations upon which arms reductions and world security must be erected.

High-altitude (25,000 miles) satellites, now invulnerable to any weapon system, Soviet or U.S., provide instant intelligence and communications. They are not only essential to military planning, but also become vital in a crisis, providing verifiable intelligence of each nation's movements. A crisis caused by cascading miscalculations, or an accidental firing of a nuclear weapon, could be wound down and the world rescued from a nuclear war only if each nation had reliable satellites providing instant information on military activities.

The problem the administration sees is that satellites are also vital if modern military forces are to be effective in combat. But the beneficial values of satellites in verifying arms treaties, building confidence among nations and otherwise preventing nuclear ambushes, far outweigh any risk of their wartime usefulness being diminished.

Further, scientists have pointed out that construction of a successful system to shoot down or destroy enemy satellites cannot be separated from the technology needed to protect against nuclear missiles. Thus ASATs, or anti-satellite weapons, smash head-on against the ABM Treaty. Any success in developing machines to destroy high-altitude satellites can be transferred to weapons useful in attacking strategic missiles in space. As such a threat emerges, the temptation to strike first, before the command controls, communications and intelligence vital to protecting a nation are lost, might prove irresistible, even for Washington.

It is vital that the Congress limit anti-satellite weapons. In a world of growing anger between the superpowers, any effort to contain the arms race to its earthly horrors is a great step and must not be missed.

Roanoke Times & World-News

Roanoke, VA, June 14, 1984

IN A TEST, a computer-guided U.S. missile has tracked down and destroyed a make-believe warhead more than 100 miles above Earth. The Pentagon hails it as an "absolutely tremendous success" and a "major breakthrough" to developing an American defense against enemy missiles.

U.S.S.R.
DUMMY MISSILE
UNITED STATES
INTERCEPTOR DEVICE COLLIDES WITH WARHEAD
Vandenberg Air Force Base California
Hawaiian Is.
Pacific Ocean
Kwajalein Atoll
Homing Overlay Experiment (HOE)

There's no doubt about the technical achievement: To locate its target, the destroyer missile used not only computers but also infrared sensors capable of detecting, in space, less than 100-degree heat more than 1,000 miles away. It caught up with the warhead-carrying missile at a speed of 20,000 feet per second and, like a kamikaze aircraft, rammed it. Result: wipe-out.

This is a step toward the "Star Wars" defense envisioned by President Reagan in a speech more than a year ago. "It would provide a way," said an Army spokesman, "to work on an enemy's ICBM attack outside of the United States, at a relatively safe distance."

So far, so good. This looks virtually foolproof as long as the Soviets can be persuaded to fire their missiles one at a time. They need to send up only real missiles, so that none of our defensive missiles are wasted by chasing decoys or clouds of chaff. Those attacking missiles also must keep their rockets burning — no fair shutting down and coasting — so that our heat-seekers can find them far out in space.

Asked about decoys and other enemy trickery, the Army spokesman said the program is "dealing with" such problems. When it has solved those when there's no prospect that sheer numbers could overwhelm the defense and get only 5 percent of the incoming nuclear warheads to their targets — then the Pentagon really will have achieved a major breakthrough. Before that happens, of course, the Soviets may have decided they'd better attack while they still have a chance to prevail. There are some strategic and political problems that technology can't overcome.

Houston Chronicle

Houston, TX,
June 13, 1984

As thousands of Houstonians flock to see the latest space adventure at the movies, the Army has pulled off a technical feat that rivals any special effects depicted on the screen. By intercepting and destroying a speeding target in space, the Army's experimental missile shows that space weaponry is no longer strictly in the realm of fantasy.

President Reagan has asked Congress to fund continued research into his proposed Strategic Defense Initiative — the popularly dubbed *Star Wars* defense against nuclear attack. The Army test does not mean that a 100 percent effective umbrella against nuclear warheads is necessarily more feasible today than it was last week. But it does demonstrate that research and development of such technology constitute more than a search for an ethereal holy grail.

The Army missile is designed to be used against incoming warheads, and must be separated from related efforts to develop anti-satellite weapons — the subject of a proposed Soviet treaty ban and a closed Senate session Tuesday. The Soviets have already developed a primitive ASAT system and are working on electronic equipment that could blind U.S. spy satellites and jam their communications. The United States is working on a small ASAT rocket that would be launched from a high-altitude jet fighter.

Administration officials have rejected various proposed treaties to ban testing and deployment of space weapons because the treaty provisions could not be verified. The underlying and reasonable assumption in the U.S. demand for verification is that the Soviets cannot be trusted. Until such verification can be attained, it makes little sense to cede this critical area of research to the Soviets.

St. Petersburg Times

St. Petersburg, FL, June 3, 1984

Scientists say that missile-destroying lasers and similar exotic weapons are technologically feasible, but it doesn't necessarily make them strategically useful. Even if only one warhead got through, Washington would be lost. Or New York. Or St. Petersburg.

The presidential candidates are no longer saying anything much that matters. Back in Congress, however, the most important debate in memory goes on with little attention being paid to it: Can the nuclear arms race be contained before it's too late? Or will it spread into the last frontier, space?

As it debates the military authorization bill, the House has voted three separate restraints on President Reagan's arms projects, hoping to force a reluctant administration to pursue possible nuclear arms control options with the Soviet Union. One cuts MX missile procurement from 40 to 15, to proceed only if Congress decides next spring that the Soviets are unwilling to return in good faith to the arms talks at Geneva. Another forbids the deployment of sea-launched cruise missiles with nuclear warheads unless the Soviets deploy similar weapons. The third, perhaps the most significant, extends an existing congressional moratorium on the testing of any U.S. anti-satellite missiles so long as the Soviets refrain from testing theirs. Mr. Reagan refuses to negotiate an anti-satellite treaty, as the Soviets have offered to do, on the premise that testing violations could not be detected. That's true, perhaps, of isolated violations — but not of the wholesale testing that either side would have to undertake to muster any confidence that its devices would work when they were needed.

BOTH SUPERPOWERS depend on satellites for early warning of ballistic missile attack so any new anti-satellite threat raises the risk of accidental war. That, alone, is reason enough to hold back. But the House vote was also a possible test run for stopping the President's so-called "Star Wars" missile defense system, which envisions space-based nuclear-powered lasers to shoot down Soviet missiles in flight.

Rep. Dante Fascell, D-Fla., chairman of the House Foreign Affairs Committee, warns accurately that it would be "costly, technically unworkable and destabilizing." Sen. Larry Pressler of South Dakota, the Republican chairman of the Senate arms control subcommittee, says the administration is hiding the true costs, which Pressler contends would top $1-*trillion.* Even if the nation's strained economy could afford the money, can the human race afford the risk?

SCIENTISTS SAY that missile-destroying lasers and similar exotic weapons are technologically feasible, but it doesn't necessarily make them strategically useful. Even if only one warhead got through, Washington would be lost. Or New York. Or St. Petersburg. The Soviet Union would simply deploy as many new warheads as needed to overwhelm the system, forcing the U.S. to position more space lasers in response. An even greater danger is that the Kremlin would launch a preemptive strike before the U.S. could acquire an advantage. Meanwhile, the arms race would have spread uncontrollably into space.

In an interview in the current *Common Cause Magazine,* astronomer Carl Sagan has explained the simple logic that apparently escapes Mr. Reagan. "Anything short of an impermeable system," Sagan warned, "is significantly more dangerous than no defense system whatever."

Each X-ray laser, Sagan pointed out, requires the explosion of a one-megaton hydrogen bomb. "Such a device would simultaneously abrogate the outer space treaty, the limited test ban treaty and the anti-ballistic missile treaty."

THAT'S WHY it's so important for Congress to stop the futuristic nonsense before it begins and to insist on the here-and-now alternative of meaningful negotiations. The Senate, of course, may not go along with anything the House has done, and even the House seems willing to indulge the administration in $1.4-billion worth of basic research, $407-million less than the Pentagon asked, into the feasibility of the star wars systems. The problem is, as always, that once the project starts rolling it could become impossible to stop. As weapons physicist Richard Garwin has said, "If all else fails, you call it a bargaining chip. That's the last step before you say you have to keep a program for the jobs it creates."

What's at stake now is not jobs but life. All life. Nothing else in the 1984 campaign matters nearly as much.

the Charleston Gazette

Charleston, WV, May 3, 1984

RESEARCH into high-energy beams — X-ray lasers, free-electron lasers, particle beams, excimer lasers — will go forward by American scientists. But it would be a blunder to make a breakneck rush to install such devices on space platforms before the Soviets do, redoubling the superpowers' race to bankruptcy.

President Reagan is proceeding with his "star wars" defense shield plan, has appointed an Air Force general to direct it, and has budgeted $2 billion for the first year.

Yet members of the Union of Concerned Scientists — including Nobel Prize-winner Hans Bethe and hydrogen bomb contributor Richard Garwin — say flatly that the shield won't work. Beams easily could be foiled by countermeasures. At best, enough missiles would get through to kill millions upon millions of Americans.

Up to $500 billion cost is projected for the shield — a price the Soviets surely would match. Why impoverish both nations, when the futile stampede to install unworkable space lasers could be avoided by a treaty banning orbital weapons?

DAILY NEWS

New York, NY, June 13, 1984

IT WAS BETTER than hitting a bullet with a bullet: The combined speeds of the missile launched from California and the missile that hit it over the Pacific was 10 times the speed of a bullet. It was a magnificent technological achievement. The question is whether it is in our interest, or humanity's, to pursue a new technological arms race against the Soviets.

This isn't President Reagan launching a new program, and damn the consequences. The successful test of the antiballistic missile (ABM) system on Sunday was the fruit of years of work, pushed vigorously under President Carter, who reproved an "inordinate fear of communism" and signed the SALT-2 treaty. Reagan is carrying on his predecessor's policies.

The U.S. and the USSR signed the ABM treaty in 1972. The only sure guarantee against nuclear war is that no one can win it. If one side attacks, it would itself be wiped out by its victim in a horrible spasm. But if either side had a workable ABM system, it might be tempted to launch an attack while the advantage lasted. Both sides agreed that it would be safer if they limited their ABMs to existing systems.

The Soviets have ABMs in Moscow. The U.S. chose to defend the missile sites in North Dakota—but when the system was installed, decided it wouldn't work and scrapped it. The treaty says ABM research can continue, and we now learn it's far advanced on our side. We are well ahead of the Russians, and we can be sure the Kremlin is anxiously reexamining its ABM research. It looks as though the 1972 treaty is dead.

Now the Soviets want negotiations to ban space weapons. These aren't ABMs. They are rockets sent aloft to shoot down enemy satellites. Since military communications, surveillance of enemy territory, and guidance systems for intercontinental missiles all depend on satellites, destroying them would mean victory. That's Reagan's Star Wars proposal.

The USSR wants to stop it before it goes so far, like the new U.S. ABM system, that it can't be stopped, and nuclear war becomes more likely in 10 or 15 years. They also are scared the U.S. may pull ahead again. The U.S. says it would be impossible to verify an "anti-Star Wars" treaty and won't negotiate. That's wrong. We should accept the invitation to talk.

It's an important issue that should be thoroughly examined—Reagan launched the U.S. into Star Wars without any public debate. Negotiating with the Russians is itself very important. They walked out of the two sets of nuclear disarmament talks and have refused to return. Antisatellite weapons negotiations, begun now, besides being good election-year politics, might bring them back to the table next year, and that will be a major objective, whoever wins in November.

ALBUQUERQUE JOURNAL

Albuquerque, NM, June 13, 1984

Next, we suppose, there will be a missile to knock down the missile that's supposed to knock down the first missile. And then, a missile to ...

Reports that the U.S. Army successfully hit an incoming ballistic missile warhead with an experimental missile inject a new element into mutual assured destruction arguments. The success offsets treaties barring anti-missile weapons. The success also may have caused Soviet leader Konstantin Chernenko to issue an urgent call for resumption of negotiations on banning new anti-satellite weapons.

The propaganda value of the Army announcement won't be lost on the Soviet Union, whose scientists no doubt are scrambling to develop a similar weapon and to devise countermeasures to protect their intercontinental ballistic missiles.

The Army missile collided 100 miles above the earth with a dummy warhead fired from thousands of miles away. The two bodies closed at a speed of more than 13,000 miles an hour and the interceptor scored a direct hit.

The new weapon took six years to develop and cost $300 million. That's a small price to pay if the missile can consistently perform as planned. The fourth and final shot closed out the development program.

The Army, by demonstrating the missile's potential, adds to the arms control debate and gives the Soviets reason to seriously consider returning to arms reduction talks. The development also adds to congressional debate about the need for deployment of MX missiles and funding for President Reagan's "Star Wars" weapons development program.

One successful interception does not guarantee reliability, does not ease first strike concerns or pressure for a nuclear weapons freeze. Instead, it adds a destabilizing element to the international arms control debate. It nevertheless offers an element of comfort for many Americans to know that — given necessary production and deployment time — the United States now apparently has the technology to shoot down enemy missiles if the enemy shoots first.

Reagan, Mondale Debate Defense; "Star Wars" Concept Argued

President Ronald Reagan and Walter Mondale debated foreign policy and national defense in their second televised 90-minute encounter of the presidential campaign Oct. 21, 1984. Mondale, the Democratic challenger, hammered hard on the issues of leadership and arms control. The incumbant Republican president accused Mondale of weakness on the defense issue throughout his career and of being opposed to major weapons programs. The President was asked by a panelist about his "Star Wars" proposal, about whether he would share "this very super-sophisticated technology with the Soviet Union" as he had proposed. "Why not?" Reagan replied. " 'Say look,' we could say to the Soviet Union," he said. " 'Here's what we can do. We'll even give it to you, now you will sit down with us and once and for all get rid—all of us—of these nuclear weapons and free mankind from that threat.' I think that would be the greatest use of a defensive weapon." Mondale disagreed "sharply" about sharing technology with the Soviet Union. That was "a total not-starter," he said. "I would not let the Soviet Union get their hands on it at all." There was nothing wrong with the theory of the "Star Wars" concept, Mondale said. "If we could develop a principle that would say both sides could fire all their missiles and no one would get hurt, I suppose it's a good idea." But the fact that the research was still in the primitive stage and commitment to a buildup of antisatellite and space weapons at this time "would bring about an arms race that's very dangerous indeed." Mondale asked, "Why don't we stop this madness now and draw a line and keep the heavens free from war?" Reagan reviewed the concept again, that we could give the Russians "a demonstration, and then say, here's what we can do." Then, if they were willing to join the U.S. "in getting rid of all the nuclear weapons in the world," the U.S. would give the Soviets the defensive weapon itself. Reagan said he did not know what the defensive weapon would be, "but if we can come up with one, I think the world will be better off." Mondale said, "Well, that's what a president's supposed to know."

The Washington Times

Washington, DC, October 29, 1984

Getting desperate, the Mondale campaign is slipping into dishonorable tactics wisely avoided up to now, the worst offender being the so-called Star Wars ad on TV — an ad so well done and so misleading that it may endanger national security. Isn't this a high price to pay for a few extra votes in a losing campaign?

The ad starts in space, with the narrator telling us: "Ronald Reagan is determined to put killer weapons in space. The Soviets will have to match us, and the arms race will rage out of control. Orbiting, aiming, waiting . . ." (Camera shows "War Games"-style computer graphics simulating nuclear war).

The narrator continues: ". . . with a response time so short there will be no time to wake a president. Computers will take control. On Nov. 6 you can take control. No weapons in space on either side. Draw the lines in the heavens with Mondale."

The distortions are awesome. The ad clearly implies, though carefully never says, that the weapons under discussion are nuclear. "Killer weapons . . . orbiting, aiming, waiting"—along with the "War Games" graphics, such words suggest that Ronald Reagan is about to orbit nuclear weapons. In fact, Star Wars weapons are designed to defend *against* nuclear weapons. "Killer weapons"? Space-based defenses would be aimed at unmanned missiles that would be destroyed in the emptiness of space.

The "no time to wake a president" and "computers will take control" lines are equally deceptive. While it may be true that space-based defenses would probably provide for computer-ordered firing — ICBMs move too quickly to allow for meetings of the general staff — these weapons are entirely defensive, aren't nuclear, and aren't aimed at any earth targets. The consequences of an accidental misfire, therefore, would be almost nil.

The widely discussed "high frontier system," for instance, would launch a small non-explosive "spear" into the predicted path of any enemy ICBM. If high frontier fired and it turned out the enemy ICBMs hadn't really been launched, we would be stuck with some extra space litter. Big deal.

As for "no weapons in space on either side" or (in another version of the commerical) "he'll spend a trillion dollars," the first is an absurd dream, the second a statistical mirage.

Space based defenses offer the first glimmering hope in the nuclear age. Using this hopeful development to convince people that Ronald Reagan is a mad bomber may be the greatest disservice Walter Mondale has ever rendered his country. Now that he must know he cannot win, could he not at least withdraw this ad and lose honorably?

THE ATLANTA CONSTITUTION

Atlanta, GA, October 23, 1984

Ronald Reagan isn't the sort to go all soft and mushy when it comes to the Soviet Union, right?

Right. And many Americans, who are a little hazy about the specifics of his original "Star Wars" proposal of April 1983, must have asked themselves after Sunday night's debate: Why in the world would the president of the United States be so eager to share our most sophisticated technology with Moscow — assuming that our scientists can ever solve the infinitely complex questions involved in defending against an aggressor's nuclear-missile attack?

One can reasonably figure Reagan's motive is not so altruistic as it is — to characterize it charitably — pragmatic.

If the United States seemed poised to attain an enormous defensive advantage, so the thinking goes, the pressure might become unbearable on Moscow to initiate a first-strike against us before our system was operative. To counter such a frightful possibility, the argument follows, we should offer a carbon-copy of our Star Wars system to Moscow, which automatically would make nuclear weapons obsolete, and everybody would live happily ever after.

Well-l-l, maybe not.

Walter Mondale spotted one obvious flaw Sunday night: It strains belief to think the hyper-suspicious Soviets would trust Washington to give them what it purported to be a totally foolproof defensive system.

But let's assume that Moscow *does* accept the offer. We can be sure that between now and that momentous date Soviet missile scientists will be under the gun to devise countermeasures to defeat our system. With the complete package in hand, the logical step for them would be to confirm their earlier findings and, possibly, find new chinks in our Star Wars armor.

In sum, all the money and effort we would have expended to develop the system would be at risk — not to mention the fact that this unprecedented transfer of state-of-the-art technology (to name a few: radars, computers, lasers and communications) would compromise our security in myriad other ways.

Sunday night the president made a couple of startling remarks: 1) that he had not even *discussed* ("roundtabled" was the word he used) with the Joint Chiefs of Staff the implications of an eventual and radical shift from mutual assured destruction to strategic defense, and 2) that we could capture the Soviets' attention with a Star Wars "demonstration" — an idea he trotted out for the first time and seemed to embrace more fervently as the debate wore on, much as Mickey Rooney might in a 1940s movie ("I've got it — let's put on a show!")

None of this sounds like the musings of a genuinely thoughtful leader. Oh, sure, the chief executive shouldn't be expected to master all the scientific minutiae involved in such a massive and expensive undertaking, but he should have at least carefully reasoned, step by step, where it would lead this country. President Reagan obviously hasn't.

THE BILLINGS GAZETTE

Billings, MT, October 1, 1984

Picky, picky, picky.

The American public is just too darn picky.

Mention the military and everybody becomes an expert.

Nobody wants the MX in his backyard.

Everybody criticises the B-1 bomber.

The stealth bomber, the one "invisible" to radar, languishes somewhere out of sight.

White trains laden with nuclear warheads are picketed as they carry their burdens to nuclear submarine nests on the west coast.

Picky, picky, picky.

The public seems bent on taking away all the generals' toys, and it just isn't fair.

It isn't any fun being a general if you don't have the latest gimmickry.

President Reagan recognizes that, so he has suggested a $26 billion "Star Wars" initiative.

And what do we hear?

More gripes.

War shouldn't be carried into man's new frontier.

We should leave at least one aspect of our lives untouched by the weapons of death and destruction.

Man should be able to cast his eyes toward the heavens without seeing the articles of the apocalypse.

Picky, picky, picky.

But give the proposal a moment of thought.

We're up to our gills in nuclear weapons down here on earth.

The Soviet Union and the United States know that if either side pushes the button, both will go up in smoke.

The fact that the rest of the world goes, too, is not important to the equation.

But if we move our cold war into space, and we certainly will, one side may be able to pick off the other side's ballistic missiles with lasers or particle beams.

That would leave the winner free to bomb the world into oblivion.

~~Anyone should be able to see the benefit of that.~~

There are other advantages, too.

The U.S. Air Force is currently building a "secret space force of as many as 50 astronauts."

That seems like a lot until you compare it with the number of people now under arms in the Soviet Union and the United States.

The point is that Star Wars are economical in regard to the number of lives at stake.

Most of the fighting could be conducted by generals standing around video screens pushing buttons to score hits on the other side's satellites.

Annual tournaments could be staged with the winners decided by a point system. The losers would have to buy the beer, or borscht.

That's cheap entertainment compared to what we pay for now.

There are a lot of reasons why the American public should be happy to pay $26 billion.

But in the days ahead, we'll likely hear more criticism of the proposal than praise.

There are always a few soreheads who can't seem to do anything but complain.

You just can't please them.

Picky, picky, picky.

Arkansas Gazette.

Little Rock, AR, November 13, 1984

Aside from the cost — up to $1 trillion — the fallacy in the proposed plan for a Star Wars defense is that it would offer false hope of security against massive nuclear destruction. Even its most ardent supporters concede that an "umbrella" against Soviet Intercontinental Ballistic Missiles would not block all of them, conceding that only about 85 per cent of these warheads could be stopped short of their American targets.

Take a more conservative figure. Say that the Star Wars defense allowed not 15 per cent but instead only 10 per cent of the land-based Soviet ICBMs to penetrate and explode on or over the United States. If the Soviets fired, say, only 20 per cent of their land-based ICBM force in a first attack, 116 ICBMs would penetrate the "umbrella" and reach their targets.

How much firepower this would represent depends upon the size of the warheads on the ICBMs. The Soviets generally are thought to have larger warheads mounted on their land-based ICBMs than the United States, but for the sake of this exercise, assume that these incoming warheads equalled the force of one million tons of TNT. (The United States' Minuteman II carries a warhead of 1.2 million tons.) A Soviet attack of this size would unload a nuclear force on the United States equal in tonnage to 7,250 bombs of the size that destroyed Hiroshima.

It was at the dawn of the nuclear age, about the time of Hiroshima and Nagasaki, that Albert Einstein admonished the world to think differently about war in the nuclear age, saying on one occasion: "The unleashed power of the atom has changed everything except our way of thinking. Thus we are drifting toward a catastrophe beyond conception. We shall require a substantially new manner of thinking if mankind is to survive." The proposal for a Star Wars defense is another illustration that political thinking hasn't changed much in the interim despite what Einstein said.

The testimony that Star Wars would not offer the United States — or the Soviet Union or the world, for that matter — a defense against nuclear firepower is strong and compelling, leading the Center for Defense Information, which is headed by high-ranking retired military officers and strategic planners, to this stark conclusion: "There will be no defense in this century, if ever. It makes no difference which side strikes first and which side retaliates. Both nations will be destroyed, utterly and completely. We are *mutually* inferior because there is no superiority in mutual destruction."

A group of distinguished American scientists also has raised its voice in recent weeks in opposition to a Star Wars defense. One of its spokesmen, Carl Sagan, the astronomer, says flatly that "It won't work, it's easily defeated by countermeasures, it would be ruinously expensive" and it could encourage the Soviets to launch a pre-emptive nuclear attack.

How expensive? The non-profit Council on Economic Priorities has estimated the Star Wars cost from $400 billion to $800 billion, but Sagan estimates that the total would be $1 trillion. Research already undertaken by the Pentagon will cost $26 billion over the next five years. It is spending $1.7 billion in the current fiscal year.

If all this could prevent a nuclear attack, saving humankind in the process, the price would be cheap, but there is another dimension to this Star Wars plan that goes largely unmentioned. The Soviet Union and the United States have massive nuclear power at their instant call that is not found in their land-based ICBM forces. That is, their arsenals are so large and so diverse and so scattered that both nations could destroy the other, if such were possible, several times over even without using their land-based ICBMs, or with perfect Star War defensive systems in place.

For example, both nations are shifting some of their emphasis to new generations of cruise missiles, which may be fired from positions on land, at sea and in the air. They hug the terrain a short distance above the earth's surface until they reach their pre-determined target. The United States alone has 28 different delivery systems bearing 26 different types of nuclear warheads. Clearly enough if only a few of them reached their targets the destructive force would surpass rational human imagination.

Any discussion of nuclear weaponry assumes a surrealistic quality, in which strategic planners endeavor to give form to nightmarish thoughts, but the essential facts about all this are stunningly plain and simple. Both the United States and the Soviet Union possess such massive nuclear firepower that they can easily overcome or circumvent a defense. To pretend otherwise, to concentrate resources of either on a space-age defensive net, merely would give false hopes while diverting attention from the real threats to human survival. The United States and the Soviet Union already are captive of their own nuclear firepower, for each is threatened as much by its own arsenal as it is by the arsenal of its enemy.

Nuclear war is a misnomer, if it should imply that one side would defeat the other, and a new order would rise from the rubble. No phoenix rises in a nuclear world of over 50,000 warheads. Life itself perishes in a "war" of this nature. Preventing it is everyone's priority. A Star Wars defensive system only distracts from prevention.

FORT WORTH STAR-TELEGRAM

Fort Worth, TX, November 1, 1984

Han Solo, who has successfully fought "Star Wars" in three epic motion pictures, would be dismayed by the Council on Economic Priorities' report on the cost of a real-life space-based defense system against nuclear attack. The council, hardly an unbiased source since it opposes the Reagan administration defense buildup and espouses an emphasis on social programs, may well be right when it says the space age defenses could cost $800 billion.

The council may even be correct that, at any price, the system wouldn't be effective. Certainly there is precedent for costly programs that don't work, though not that costly.

But when President Reagan first advanced the idea of such a space missile defense program more than 19 months ago, he was not calling for a crash program. He has sought research into the program, with the thought that in 15 or 20 years such a defense system might replace the current "defense by deterrence" philosophy and might be safer because it would not rely on offensive weapons.

With that in mind, a panel of scientists has been formed to conduct research into the technical feasibility of such a defense system. Congress compromised, before its 1984 adjournment, on a $1.4 billion appropriation for this effort — not to build such a system, but to begin the research.

No one has claimed that any "Star Wars" defense system would be a model of speed or economy. Almost a year ago, Pentagon research head Richard D. LeLauer testified that it would take two decades to develop any kind of nationwide "shield" against nuclear attack and that the cost would be "staggering."

In short, such a defense might not work. It might not be practical. And it might be too costly.

But the Soviet Union, our prime adversary and the only other nation capable of devising and employing such an effort in space, is known to be working on "Star Wars" technology of its own.

The Soviets, too, may find it less than a workable idea.

While there are reasons to go slowly toward such a program, and to resist the expenditure of hundreds of billions of dollars, there are also reasons to at least go ahead with the research. In a dangerous world, which could expand into a dangerous cosmos with or without us, the United States must know all it can about the possibilities, the problems and the potential for space-oriented nuclear defense.

The research should go on, even if it costs a few billion dollars over the next few years and even if it only serves to determine that such a system isn't practical.

Defense in space is an area where what we don't know could hurt us, while what we learn might well save us from something worse.

Boston Sunday Globe

Boston, MA, October 28, 1984

The exchange over the Star Wars space-based missile-defense system was a curious part of last Sunday's debate between President Reagan and Walter Mondale.

Reagan waxed visionary: He described a weapons-defense system, shared with the Soviet Union, that would allow both countries "once and for all to get rid – all of us – of these nuclear weapons. And free mankind from that threat."

Mondale's response was blunt: Sharing "the most advanced, the most dangerous, the most important technology . . . with the Soviet Union is, in my opinion, a total non starter. I would not let the Soviet Union get their hands on it at all."

Regrettably, Mondale did not go further and expose for the benefit of millions of Americans the technologically loony and militarily dangerous fraud that is Star Wars.

□

The notion of replacing the escalation of ever more destructive weapons systems with the acquisition of a perfect defense system is a beguiling one.

As advanced by President Reagan in March 1983, the system would involve placing into space an antimissile weapon that would have the capability of destroying any missiles launched in attack against the United States.

One billion dollars was spent on research and development last year, and $1.4 billion is budgeted for the current fiscal year. The Defense Department has estimated that it could take 30 years and more than $400 billion to perfect the system.

As outlined by the Union of Concerned Scientists in its recently published primer, "The Fallacy of Star Wars," the task is to design a foolproof defense that can shoot down 90 percent of any nuclear-armed missiles in the first 180 seconds after launch. Once past that initial booster stage, a missile will release a number of warheads – its MIRVs – as well as a large number of decoys, vastly complicating the task of defense.

The technological feats necessary to accomplish the task are incredible, involving orbiting mirrors 20 times the diameter of the one at Mount Palomar, perfecting laser beams that can focus on a fast-flying missile from hundreds of miles away for long enough to destroy it, and designing and constructing submarines capable of popping the laser weapons into firing range.

Even if this technology can be perfected, the countermeasure technology is already well developed – space mines that could be planted in the path of an orbiting laser platform – and could be operational long before the United States could perform the Star Wars demonstration that Reagan promised during the debate.

□

Twice during last Sunday's debate, Mondale referred to Strobe Talbott's "Deady Gambits" – the gripping account of arms control under the Reagan Administration. The record of the past four years which Talbott details – the stalling, the sidetracking, the outright rejection of Soviet overtures – must raise serious questions about whether the Reagan Administration even wants a mutual control of nuclear arms.

Reagan's advocacy of Star Wars takes that grim scenario a step further. Rather than promising a "mutual deterrence," the curious technology of Star Wars actually increases the pressures for a preemptive first strike. The system's only possible value is against a retaliatory attack, to shoot down the few remaining Soviet missiles left after an American first strike. Realizing that, Soviet policy-makers will have to consider the possibility that the United States is plotting a first strike and thus begin to consider whether they should not launch their own preemptive strike.

In the week which remains in the presidential campaign – while the Star Wars discussion still lingers in the minds of American voters – Walter Mondale should make clear the fraudulent nature of President Reagan's proposal and show it up as not just loony, but infinitely dangerous.

THE WALL STREET JOURNAL.

New York, NY, October 24, 1984

To commit this nation to a buildup of anti-satellite and space weapons at this time in their crude state would bring about an arms race that's very dangerous indeed. One final point: The most dangerous aspect of this proposal is for the first time we would delegate to computers the decision as to whether to start a war. That's dead wrong. There wouldn't be time for a president to decide. It would be decided by these remote computers. It might be an oil fire, it might be a jet exhaust, the computer might decide it's a missile and off we go. Why don't we stop this madness now and draw a line and keep the heavens free from war?

I remember the night before I became vice president. I was given the briefing and told that any time, night or day, I might be called upon to make the most fateful decision on earth–whether to fire these atomic weapons that could destroy the human species.

–Walter Mondale, in his debate with President Reagan.

At this point the American people needed a few basic facts about the Strategic Defense Initiative, or the "Star Wars" missile defense. The great communicator didn't give them much help. Someone ought to. So let's start with one little detail: None of the weapons being seriously studied would be armed with nuclear warheads. To repeat: No nuclear warheads.

The object of the proposed defense is to destroy nuclear weapons headed toward the U.S. Many of the systems being studied would be totally incapable of doing any damage to anything in the Soviet Union. Put aside the question of whether a computer is more likely to make an error than a president being awakened in the middle of the night. If the computer makes an error with Star Wars, you may get a meteor shower. At worst, assuming successful development of the most visionary proposals, you would get a laser beam striking somewhere in Russia, and the systems could be designed to avoid even that. Under the present offense-only strategic posture, if the president makes an error the result might or might not be the end of the species but would clearly be a catastrophe of unimaginable proportions.

Mr. Mondale surely knows all this. At least, he is running around the country bragging about his superior mastery of the details of strategic questions. Of course neither his debate statements nor his even more strident TV ads ever explicitly utter the implied falsehood–that the Star Wars proposal consists of stationing nuclear weapons in space and letting computers decide whether to explode them. Yet this is precisely the impression both are calculated to leave, and this is sheer demagogy.

The danger here is not to the Reagan campaign. The president can take care of himself perfectly well; we only wish he could do the same for the issues involved. The next president and next Congress will have a historic decision to make about strategic posture, and in this election the American voters have the chance to make their contribution. They deserve better from both candidates.

The technological facts are these: With the improvement in missile accuracies, and particularly with the ongoing Soviet deployment of more and more missiles, a first nuclear strike becomes militarily more attractive. By striking first, the Soviets could destroy most of our land-based retaliatory missiles and probably also most of our nuclear-armed aircraft. Missiles aboard submarines would be far more difficult to attack, but will not be invulnerable forever. A president conceivably might be faced with deciding to retaliate with a vastly crippled force, knowing that the Soviets retain more missiles to launch against civilian targets if they so choose.

At the same time, technological advances make it possible to blunt the first strike by deploying missile defenses and without using nuclear weapons. Most concepts rely on a tiered defense. Space stations that detect missiles and throw steel nets or showers of rocks in their way (or in more advanced versions, destroy them with lasers), ground-guided missiles with super accuracies that can actually collide with missiles as they reenter, Swarmjet systems near our missile silos that shoot a shotgun blast of rockets in the way of a descending warhead.

Thanks to the advance of computers and information processing generally, these systems have now become feasible, though of course further testing and development are required. Missile defenses will not soon or perhaps ever be leakproof; no weapon has ever been either an irresistible force or an immovable object. But even a limited strategic defense could do much to make any first strike far more uncertain and less attractive, and thus help keep any president from being awakened in the middle of the night and asked to decide whether to blow up the world.

Why is this a historic decision, then, or even a particularly close one? Because of a strategic doctrine called mutual assured destruction, which holds that the way to avoid war is to make it unthinkably horrible. Thus offensive missiles are good, especially if not very accurate and targeted at masses of civilians. And defense is immoral. Incredible as it may seem to the man in the street or for that matter to many military officers, MAD has for years been the dominant strategic thinking among defense intellectuals. By now it is deeply embedded in the arms-control treaties we've negotiated with the Soviets.

The real meaning behind Mr. Mondale's statements and commercials is that he prefers to rely on the MAD philosophy; otherwise we will have an "arms race." While President Reagan's position could be more cogent, clearly it does point away from MAD. There is a real difference here; we hope the voters will understand it well enough to decide which approach promotes peace and which increases the chance of war.

The Idaho STATESMAN

Boise, ID, October 26, 1984

Forget, for a minute, the rest of Sunday's debate between President Reagan and Walter Mondale. The crucial dialogue, the issue that will determine the future of the arms race and of the earth itself, was sliced clearly and unmistakably for the American voters.

The president wants to move the arms race into space by spending $26 billion in the next five years to research a "Star Wars" defense of space stations supposedly capable of blasting Russian nuclear missiles from the skies. The improbable idea is that by constructing a network of space defenses, we can make ourselves immune to nuclear attack, end the arms race and live happily ever after.

If only it were so.

The grim truth is as Mr. Mondale said. The nuclear genie has been unbottled since Hiroshima. There is no forseeable way to cap it with technology Deploying laser-equipped satellites into the heavens will force the Soviets not only to do likewise, but to find ways to disable our system. It wouldn't be long before the whole thing leapfrogged out of control. Besides, a "Star Wars" system would be effective against ICBMs. It would not halt cruise missles, backfire or stealth bombers. Nuclear war still would be possible.

The "Star Wars" plan also is destabilizing and easily could bring on Armageddon, not prevent it. If the Russians have no comparable system to protect themselves, they face the choice of a preemptive strike before our system is in place or living under our nuclear shadow without a retaliatory capability. While we might wish to have that clear-cut nuclear superiority, we should understand the Soviets' fears. We would never tolerate a reverse situation.

The president's offer to mitigate this problem by giving such a system to the Russians is untenable because we also would be giving away vital defense secrets. If Mr. Mondale had made the giveaway proposal, he would have been laughed off the debate stage Sunday.

The president, by his failure to negotiate a nuclear arms agreement and by his commitment to an arms buildup to force the Russians to the negotiating table, wants to dictate treaty terms from a position of strength. We wouldn't allow the Soviets to do that, and neither will they allow us to do that.

Clearly, the only answer to the nuclear nightmare is a more realistic solution of negotiation. And for negotiations to be successful, both countries will have to reach out in a spirit of trust, not one-upsmanship.

U.S. to Discuss "Star Wars"; Weinberger Backs S.D.I

An unidentified Reagan Administration official December 20, 1984 said that United States plans for space-based Strategic Defense Initiative (SDI) defenses against nuclear attack, widely know as "Star Wars", would be "on the table" at the U.S.-Soviet talks scheduled for January, 1985. The official said that since the defenses did not yet exist, "it is intellectually obscure" to speculate over what the U.S. could gain in return for canceling them. However, he added, the program would be the subject of bargaining. "The research program and our intentions for it have to be on the table an a matter for discussion and negotiation between us [the U.S. and U.S.S.R.]," the official said. "And they surely would be," he added.

Defense Secretary Caspar Weinberger Dec. 19 attempted to clarify the Administration's position on its SDI plan. Weinberger defended the initiative, saying that the current policy of nuclear deterrence hinging on "mutual terror" condemned the nation "to a future in which our safety is based only on the threat of avenging aggression." He moved to assure nervous U.S. allies that the U.S. would not create a protective system at the expense of its overseas commitments. "The security of the U.S. is inseparable from the security of Western Europe," Weinberger said. The news conference came a few days after Mikhail Gorbachev, the second secretary of the Soviet Communist Party, had said in London that U.S. plans for space weapons were a potential barrier to nuclear arms reductions by the U.S.S.R.

The goal of the Star Wars program had been scaled back to protecting only U.S. land-based ballistic missiles, according to the New York Times Dec. 23. George A. Keyworth 2d, President Reagan's science adviser, told the Times that the original goal of SDI—protecting the entire U.S. population—had been set aside for the present. The Times story drew denials by the Administration. President Reagan said that the space-defense program "isn't going to protect missiles." Defense Secretary Weinberger admitted that there might be a "transition phase" in the program, but maintained that its goal was not to "protect any particular target."

WORCESTER TELEGRAM.
Worcester, MA, December 24, 1984

The campaign against President Reagan's Strategic Defense Initiative is getting frenzied. Critics ridicule it as "Star Wars." The big opinion-makers of the media are in full cry against it. Democratic congressmen are speaking out. Four former government officials assert that the thing won't work and, if it does, it might upset the delicate balance of peace.

But all of that doesn't hold a candle to the frantic cries of the "Union of Concerned Scientists." A few months ago, that curious group said that the space defense envisaged by Reagan would require "thousands of satellites," each one weighing 40,000 tons. The system would take 60 percent of the total power output of the United States, said UCS.

That moved Robert Jastrow, a real scientist, to do some computations. He calculates that the job could be done with between 100 and 200 satellites. He later found the Livermore Laboratory in California has run a computer analysis showing that the actual number would be somewhere between 45 and 90. His article is in the current issue of Commentary. When the Union of Concerned Scientists looked at Jastrow's calculations, they lowered their figures, first to 800 then to 300.

The Union of Concerned Scientists is a strange hybrid. Stanley Rothman and Robert Lichter are writers who tried to find out who is behind the organization. They found that it doesn't include many scientists — maybe 200 out of the 130,000 listed in American Men and Women of Science. Anyone can join; all it takes is a contribution of $15. The organization is an anti-nuclear sounding board for all sorts of fringe groups. It seldom endorses any measures to enhance the defense and security of the United States.

In the coming storm over "Star Wars," there will be all kinds of claims and counter-claims. Maybe it won't work. But just what is wrong with the idea of trying to defend America by putting satellites in space capable of shooting down hostile enemy missiles?

The Washington Times
Washington, DC, December 25, 1984

Richard Lugar, the Indiana senator who was elected by his Republican peers to head the crucial Foreign Relations Committee, got off on a good foot the other day by urging President Reagan to stick to his guns on "Star Wars." Mr. Lugar's encouragement comes at a particularly critical time, with some members of the next Congress already whetting the budget ax for defense cuts.

"Star Wars" backers in Congress, who are sure to have their hands full fighting off those who want to slash research and development expenditures for the program, generally fall into two groups: those who see it as a plausible means of defending the United States from a ballistic missile attack, and those who grudgingly back it as a bargaining chip at some future arms-control negotiations. Mr. Lugar, it is comforting to note, counts himself among the former.

It is difficult to imagine a nation as rich in know-how as the United States being unable to refine the technology necessary to mount a credible ballistic-missile defense. Indeed, a significant number of physicists, engineers, and other scientists believe it is altogether possible. For the United States to refrain from discovering whether a "Star Wars" technology were militarily and financially feasible would be folly, especially since it would effectively terminate the nuclear blackmail with which each superpower holds the other hostage.

Similarly, it would make no sense to proceed with "Star Wars" after having announced to the world that it was nothing more than a bargaining chip. That is not to say, however, that given the proper quid pro quo from the Soviets the administration should refuse to lay off "Star Wars" against some future Kremlin offer. As Mr. Reagan has repeatedly emphasized, all things are possible if the Soviets are willing to bargain in good faith.

For the time being, "Star Wars" will need every ally it can muster in the Senate and House, in addition to the president's not inconsequential persuasive abilities. It is a good omen indeed that Mr. Lugar has committed himself — along with, one hopes, the power and prestige of his committee — to completing the research into a program that very possibly could bring an end to the nuclear arms race.

THE PLAIN DEALER

Cleveland, OH, December 27, 1984

"2001, A Space Odyssey." A baboon flings a high-velocity bone past a huge black Monolith. The bone turns into a spaceship. Is that gigantic allegory: (a) a heavy nod at the advent of weapons, or; (b) enigmatic and confusing and so who cares, or; (c) an astounding display of prescience with regard to the Strategic Defense Initiative (SDI).

Answer (c) looks good to us. Indeed, the baboon's bone is a highly suitable metaphor for the SDI (the Star Wars program) currently being defined and redefined by the administration.

Be it killer bone or killer satellite, SDI is a dubious idea, made more dubious by the vacillation of its proponents. One day SDI is a "bargaining chip," the next it is non-negotiable. One day it is an abbreviated project to defend the nation's land-based ICBM arsenal; then it is a national defense structure; then a semi-global one extending all the way to Western Europe. It is possible, impossible; short-term, long-term, long-term with limited short-term applications—all depending on which Washington source is tossing the bone. Compared to this, the Monolith reads like an open book.

The problem is that, regardless of how it is defined, SDI is still fanciful, destabilizing and dubious. If it is used as a bargaining chip, then it risks the ambivalence of "dual-trackism"—that is: It might be attached to negotiating conditions so vague or so unrealistic that they cannot be fulfilled, thus making the system's ultimate dispensation irresolute. Even if it is not a bargaining chip, its high-tech implausibility (Lasers? Particle beams? Where's the X-ray vision?) makes it an enormous political and scientific gamble.

If SDI is an abbreviated system designed to protect the nation's ICBM fields, then it perpetuates the risk of escalation to global thermonuclear war. The system cannot be utterly impenetrable, so any strike will have at least limited success, thereby requiring counter-strike. Worse though, SDI could therefore destroy international confidence in the nuclear balance, conceivably prompting a pre-emptive attack.

What's more, both a limited system and a national one will spark new Atlantic Alliance fears of "decoupling"—the theory that the United States, feeling secure, might not honor its commitment to the security of Europe.

Lastly, if an SDI defense umbrella is extended to Europe, the Soviets have no choice but to enter the space-war race full bore. That ultimately could mean that nuclear conflict would not only end with space weapons but, hideously, begin with them.

All of which is made even more parlous by the project's legal rationales. President Reagan says SDI abrogates no treaty because it is only a research project. British Prime Minister Margaret Thatcher agrees, saying, essentially, that SDI should not be considered a component of the arms race until it is a reality.

What the president and the prime minister ignore is that, after sinking billions of dollars into the project, it will not be abandoned. Rather, the project will continue unabated. Thus accepting the legitimacy of current research is tantamount to accepting the system's inevitability.

There are plenty of reasons to oppose the SDI project apart from the administration's inability to define it. It could aggravate the arms race, cost incalculable sums, lower the threshold of nuclear war, weaken the Atlantic Alliance. It might also delay or thwart shifts to strategies of de-escalation and away from the huge, land-based globe-killers.

But those are future contentions, future baboon bones. For the moment, the most compelling argument against SDI is that nobody seems to know what to do with it. Not even in theory.

THE SUN

Baltimore, MD, November 26, 1984

Darth Vader and the evil empire he represented in the popular "Star Wars" films had handy-dandy ray guns that could blast not just missiles but whole earth-sized planets out of the sky. Real life is different. Real ray guns, whether they project laser or particle beams, operate according to the principles of physics. That means that the immense output of energy needed to destroy missiles with ray guns would require an even greater input of energy to make the ray guns work.

An estimate of the amount of energy needed is provided in a *Scientific American* article written by the eminent nuclear physicist Hans Bethe and three other U.S. scientists with solid credentials. A system to defend against the 1,400 Soviet inter-continental ballistic missiles would, they say, have a power requirement of "more than 60 percent of the electrical generating capacity of the entire U.S. . . . [T]he outlay for the power supply alone would exceed $100 billion."

The above may not be the single most compelling argument against President Reagan's Strategic Defense Initiative, otherwise known as "Star Wars." Just as compelling is the strategic argument. The four scientists say that for the United States to begin serious work on the SDI would be profoundly dangerous to ourselves and the world. First, to undertake such a project would be in clear violation of an existing treaty against anti-ballistic missiles (ABMs). Second, one of the best counters to a Star Wars defense would be large increases in numbers of intercontinental and other ballistic missiles (and dummy missiles designed to confound the defensive systems). Thus, they say, an arms race in space would greatly accelerate the existing arms race between the superpowers.

A defense system effective against a large-scale missile attack is still far beyond human technological capability for dozens of reasons beyond the energy requirements. Problems range from inability to protect delicate equipment in orbit to the ease with which missiles could be hardened against — or built to confound — the proposed defensive systems. The scientific community is in broad agreement that the hope for a workable defense is remote.

But even if Star Wars weapons would not be very effective against missiles, they would be effective against an adversary's *satellites*, including military communications satellites in high geosynchronous orbit. For this reason, and because they might be *partially* effective against missiles, they would generate counter-efforts to checkmate this destablizing technology. Even though the administration is not yet committed to deployment of a such a military system, it is involved in a research effort of such magnitude that a threshold is foreseeable. The Soviets also are working on particle-beam space weapons, and both sides already have weapons that can take out low-orbit satellites. A U.S.-Soviet agreement to halt work on space weapons is needed soon.

Los Angeles, CA, December 26, 1984

If the Reagan administration ever believed its own rhetoric about "Star Wars" space defenses against a nuclear attack on the U.S. that belief seems to be fading. According to a New York Times News Service story in the Herald, scientists at work on the system have concluded, at least for now, that the construction of an impenetrable shield is impossible — and that the aim of "Star Wars" therefore should be scaled back from protecting America's cities to protecting its land-based nuclear missiles.

There are some political benefits for the administration in this. Less grandiose defenses are sure to be cheaper defenses, and the White House may thus appear to be meeting critics of its defense-spending policies halfway. But even if a space-based defense could work, and even if the U.S. can afford one — two big ifs — using it to protect only land-based missiles would create new problems.

Most important, it might require the abrogation of the anti-ballistic missile pact of 1972, which sharply limited the ways in which the U.S. and USSR could protect their ICBMs. If the "Star Wars" system couldn't destroy enemy warheads during their flight from the Soviet Union, it would have to do so when they re-enter the atmosphere, which probably would require dusting off the ABM technology we mothballed a decade ago.

In other words, the new, limited "Star Wars" would not shield Americans against nuclear attack; it would simply give new life to deterrence by making it harder for the Soviets to destroy our ICBMs in their silos. Increased deterrence is not bad in itself, of course, but we wonder in this case if we're seeing another attempt to sell a recaciltrant Congress on the plan to deploy MX missiles in highly vulnerable Minuteman silos. If so, rather than forcing the Soviets to give up all thought of attack, it would almost certainly prompt them to deploy more and bigger, silo-busting warheads. Says former Defense Secretary James Schlesinger: "One might be willing to upset the ABM treaty to protect every man, woman and child in North America. But to upset it for possible improvements in deterrence is a much more difficult decision."

The appeal of "Star Wars" is that it promised, accurately or otherwise, to make nuclear attack impossible. A modernized ABM system would simply make any attack more overwhelming. Either way you look at it, it's a dangerous gamble — at long odds. ■

ST. LOUIS POST-DISPATCH

St. Louis, MO,
December 24, 1984

In a speech to foreign journalists in Washington, Defense Secretary Caspar Weinberger has argued that unless this nation goes ahead with the development of a full-scale space-based "Star Wars" missile defense system and strategy, the U.S. will be condemning itself "to a future in which our safety is based only on the threat of avenging aggression." To Mr. Weinberger, the policy of mutual assured destruction is a "disproven strategic dogma." He dismissed the argument that a Star Wars system is technologically unworkable by saying, "History is filled with flat predictions about the impossibility of technical achievements that we have taken for granted."

There is something to that argument, but at its heart it is fundamentally flawed and historically wrong. First, mutual assured destruction — or the balance of terror, as it is also called — is not disproven dogma; it has worked. Even if a Star Wars anti-missile system can destroy 99 percent of all incoming Soviet warheads (a wildly optimistic percentage), dozens of nuclear warheads would still get through to their targets; more than enough to kill 100 million people. Nuclear terror will remain.

Mr. Weinberger also takes an ahistorical view of weapon development. Yes, there are always new weapons coming down the pike that make established weapons obsolete — such as steam-powered ironclads outclassing wooden sailing warships. But arms are never developed in a vacuum. Any U.S. Star Wars system is certain to be met by a counter-Soviet system (or more likely, systems) designed to destroy it. There is no such thing as an ultimate weapon.

Mr. Weinberger is trapped in the fallacy of the ultimate move: If we can just do "X" then all will be right. It doesn't work that way in business — or in nuclear warfare.

BUFFALO EVENING NEWS

Buffalo, NY, December 5, 1984

FOUR FORMER high-level officials in both Republican and Democratic administrations have sounded a timely warning about President Reagan's plan for a space-based nuclear defense — the so-called "Star Wars" program.

Writing in Foreign Affairs magazine, the four former officials — Robert McNamara, George Kennan, McGeorge Bundy and Gerard Smith — emphasized that progress in arms control negotiations with the Soviet Union is unlikely if Mr. Reagan proceeds with the development of a Star Wars system.

Mr. Reagan proposed the Star Wars defense in 1983, saying that it held hope for rendering all nuclear weapons "impotent and obsolete." But a distinguished panel of scientists concluded after a year's study that such a defense system, based on the use of laser beams and particle beams in space, was technologically "unattainable."

The four former officials have added to that argument the certainty, as they put it, that the Star Wars defense would lead to "a large-scale expansion of both offensive and defensive systems on both sides." Such an expansion would come because, as any anti-ballistic missile system develops the capability of filtering out incoming missiles, an adversary would seek to build more missiles in order to maintain its offensive power. And this would obviously work against the goal of reducing the number of offensive missiles.

These considerations lend support to the recommendation of the four former officials that the Star Wars program be cut back and confined to research.

Mr. McNamara, a former defense secretary, and Mr. Kennan, a former ambassador to Moscow, also took part recently in a study by a group of many world figures, including former Canadian Prime Minister Pierre Trudeau, former British Prime Minister James Callaghan and former West German Chancellor Helmut Schmidt. Written before the recent agreement for the resumption of nuclear talks, their report stressed the importance of continued negotiations to achieve a "stable balance" in nuclear weapons.

The report noted the dangers in NATO's reliance on nuclear weapons for the defense of Western Europe and urged the strengthening of conventional forces. It urged that NATO's battlefield nuclear weapons be pulled back from the front lines so that they would not be overrun in a conventional attack.

The report did not call for "no first use" of nuclear weapons, since, until the West's conventional forces are strengthened, this would only invite attack. But it did urge "no early use," recommending the pullback of nuclear weapons that would make them available as a last resort.

The Dallas Morning News

Dallas, TX, December 17, 1984

MEET another Democratic congressman who during the '80s has learned nothing and forgotten nothing — Joseph P. Addabbo of New York. He wouldn't be a problem if he weren't chairman of the House military-appropriations subcommittee and an ardent foe of strategic defense, known to its detractors as "Star Wars."

Addabbo, in an interview with the *New York Times*, promised the other day to freeze and maybe even cut funds for strategic defense. The Soviet military will no doubt be delighted to learn of his resolve.

Walter Mondale tried and failed to make a campaign issue of "Star Wars," the Reagan administration's plan for a space-based defense system that would blunt a Soviet nuclear attack on the United States by destroying incoming missiles, chiefly with lasers.

Mondale called on his countrymen to "draw a line and keep the heavens free from war." Nice cosmic touch there — except that the purpose of stategic defense is to keep the earth free of nuclear war by making such a war unwageable and unwinnable.

If strategic defense were intended as a tool of U.S. conquest and domination, President Reagan would hardly have suggested that his successor could give the new technology to the Soviets. Having, in other words, neutralized their nuclear missiles, we would proceed to neutralize our own.

Reagan's successor, according to Reagan, then could say to the Soviets: "I am willing to do away with all my missiles. You do away with all yours."

This achievement wouldn't of course turn the Soviets into nice guys: Leninist lions ready to lie down with capitalist lambs. The Kremlin would go on pushing its objectives, but without its nuclear ace in the hole.

It is easy to see why the Soviet Union, preeminent in nuclear weaponry, strenuously opposes strategic defense. Why American congressmen, and presidential candidates, should oppose it is less clear, except that cobwebbed cliches are hard to sweep from the mind. Such a cliche is the notion that new weapons always endanger rather than enhance prospects for peace.

Strategic defense doesn't cost much — probably no more than $5 billion in a nearly $1 trillion federal budget and about $25 billion over five years. It should be interesting to hear Addabbo and his friends explain why that's too much to pay for security from nuclear destruction.

AKRON BEACON JOURNAL

Akron, OH, November 12, 1984

NOW THAT THE election is over, the Arms Control and Disarmament Agency should release a study it has been holding that concerns antisatellite weapons.

The study, by a Harvard researcher on contract from the agency, challenges the Reagan administration position on controlling space weapons by concluding: "Arms control for antisatellite weapons can support the United States security interests."

The report urges work toward arms agreements on space arms technology now, rather than after further development of such technology because "arms-control agreements, historically, have been less difficult to apply as preventive measures than as remedial measures."

The world is seeing the truth of that statement in the present problems the United States and Russia are having in limiting their stockpiles of nuclear weapons. Once that arms race is exported into space — through President Reagan's *Star Wars* missile defense system or some other weapon — control will be more difficult, if not impossible.

The arms control study was limited to antisatellite arms — weapons that can knock out the satellites we depend on for weather and defense surveillance and other uses. Control of antisatellite weapons can be of great benefit to the United States because this nation depends so heavily on those systems. The discussion also should include other space weapons systems, however, since very similar technology is involved.

Officials at the arms-control agency said the report was being held up until after the election, not because it disagrees with the President, but because the agency did not want the study to become a "political football." Fine. Now the election is over, and this report can add to the important debate that must take place soon on the issue of arms in space.

THE INDIANAPOLIS STAR

Indianapolis, IN, December 30, 1984

"Star Wars" is a derisive label fastened by hostile propagandists onto President Ronald Reagan's proposal to develop a defense technology capable of destroying incoming enemy nuclear-warhead missiles.

The same idea had been proposed by the famed nuclear scientist, Dr. Edward Teller, "father of the hydrogen bomb," and Lt. Gen. Daniel O. Graham, U.S. Army (ret.), leading proponent of the High Frontier Program, and others.

The anti-defense lobby immediately attacked the president's proposal saying that (1) it should not be tried and that (2) it would not work, a pair of odd charges considering that the president had not said specifically what "it" was, his proposal being of a general nature.

Although technology that would contribute to a radical space defense system is moving ahead by leaps and bounds, opponents of the "Sky Shield" proposal are still sneering.

They are still deriding the idea by labeling it "Star Wars," using the old poisoned propaganda gimmick summed up in the saying, "Give a dog a bad name and hang him."

"Star Wars" gives it a Hollywood flavor, and that helps to undermine it. Who wants to undermine it? The Soviet Union does. So do the same U.S. groups that favor unilateral arms freezes and the halt of U.S. support to any country that is under attack by Soviet-backed military forces.

What is the alternative to developing a "Sky Shield" defense that would destroy U.S.-bound missiles harmlessly in flight? The alternative is continuation of the mid-'60s balance of terror doctrine of Mutual Assured Destruction (MAD) in which the sole deterrent against all-out nuclear war is the threat that any side launching an all-out nuclear attack will in turn be annihilated by an all-out nuclear counterattack.

As anyone can see, the flaw in this outmoded and immoral doctrine is that a mistake, a miscalculation, a sudden military move or possibly the action of some fanatical terrorist group could lead to a global nuclear conflict.

President Reagan and Gen. Graham have both said they would not object to making "Mutual Assured Survival" a joint U.S.-Soviet effort, and though critics have taken them to task for this, no doubt millions of Americans would agree.

President Reagan, in calling March 23, 1983, for a missile defense program, said that "we are launching an effort which holds the promise of changing the course of human history. There will be risks, and results take time. But with your support, I believe we can do it."

And although there remains an influential lobby which, for some unfathomable reason, demands that we go on living under the threat of nuclear Doomsday, no doubt we *can* succeed in making nuclear weapons obsolete.

U.S. to Test "Star Wars" Weapons; Soviets Suggest A-Test Freeze

The U.S. Defense Department April 16, 1985 revealed a plan to test elements of the Strategic Defense Initiative (SDI), the so-called "Star Wars" program, without violating the 1972 ABM (antiballistic missile) treaty. The Pentagon plan, contained in a report sent to Congress, was to undertake activity "short of field testing of a prototype system or component." The report listed 15 "major experiments" that, according to the Pentagon, would fall within a loophole in the treaty. For example, the Defense Department noted that the treaty did not precisely define "component." Although the experiments were to be conducted in ways that would forestall criticism that the United States was violating the treaty, the report stated that the U.S. "reserved the right" to disregard provisions of ABM in response to continued Soviet violations of the treaty. The plan included the testing of space weapons designed to intercept Projectiles. Paul C. Warnke, the head of the U.S. Arms Control and Disarmament Agency under President Jimmy Carter, characterized the Pentagon's stand as "a total fraud...This is the kind of reasoning that brings the total arms control business into disrepute." Defense Secretary Caspar Weinberger, speaking at a gathering of the American Society of Newspaper Editors, however, contended that SDI was "necessary as a prudent hedge" against "the ominous possibility of a deliberate, rapid, unilateral Soviet development of strategic defenses."

Meanwhile, in related developments, the Soviet Union appeared anxious to establish a ban on the testing of nuclear weapons. Anatoly Dobrynin, the Soviet ambassador to the U.S., April 12 said that it was "high time for a comprehensive test ban treaty." His remarks came in an address to a conference on arms control at Emory University in Atlanta. The ambassador said that the U.S.S.R. would ban some nuclear tests within two months of ratification by the U.S. Senate of the so-called threshold bans negotiated by the two parties in the early 1970's. (The Senate had never ratified the bans.) And the Soviet Union, through the news agency Tass, April 17 indicated a willingness to suspend the testing of nuclear weapons as of August 6, the 40th anniversary of the atomic bombing of Hiroshima by the United States. The report did not say that the U.S.S.R. would definitely halt testing, nor did it indicate whether a unilateral moratorium would be possible if the suggestion were rejected by other nuclear powers.

DAYTON DAILY NEWS

Dayton, OH, April 18, 1985

Defense Secretary Caspar Weinberger's reassurances to the American people about the Star Wars program are not in the least reassuring.

Mr. Weinberger told a convention of American newspaper editors in Washington last week that a space-based defense "will not occur unless a defensive system is developed that would better contribute to deterrence."

History provides a lot of cases of American administrations producing new weapons systems that do not carry out the purposes for which those systems were originally intended.

The MX missile is only the most recent example. This missile was conceived mainly as a replacement of the United States' other land-based missiles, which were becoming more vulnerable in their silos. Once the MX was developed, the constituency for deploying it was revved up. Now the MX is being deployed — but in *existing* silos.

The best argument for Star Wars research is that the Russians are conducting such research. Research cannot convincingly be limited by treaty. Testing can be limited, however, because it can be verified. The United States ought to be willing to negotiate bans of space-based weapons.

Perhaps the Reagan administration eventually will agree to limit such weapons. So far, though, it is still playing with the peculiar moral argument that Star Wars weapons would be cleaner and nicer than nuclear weapons, especially since they would only shoot down bad nuclear missiles and not hurt people.

In other words, the administration is telling the American people not to criticize Star Wars development because no one knows what will be developed — while at the same time saying that it *does* know what kinds of weapons will be developed (non-nuclear, defensive only, etc.).

The President and secretary of defense further expect the American people to believe that, in order to allay Soviet anxieties over Star Wars developments, the Americans will share all this technology with the Soviets once it is proven solid. This is a joke, right? No — the President is keeping a straight face. So is Mr. Weinberger.

Los Angeles Times

Los Angeles, CA, April 16, 1985

An extraordinary conference at the Carter Center at Emory University in Atlanta last week could make an important contribution to the Geneva talks on nuclear arms control. It depends on how the Reagan Administration reacts to some things that were said, and whether Soviet representatives were sending serious signals or playing propaganda games.

The meeting, co-chaired by former Presidents Jimmy Carter and Gerald R. Ford, also featured three former secretaries of state, several key members of Congress, two former defense secretaries, a former chairman of the Joint Chiefs of Staff, the Soviet ambassador and a member of the Soviet general staff, and other military strategists—foreign and domestic.

There was a loose consensus among American participants on several key issues:

—The United States and the Soviet Union should continue honoring the 1972 anti-ballistic missile treaty and avoid moves that undercut provisions of the SALT II and other arms-control accords that have been negotiated but not ratified by the Senate.

—The two governments should proceed with a threshold test-ban treaty, with progressively lower limits on nuclear tests until a comprehensive test ban is in effect.

—The Geneva negotiations should take account of the relationship between offensive and defensive strategic weapons, and both should be on the bargaining table.

—Research into anti-missile defense systems cannot be reliably monitored and should be continued by the United States, in light of the "massive" effort by the Soviet Union in this field. However, the two governments should negotiate an understanding of the gray-area difference between research, which is allowed by the anti-ballistic missile treaty, and developmental testing, which is not.

—It is crucial to the arms-control process that each side have confidence that agreements are being complied with by the other. Provision should be made for on-site verification of compliance when other means of resolving accusations of violations have failed.

The public remarks of the Soviet representatives were mostly boilerplate repetitions of their long-standing positions, except on two points:

Soviet Ambassador Anatoly Dobrynin suggested, albeit vaguely, that Moscow might no longer insist on an agreement to halt the Strategic Defense Initiative, or "Star Wars" program, before taking up reduction of offensive missiles. If he meant it, that is a major shift in the Soviet position.

Dobrynin also said that U.S. representatives might be invited to visit a big new radar installation at Krasnoyarsk, which most American experts consider a violation of the ABM agreement—another possible big change in the Soviet position concerning on-site inspections.

The Reagan Administration is entitled to suggest that the best place for the Soviets to make helpful shifts in their negotiating posture is in Geneva. Indeed, the Russians may have been less interested in sending serious signals than in exploiting a major American media event to their own advantage by making vague formulations for which they cannot be held accountable.

For its own part, however, the Administration may find it hard to ignore the unease over Washington's present arms-control policies that was clearly reflected in Atlanta by highly respected American experts in the field, including some with service in Republican administrations. President Reagan could do himself and the country a favor by taking the implied criticism to heart.

Herald News

Fall River, MA, April 25, 1985

The first round of the resumed arms negotiations in Geneva concluded with no sign of any agreement on any substantive issue.

On the other hand, they also concluded on a note of apparently unforced good will between the chief negotiators.

The talks are scheduled to resume on May 20, and in their second round, it should become clearer whether they hold any promise of eventual success.

Meanwhile Mikhail Gorbachev, the Soviet leader, has expressed the opinion that the United States has already proved it is not negotiating in earnest by refusing to discuss space-based weapons.

The charge is hardly worth answering.

The Reagan administration made its policy known well in advance of the Geneva meeting. The space-based weapons do not exist, and when and if they are developed, they will form a defensive system rather than a potential threat to any one.

They do not belong in the range of discussion of the Geneva talks, which concern the limitation of nuclear weapons designed and constructed for rttack.

Whatever one's view of the administration's plan to construct space-based weapons, it is clear that the weapoons are meant to repel attack, not to initiate it.

The Soviet Union's desire to include discussion of them in the Geneva talks is an attempt to dismantle a possible method of defense against nuclear attack.

It has nothing to do with the limitation of the nuclear weapons built for possible use in attack by both sides. The United States and the Soviet Union have both sought deterrence by constructing roughly comparable systems of offensive weapons.

The space-based weapons this country now has in mind would, at least theoretically, create a real defense against nuclear weapons apart from the threat of retaliation.

They would not, however, constitute a threat in themselves.

It is possible to think the space-based weapons system would not work or that it would be too expensive even for the United States to mount.

But it does not fall within the categories legitimately under discussion at Geneva, and the efforts of the Soviet Union to introduce it are, to say the least, counter-productive.

The Geneva conference is about limiting offensive weapons which both sides possess and have built up to deter one another.

The intention of the conference is to find ways of reducing the number of those weapons without reducing the effective deterrence that the present nuclear stockpiles provide.

There is no reason why the conference cannot proceed without reference to proposed space weapons that may or may not be developed for defensive purposes in the future.

It remains to be seen whether the Soviet Union will permit it to proceed, but Gorbachev's statement is not encouraging.

In any event the administration is right to stick to its guns and not permit the Soviet Union to dismantle its effort to develop a safe way to destroy nuclear missiles before they reach their targets rather than proceed indefinitely with the construction of still more offensive weapons.

The Geneva conference deserves to succeed, but not at the sacrifice of a possible end to the present system of deterrence by fear of retaliation.

THE WALL STREET JOURNAL.

New York, NY, April 23, 1985

Release of a Pentagon report on the Reagan strategic-defense initiative has uncovered a theological rhubarb over arms control. The narrow issue is whether testing for the U.S. "Star Wars" program will violate the 1972 Anti-Ballistic Missile Treaty, which by now nearly everyone agrees has already been broken by the Soviet Union. The argument illustrates the double standards that dominate the arms-control discussion. And it raises the question of how you get out of a treaty that threatens the strategic balance and national security.

With Star Wars, Mr. Reagan wants to actually defend against nuclear missiles, which is what the ABM treaty seeks to prevent. The treaty bans deployment of certain ABM components, does not seek to prevent research, but does have some provisions limiting testing. So far, Mr. Reagan has asked for research and testing but not deployment. The issue is what tests are allowed by a treaty that bans tests of ABM "components" without exactly explaining what is a "component."

The Pentagon report describes the tests and avers that while testing involves "gray areas," it plans to "make certain" that "the U.S. is in compliance." A detailed section explains the difference between a "component" and a "sub-component," and how U.S. testing involves the latter.

Arms controllers warn that breaching the "gray areas" may wreck the treaty. Yet almost all factions now concede the Soviets have already violated it, apparently without wrecking it as far as U.S. tests are involved. For years some of us have been complaining about Soviet testing and deployment of surface-to-air missiles, some reloadable; mobile radars tested in an "ABM mode"; and all the other components needed for a breakout into a nationwide defense. Arms controllers dismissed these isolated developments as strategically insignificant, only "gray-area" violations.

With construction of an ABM radar at Krasnoyarsk, even die-hard apologists concede the Soviets are in violation. (Though we are now being told that while the radar obviously violates the restrictions in the treaty it will not be used for the purposes those restrictions were meant to prevent.)

With the ABM treaty—which tries to limit technology, an ambiguous and changing thing—nearly everything is a gray area. So under the double standard that arms controllers seek to apply, Soviet activity right up to the point of a nationwide ABM capability is "gray" and therefore allowable. But at the same time U.S. research is also gray—but therefore not allowable. That there can even be a heated debate on whether this-or-that test is a violation point illustrates the inherent, abject flaw of the ABM treaty.

The Pentagon's report rightly questions this "double standard," but it also embodies a double standard or two of its own. For example, if the Soviet research and testing really is aimed at a large capability to break out of the treaty quickly, why should we worry about whether our research and testing comply? More fundamentally, why are we spending all this money on research if we are going to abide by a treaty that outlaws deployment of the weapon if the research is successful?

There is a strong case to be made for a ballistic-missile defense program, but it cannot be made hiding behind contradictory rationales. A serious case would probably start from the premise that, even if not violated, the ABM treaty is a bad thing, for us and the Soviets. It seeks to limit defense; real arms control ought to allow for unlimited defense and try to limit offensive forces.

Likewise, the Reagan administration has been bold enough to point the finger at the Soviets, at Geneva and elsewhere for violating the treaty. But if the administration wants Americans to take such charges seriously, it will have to act as if it believed this were true. The Pentagon report at least notes that the treaty does have a withdrawal clause, and that when it was signed, negotiator Gerard Smith said that the U.S. would consider its "supreme interests" jeopardized if further limits were not placed on offensive arms. And the Pentagon further remarks: "We do reserve the right to respond to those violations in appropriate ways, some of which may eventually bear on the treaty constraints as they apply to the United States."

In this oblique and bureaucratic way, the Pentagon report does start to open the right issue: If a treaty is built on wrong principles in the first place, if the Soviets are already violating it, at what point does the U.S. stop twisting its own programs to comply, and simply and honestly say the treaty is void?

The Washington Post

Washington, DC, April 28, 1985

THE ADMINISTRATION is taking a certain amount of heat for saying it intends to stick within the terms of the 1972 anti-ballistic missile treaty as it continues its work on a missile defense system in space. From the left, it is accused of cynically planning to exploit technicalities and loopholes in the treaty in order to be able to say it is testing within its terms. From the right, it is faulted for failing simply to renounce the treaty, which, its critics maintain, was either flawed from the start or has been effectively trashed by Moscow's violations of it.

These grumbles are predictable: This is an administration that believes that the old arms control agreements undercut American interests and that the Soviets are untrustworthy negotiating partners. On this basis, President Reagan set out to search for a defense that would render obsolete not only nuclear weapons, as he has declared, but also the very need for negotiated arms agreements. Meanwhile, however, Mr. Reagan also entered ambitious negotiations. So he was bound to have to answer to arms control's traditional friends and foes alike.

Traditional arms controllers regard the ABM treaty as the high-water mark in the attempt by the two superpowers to master jointly their nuclear destinies. It helps to recall, however, that the treaty was never regarded as the be-all and end-all of American security. The text provided for research, the sure engine of change, and for amendments, periodic review and even withdrawal. The two powers were serious about the treaty, but they made it warily. They wanted restraints but not a straitjacket.

It is no surprise, then, that there is heavy pressure on the treaty now. It comes from sources well foreseen: technology and distrust. For years the two countries have been conducting research on space defense. They have also accused each other of violations.

The matter of violations is key. The Kremlin can go so much further than any American administration in pressing beyond the terms of treaties: it has no public or opposition to call it to account. This puts a special burden of policing the ABM treaty on Americans. Here it must be said that American conservatives, though they can go too far, have been attentive to issues of Soviet treaty compliance. The Pentagon is right to be troubled by the emergence of a double standard that forces Americans to be faithful to agreements that Soviets compromise. The traditional friends of arms control need to be no less attentive. It could not fail to give the Kremlin extra incentive to satisfy American anxieties about, for instance, its Krasnoyarsk radar —a very large and troubling violation—if the traditional arms controllers took the lead in complaining about it.

Meanwhile, it is better that the Pentagon reshape its testing to stay inside the ABM treaty than that it plan to test outside. Americans who think it's twisting words can go to the political arena, as they are. Soviets who believe the Pentagon is stretching the terms of the ABM treaty can go to the consultative body set up to handle these issues. There they can raise their questions about suspected violations—and they can also address the many questions about their own enterprises that are on Americans' minds.

The Dispatch

Columbus, OH, April 17, 1985

The Soviet Union may have signaled a major change in its arms negotiating position when it indicated over the weekend that it would now permit on-site verification of arms-control agreements. This issue has been a major stumbling block in U.S.-Soviet talks and could open the way to significant progress in the search for an arms reduction accord.

But a few things have to be checked out first. The signals came during a foreign policy conference sponsored by former President Jimmy Carter in Atlanta on Saturday when Soviet Ambassador Anatoly Dobrynin said that the Kremlin would consider on-site inspections of strategic installations when a need for such an inspection arises. Dobrynin had been discussing a radar site which U.S. experts think may violate an existing anti-ballistic missile treaty between the United States and the Soviet Union. The ambassador said that the installation is a non-military site and if the dispute can't be resolved in any other way, a tour could be arranged.

This statement caught Carter by surprise since it represents a significant change in Soviet policy. That policy has precluded any inspections by outsiders. On-site inspections have been demanded by the Reagan administration which feels they are the only way to guarantee compliance with treaties. When informed of Dobrynin's comments, the White House said it would await a review of the entire text of the Soviet's remarks before issuing a statement.

It is interesting that the Soviets announced this policy change in a non-official, almost informal, setting and that they did it in a rather round-about manner. It may be that new Soviet leader Mikhail Gorbachev likes to do things this way, or it may be that the Soviets are merely trying to keep Washington guessing about its policy decisions.

Whatever the case, if the switch is real, it could prove very important in the effort to reduce the nuclear threat around the world.

SEE THE DIFFERENCE?

The News and Courier
CHARLESTON EVENING POST
Charleston, SC, April 14, 1985

Recently, the Defense Department released its fourth annual report on Soviet military strength, a sobering, timely and pertinent document. Predictably, those who criticize the 1985 edition of Soviet Military Power suggest it is a Pentagon lobbying effort to support even bigger defense budgets. In one respect it might be but, viewed in toto, it presents an overwhelming case in support of the Reagan defense build-up.

For instance, the report goes into depth about Soviet research on anti-missile defensive systems, space weapons and laser programs. Ironically, at a time when the Kremlin is constantly denouncing America's "militarization of space," there have been more than 10,000 Russian scientists and engineers assigned full time to space weapons, lasers and particle-beam programs. At this point, the Soviet Union still remains the only nation with an operational anti-satellite weapons system.

As Defense Secretary Caspar Weinberger testified last week, "these are systems that the Soviets are attempting to keep the Americans from achieving. They apparently want a monopoly." That becomes obvious when you consider the top Soviet agenda item at the current Geneva arms control talks is blocking the president's Strategic Defense Initiative ("Star Wars") program.

The report contains other tidbits of information that should prove sobering reading on Capitol Hill where even some Republicans are contemplating deep cuts in the Reagan defense budget, a budget by the way that is already below the levels proposed five years ago by the Carter administration. Items:

— The first of a new class of Soviet nuclear-powered attack submarines is at sea.

— Two new intercontinental-range ballistic missiles (ICBMs), the single-warhead SS-25 and the multiple-warhead SS-24 are nearing deployment.

— Lastly, they seem to be working on a more accurate successor to the awesome, 10-warhead SS-18s, the largest ICBMs ever built.

You'd think that if Congress lived in a world of reality, such details would leave some impression, some worry that if the Soviet military spending rate continues to grow every year, we should do something to counter it. The congressional track record shows something else. Consider what the Kremlin purchased in 1984 alone compared to American production numbers Congress agreed to fund (in parenthesis). Tanks, 3,200 (840). Artillery pieces, 2,650 (850). Tactical combat aircraft, 900 (400). Military helicopters, 800 (160). Surface-to-air missiles, 28,000 (500). Submarines, 6 (4). ICBMs, 100 (21 MXs). Long-range bombers, 50 (34). The Soviets are even now preparing their first attack aircraft carrier for sea trials.

Soviet military power must continue to be the yardstick against which we measure the adequacy of our own defense budgets. This timely publication demonstrates once again how sadly we match up in that comparison.

The Globe and Mail
Toronto, Ont., April 20, 1985

In October, 1962, the Soviet Union and the United States came to the brink of nuclear war over the Cuban missile crisis. Ten months later, however, the two superpowers ratified the Limited Test Ban Treaty — the first arms control pact of the nuclear era.

The rapid shift from apocalyptic confrontation to limited co-operation in the Khrushchev-Kennedy era provides a precedent for a similar improvement in relations today. U.S.-Soviet ties were badly frayed by Moscow's walkout from the Geneva arms control talks in late 1983. While the talks have recently resumed, early indications point to a continued impasse.

Yet the possibility for at least a small step forward in mutual nuclear restraint has appeared in the Soviet offer to extend the 1963 limited test ban and make it comprehensive. The treaty currently covers tests on land, in the atmosphere and under water, but exempts underground tests. It is time to seal off that loophole.

The first U.S. response to the Soviet proposal has been a deplorable lack of interest. The Americans, however, have some reason to doubt Soviet sincerity in this area. In 1958, the USSR, the U.S. and Britain reacted to public disquiet over the levels of radioactive strontium 90 in mothers' milk by entering into a non-binding agreement to refrain from atmospheric tests for an indefinite period. Moscow was the first to renege — in late 1961.

Similarly, the United States offered to observe a comprehensive moratorium after 1963, so long as the Soviets also abided by it. But Moscow once more resorted to atmospheric tests after a short interval, and all bets were off. Clearly, any extension of the test ban will have to be done by a formal amendment if the Soviet Union is to be seriously constrained from further detonations whenever they would suit its purposes.

A comprehensive test ban should not be overrated as a contribution to arms control. While it would ban the explosion of nuclear warheads, it would not rule out tests to perfect the missiles used to deliver them. Nor would it impede tests for the development of space weapons based on lasers and particle beams. Even so, it would create a helpful political climate for the negotiation of more sticky arms control questions.

That climate is not improved, however, when Soviet peace activists visit Canada in order to spread deceit. Dr. Galina Savelieva of the Soviet Academy of Medical Sciences insults the intelligence of her Canadian hosts when she claims that there is less fear about the possibility of a nuclear war in her country because the Government and the people see eye-to-eye on how to check the nuclear spiral.

If the Soviet Union really is more serene about the nuclear threat, why has it developed a massive civil defence program, complete with fallout shelters, on a scale that dwarfs such preparations in the West? If, as Dr. Savelieva notes, there is agreement between the Soviet Government and the Soviet anti-nuclear movement, how could it be otherwise? The anti-nuclear "movement" is a state-approved purveyor of the Party line.

Soviet support for a comprehensive test ban is welcome, but Soviet resort to a selective talk ban is not.

The Cincinnati Post
Cincinnati, OH, April 18, 1985

Not long ago a group of liberal and moderate Democrats joined forces to proclaim that the Eastern emperor of arms control has no clothes.

The emperor is the Soviet Union, that perennial champion of a nuclear freeze, cutbacks in satellite-based defense research and other causes dear to the hearts of so many. As it happens, however, Soviet compliance with existing treaties has become something of a scandal even to big boosters of arms control. Some of them responded with a warning to Soviet leader Mikhail Gorbachev of "serious consequences" if his regime didn't shape up.

For the first time, in short, almost the entire spectrum of American political opinion was insisting that Soviet violations had become intolerable. Your serve, Mikhail.

Quietly and unofficially, the Soviet attitude appeared to change. Officials even conceded that a radar station in Siberia—their most flagrant violation—perhaps wasn't what they'd said it was. It did have a military purpose after all.

That's what U.S. experts have been saying for two years, while pointing out that the station therefore violates the Antiballistic Missile Treaty of 1972.

There remains the question of what to do about the radar—short of the unlikely option of getting it torn down. At least, however, the two countries are quietly exploring solutions, which is more than they were doing before U.S. opinion hardened.

The lesson here is as old as the Cold War, but worth repeating. So long as an adversary is divided over tactics or the interpretation of a conflict, the Soviets rarely concede anything.

When confronted with consensus, however, the Soviets tend to respond differently. In fact, they resort to the sort of give and take generally expected in serious negotiation. Soviet officials are proving again they are not at all unreasonable—so long as they aren't given a chance to be.

Moscow Condemns U.S. Plans for Antisatellite Missile Test

The White House August 20, 1985 announced that the United States was preparing for its first test of an antisatellite missile against a target in space. The test, to take place on an undisclosed date in the fall, would not, the announcement said, violate the United Nations Charter on any existing treaties with the U.S.S.R., including the 1972 ABM (anitballistic missile) treaty. Congress in 1984 had released $19.4 million for three ASAT tests, upon these conditions: that it be given 15 days advance notice of the tests, that it receive presidential assurances that the tests were necessary for national security, and that the tests not violate the treaties or interfere with ongoing efforts to negotiate limits on space weaponry.

Both the U.S. and the Soviet Union had begun developing ground-launched ASAT missiles in the 1960's. The systems were designed to destroy reconnaissance satellites, which flew in relatively low orbits. The U.S. ASAT program was aimed at developing air-launched missiles that could destroy early warning and communications satellites, which orbited high above the Earth. The Air Force in 1984 had conducted its first, and only, test of an air-launched ASAT missile. The 1985 U.S. test was to involve an ASAT missile carried by an F-15 jet fighter. The target was to be an obsolete U.S. satellite that was still in orbit. The plan called for the plane, guided by radar, to release the missile at a high altitude. The missile, guided by on-board infrared sensors, was to destroy the target satellite by high-speed collision rather than by explosion.

Tass, the Soviet news agency, Aug. 21 warned that the U.S. decision could force the U.S.S.R., which had announced a unilateral moratorium in 1983, to resume its own ASAT tests. The U.S. move, Tass asserted, was in effect a negative response to a Kremlin proposal that the United Nations sponsor an international conference on curbing space weapons. Tass restated the Kremlin's view that U.S. ASAT development was tied to the U.S. Strategic Defense Initiative; the Reagan Administration maintained that its antisatellite and antimissile programs were separate entities.

The Miami Herald

Miami, FL, August 28, 1985

ASAT Weapons

SOMETIME in September, an Air Force F-15 pilot will launch a small, complex, nonexplosive rocket at an older, low-orbiting U.S. satellite and attempt to disable it. One hopes that the pilot's timing is better than that of the Reagan Administration, which scheduled this test only two months before President Reagan and Soviet Premier Mikhail Gorbachev are to hold their November summit meeting.

The Administration's rationale for testing the anti-satellite (ASAT) system now is that the Soviets already have an operational ASAT system and the United States doesn't. That's true — as far as it goes. The Soviets' ASAT system is old, ground-based, and is believed incapable of striking at U.S. satellites in orbits higher than about 22,000 miles above the Earth. The Soviets unilaterally halted their own ASAT testing in 1983.

The U.S. test, by contrast, involves a system that — if it works — would be immensely more flexible because its smaller rocket could be launched from aircraft. The Air Force has had lingering problems with the ASAT device. It packs 235 parts into a container about the size of a gallon of milk, and neither its design nor its cost has been fixed.

In Washington, armaments experts on both sides of the issue concur on one factor: The ASAT system's technology is inseparable from that involved in research on the Administration's Strategic Defense Initiative (SDI), popularly called "Star Wars." The 1972 anti-ballistic-missile (ABM) treaty between the United States and the Soviet Union permits research on SDI-type systems but bans their actual deployment. The treaty doesn't cover ASAT systems.

Clearly, however, lessons learned in research on anti-satellite weapons would be applicable to systems designed to down ABMs. Given the level of U.S.-Soviet mutual intransigence and each nation's determination to match the other's excursions into new weaponry, the Administration's planned ASAT test is both a provocation and an invitation to escalate the arms race.

Such a provocation might be defensible if it involved merely the sequential next step in an existing program of testing or if its timing required unchangeable conditions, such as the "window" for launching space missions. Neither condition obtains here, however. This ASAT test's timing seems based on little more than a spurious "we gotta match the Soviets" argument and the President's stated belief that ASAT treaties are unverifiable.

Even while saying that, however, the President has told Congress that his Administration will make every good-faith effort to negotiate a treaty curbing ASAT devices. Then why time this test so near the summit meeting? Why not, instead, wait until after the summit and, if those talks produce no progress on an ASAT agreement, determine whether to proceed with this test?

The Grand Rapids Press

Grand Rapids, MI, August 23, 1985

President Reagan's plan to launch a satellite-destroying missile into space this fall could knock down more than just its target.

The proposal is likely to cripple prospects for advancement in the November summit between Mr. Reagan and Soviet leader Mikhail Gorbachev. It could also set back Geneva progrss toward an arms control treaty.

National security adviser Robert McFarlane justifies the launch on grounds that this country needs to "play catch-up ball" with the U.S.S.R., which has a monopoly on space weaponry. Mr. Reagan told Congress the testing is crucial to the nation's security. Neither, however, makes a strong enough case to justify the military and diplomatic risks.

Although the Soviets have an anti-satellite (ASAT) system which they have repeatedly tested in space since the late 1960s, their missiles amount to something of a blunderbuss when contrasted with American weaponry. The U.S.S.R.'s ground-based satellite slayers require at least 24 hours between their firing and the time they find their target.

Additionally, the Soviet missiles are capable of hitting only relatively low-flying targets. They can strike American spy satellites which fly in low orbits but can't touch communications satellites, upon which the nation is so dependent. The Soviet space tests of their weaponry, which concluded three years ago, were successful only half of the time.

By contrast, this country's ASAT missile is one-eighth the size of the Soviet weapon and 24 times faster in finding its target. The American ASAT, whose components have all been successfully tested on the ground, is believed to be faster, more flexible, more reliable, and considerably more effective than the Soviet counterpart. Once launched from the fighter jet, the American weapon can travel at 500 miles a minute, while the Soviet arm's pace is 13 miles a minute from ground to target. Even the Central Intelligence Agency agrees that the Soviet system is no threat to U.S. satellites.

Little if anything is to be gained by testing this nation's ASAT system. If there is hope of negotiating a ban on such weapons, it will be hurt by the fall launching.

The White House is engaging in bureaucratic doublespeak on this subject. Mr. Reagan says the launch is necessary because of the "growing threat" of the Soviet system to American satellites — despite the contrary findings by the CIA.

White House spokesman Larry Speakes lauds the launch plan as a potential incentive for the Soviets to begin serious bargaining with the United States. In his next breath, he says the test would have no bearing upon the Geneva talks.

Rather than propel the Soviets to the negotiating table, the American anti-satellite missile test is likely to drive them back to the missile drawing boards. That is the way arms races go and that is how the United States and the Soviets have accumulated enough firepower between them to blow up the world several times over.

President Reagan through most of this year has made a point of his desire for a mutual U.S.-Soviet reduction in weapons. The ASAT missile can be helpful in advancing that hope. All he needs to do is return it to the shelf. In the absence of a real Soviet threat, the risk of such a hold is insignificant. The chance of avoiding another round of expensive and dangerous arms development, on the other hand, is great.

The Boston Globe

Boston, MA, August 23, 1985

The Reagan Administration's decision to test its antisatellite weapon against a target in space will be read round the world as a signal that the president is probably not sincere about negotiating arms reductions with the Russians.

The decision to press ahead with ASAT testing on the eve of an important summit meeting between the president and Soviet leader Mikhail Gorbachev is particularly unfortunate. The testing will cast a shadow over the stalled arms-control process that a successful summit might unlock

In looking at possibilities for a deal at Geneva, the key issue is President Reagan's intent on space weaponry. The administration wants the Soviets to agree to mutual cuts in offensive weapons. The Soviets reply that they have no problem in principle with such cuts if the United States stops the arms race in space.

The space race falls into two categories. The first is ASAT weapons, to black out surveillance, communication and early-warning satellites. The second is the futuristic Star Wars technology which, in theory, will shoot incoming ballistic missiles.

The United States would do well to avoid satellite killers because satellites are vital for both sides in reducing fears and managing crises. But more to the point of Geneva, ASAT also happens to be the only category of space weaponry where the administration could make a meaningful trade-off in return for Soviet offensive cuts.

Star Wars concepts are useless as bargaining chips because they barely exist even on paper. The hardware will not be operational for decades, if ever. The administration and the Soviets agree Star Wars research cannot be restrained, because a halt on research cannot be verified. The administration cannot offer anything but unverifiable promises not to deploy the technology. Promises about Buck Rogers gadgets will not get Soviet concessions on offensive missiles.

Unlike Star Wars, ASAT weaponry consists of working hardware in the final stages of testing. The US system will undoubtedly work, and will be fast, concealable and accurate, far superior to the crude, existing Soviet system.

That means two things. If it becomes operational, the Soviets will have to match it, putting US "space assets" at far greater risk. On the other hand, a US offer to stop testing would have real value at the bargaining table.

The basic philosophical problem of this administration in arms control is that it counts entirely on a coercive model of diplomacy, relying solely on pressure rather than compromise to win concessions. The result so far has been a barren record in arms control.

There had been some hope that in his second term Reagan would come seriously to grips with arms-race issues. The Soviets have made significant gestures – including their recent declaration of a unilateral moratorium on underground nuclear tests. They have had an ASAT moratorium in effect for three years. They are in an increasingly strong position to argue that they, not the United States, are the peace-makers.

The administration claims it must proceed with ASAT tests "to restore a military balance." It says it must demonstrate determination to give the Soviets an incentive to bargain at Geneva. This is false. What President Reagan has lacked all along is not bargaining leverage but the will to compromise.

The United States will have to grant something important to the Soviets to win anything of importance in return. Space happens to be the only game in town. Without US restraint in space – and pressing ASAT testing is the antithesis – the chance of getting Soviet concessions and ending the arms race is vanishingly small.

ANTI-SATELLITE TEST

Buffalo Evening News

Buffalo, NY, August 29, 1985

IT IS QUESTIONABLE whether the Reagan administration's decision to proceed with the controversial testing of an anti-satellite weapon will ultimately help or hurt the national security. The danger is that this action may lead to a further escalation of the arms race in such weapons, which threaten satellites used for communications and early warning.

The administration points out, quite correctly, that the Soviet Union already has the world's only operational anti-satellite system. But the Soviet model is a rudimentary weapon that doesn't operate at the high altitudes where the bulk of the U.S. satellites are found.

The new U.S. system would, by contrast, be highly advanced — and that's the problem. The result could be a spiraling space arms race in which each side seeks to leap frog the other, with neither achieving any net gain in nuclear security. The real need is to try to negotiate a mutual ban on satellite-destroying weapons. President Reagan has committed himself to such a goal, although the administration continues to question whether it would actually be feasible to verify compliance with anti-satellite control measures.

There is no doubt that the United States can't afford to fall behind in anti-satellite technology or permit the Soviets to threaten our military satellites. However, congressional critics argue that the coming test will make it harder to get an agreement on anti-satellite weapons.

Senator Larry Pressler, R-S.D., contends that the test is ill-timed — that "we should have waited until after the November summit in Geneva to see if the Russians would be willing to take their anti-satellite system down." That sounds like advice worth heeding.

AUGUSTA HERALD

Augusta, GA, August 22, 1985

Does anyone seriously believe that the Soviet Union would not put a ground-based laser system in space if it could? Especially if the United States could not.

That's the point to keep in mind as the faint-of-heart crowd rains criticisms down on President Reagan for deciding to test an anti-satellite weapon in space.

No sooner did administration spokesman Larry Speakes announce that the government would test its anti-satellite (ASAT) weapon than critics – both domestic and foreign (including spokesmen for the U.S.S.R.) – were all over the networks denouncing the U.S. for "spoiling the atmosphere" for the Geneva arms talks and the fall summit between Reagan and Communist boss Mikhail Gorbachev.

Yet history shows, as Speakes pointed out, that the Kremlin negotiates in good faith to control new weapons only when it's convinced the U.S. is totally committed to development.

The Russians have already tested their own ASAT system and, according to the man the American people have twice elected president by landslide majorities to make these kinds of determinations, it constitutes "a clear threat" to the United States and its allies. Reagan would be trifling with the security of his country if he didn't move to play "catch-up ball."

The Soviets, of course, would be delighted to see us give them an open field to score an ASAT touchdown. But in the arms race, as in football, deserting the field would not be interpreted as an act of generosity, but as proof of weakness and lack of will.

In recent weeks we've chided Mr. Reagan for not matching his tough rhetoric with strong actions in areas of defense spending, budget cuts, and retaliation against terrorists.

We're pleased to note his ASAT decision indicates there are limits to how far he'll go to avoid controversy with his critics on the left. More than that, we hope it portends a change of mood in which he will not back off from making similar hard-headed, common sense decisions in the future.

DAILY NEWS

New York, NY, August 22, 1985

THE WHITE HOUSE has announced the U.S. will test an antisatellite weapon, as it had promised long ago. The world would be safer if this next step in arms escalation could be avoided. But the Reagan administration has no responsible alternative.

The U.S. and the Soviet Union depend heavily upon satellites to direct their armed forces—especially the strategic forces—and to spy. At the extreme, if the U.S. launches an intercontinental missile at the USSR, its flight is directed by a satellite in orbit 20,000 miles or so over Siberia, which communicates with American military bases through a series of other U.S. satellites flying at much lower altitudes, and ground stations in the U.S. and around the world.

If the USSR knocked down the U.S. satellites, the missile would be in peril. It works both ways: If the U.S. can destroy Soviet satellites now orbiting silently above the Earth, the USSR Rocket Command will be blinded. Tuesday's announcement means the Pentagon believes it can do it.

The Soviets have protested—but they have already tested their own ASAT system. Their protests must be understood in that light. The Pentagon claims the Soviet system is far inferior to its own—but that won't last. The Soviets will invest whatever it takes to catch up. In a few years all satellites will likely be vulnerable to sudden destruction.

That would improve the chances of a sneak attack, and make the world a more dangerous place. One side could simultaneously launch missiles against its enemy's defenses and against the satellites that would guide any retaliation.

It's not the Strategic Defense Initiative—Star Wars. That's still years off, even supposing it ever works. And the ASAT test doesn't violate the provisions of the anti-ballistic missile treaty. This is today's technology, and it will work.

There are negotiations with the Soviets on curbing ASAT forces, as well as on short-range missiles, long-range missiles and troop levels in Europe. None of the talks is getting anywhere. It's increasingly apparent that the arms race is a seamless web that can only be curbed by a comprehensive agreement between the U.S. and the USSR.

There's precious little sign that such an agreement is possible. The Gorbachev-Reagan summit might, just might, prove the contrary—but in the meantime it's clear the U.S. must prepare for the worst.

The Times-Picayune
The States-Item

New Orleans, LA, August 23, 1985

President Reagan's decision to order field tests of the U.S. anti-satellite weapon is sound both as military strategy and as diplomatic tactics.

Strategically, it represents a "We can, too" response to an already tested Soviet satellite killer. And in addition, it is superior to the Soviets' space grenade. It does its work without even being explosive — it destroys by simple impact.

Tactically, the move comes during the sparring before Mr. Reagan's preliminary bout with new Soviet leader Mikhail Gorbachev in November and during a recess in the U.S.-Soviet arms-control talks at Geneva. It shows that we are determined to look to our own defense in the absence of a realistic agreement on mutual arms restraint.

The diplomatic tactics are probably more important. Militarily, anti-satellite weapons are no more a first line of defense than are our two countries' vast arsenals of nuclear weapons. Like nuclear missiles, they are basically a deterrent force.

The weapons are now primarily counters in the arms-control game being played in Moscow, Washington and Geneva. The Soviets have predictably risen in outrage at the coming tests, though they do not violate any treaty and have been specifically authorized by Congress. Domestic opposition has risen as well. With typical reverse logic, it charges that a display of strength and determination makes an arms agreement more, rather than less, difficult to achieve.

The arms talks in Geneva have not yet gotten down to serious business. Messrs. Reagan and Gorbachev have not yet met face to face, and Secretary of State Shultz has met new Soviet Foreign Minister Shevardadze only briefly. The world is watching — somewhat myopically — so let us survey the sparring before the main events.

After a concentrated period of underground nuclear testing, Mr. Gorbachev announced a unilateral moratorium on nuclear testing and invited the United States to join. The United States made public note of the previous Soviet activity, and held an underground nuclear test. The Soviets howled that that proved U.S. perfidious militarism.

The United States then scheduled its anti-satellite tests, which both Soviet and popular imaginings tend erroneously to equate with the controversial "star wars" space defense. The Soviets are obsessively anxious to stop even U.S. research on "star wars" in the arms-control negotiations. The Soviets et al again howled.

In view of these U.S. moves, the Soviets can hardly expect to intimidate Mr. Reagan when he meets Mr. Shevardnadze in September and Mr. Gorbachev in November. They may, indeed, at last get the message: If they want to reduce their own risks, they had better sit down and talk seriously about reducing our common risks.

And if they will not do this, we will not have skipped a step keeping up with them.

Los Angeles Times

Los Angeles, CA, August 22, 1985

The White House intends to shatter the sanctuary of space with a satellite-killing missile sometime in September. We hope the equipment is not as flimsy as the case that the Reagan Administration makes for doing it.

The shattering device, known by the Air Force as an ASAT system, will be a missile launched from racks under a high-flying F-15 fighter plane and programmed to cripple satellites by colliding with them. In the test, the target will be an exhausted American satellite, still wandering around the Earth in orbit.

In a written notice to Congress, President Reagan argues that final testing of the satellite-killer has an urgent bearing on the nation's security. White House aides try to reinforce the argument, claiming that the Soviet Union has a monopoly on satellite-killers that must be broken. The truth is more complicated than that.

The Soviet Union has a ground-based satellite-killer that takes as long as 24 hours to get ready to fire and has, at best, proved effective less than half the time in tests—he last of which was conducted in 1982. Robert C. McFarlane, the President's national-security adviser, calls that a monopoly. McFarlane, a former Marine Corps colonel, surely would hesitate to order men into action carrying rifles that would misfire half the time. A Soviet system with the same track record can hardly be called a monopoly.

In one breath White House spokesman Larry Speakes said this week that the September test might give the Soviets an "incentive" to bargain seriously on a treaty limiting or banning satellite-killers. In the next he said that the test would not affect Geneva arms-control talks. As for incentive, the Soviets have tried for months to interest the United States in such a ban, without apparent success.

In his letter Reagan told Congress that the United States must move, in part, because of a "growing threat from present and prospective" Soviet systems to American satellites. The CIA has told Congress that the Soviet systems pose no threat to American satellites. A prospective threat can be dealt with by treaty.

The White House arguments do not discuss the fact that America depends far more on its satellites to link this country's global forces than the Soviets depend on theirs. They are misleading in their discussion of using American satellite-killers to "deter" Soviet systems. Satellite-killers cannot fend off attacks on satellites; they can only destroy satellites. White House statements equate weapons in space with militarization of space—also misleading. Spy satellites long ago "militarized" space, and the world is better for it. Satellites make it impossible for the Soviet Union to move forces around without America and its allies seeing them. Deploying weapons in space would provide no such stabilizing influence.

The White House should call off the September test. If not, it must at least level with Americans about the implications of the test. The risk that the test will stampede the Soviets into trying to build a better batch of satellite-killers is far higher than the chance that it will make them jump to negotiate.

The Honolulu Advertiser

Honolulu, HI, August 23, 1985

War in space has been a reality for at least 14 years. That was when the Soviet Union developed the first operational weapons system to knock out satellites.

Since then the United States has played catch-up. Now the U.S. is scheduled to conduct the first test of its own system after Labor Day. A high-flying F-15 fighter will fire a rocket that is to home in on a satellite in low orbit.

GIVEN THE Soviets' established space weapons program, Moscow's cries of foul are hollow indeed. Had the Soviets been concerned with maintaining peace in outer space, they would not have spent so much on their anti-satellite system.

But now that the U.S. is approaching equality in anti-satellite weapons (and, in fact, may soon shoot ahead of the Soviets), the possibility of negotiating a treaty to control a space arms race has improved.

Moscow is certainly eager to discuss the topic; it had been cool to the idea before the U.S. announced progress with its weapons.

A key factor will be what happens with the Reagan administration's Star Wars program: the plan to have a network of lasers or other weapons to knock out ICBMs in flight.

And drawing up a treaty would also be dependent upon progress in the intermediate-range and strategic nuclear weapons negotiations that have resumed in Geneva. At the moment those talks are moving slowly.

THE ANNOUNCEMENT of the upcoming U.S. "satellite killer" test is an escalation of sorts in the arms race. But in catching up with the Soviet Union's capabilities, Washington has set the stage for real progress in controlling the spread of weapons into space.

The test is now how willing each side is to take advantage of this opportunity.

Reagan, Gorbachev Talks Collapse; "Star Wars" Blocks Arms Accord

U.S. President Ronald Reagan and Soviet leader Mikhail Gorbachev met in Reykjavik, Iceland Oct. 11-12 for what were to have been discussions to set the agenda for a true summit in the U.S. Instead, the talks turned into a intense and detailed negotiations over arms control. The parley fell apart Oct. 12, in a vehement disagreement over the U.S. Strategic Defense Initiative (SDI), the so-called "Star Wars" program. The stalemate came as the superpowers appeared on the verge of concluding a pact on the substantial reduction of offensive nuclear weapons. No date was set for a Gorbachev visit to the U.S., one of the original purposes of the Reykjavik talks. In the aftermath, each side accused the other of intransigence, while observers throughout the world struggled to understand what went wrong. The talks were conducted under tight security and, under a mutual agreement, a news blackout.

In the morning and afternoon sessions Oct. 11, Reagan and Gorbachev met for nearly four hours in a room on the first floor of Hofdi House, a small 75-year-old mansion that overlooks Reykjavik Bay. The only others in the room were their foreign ministers, George Shultz and Eduard Shevardnadze, their translators and note-takers. Officials on both sides described the first day's meetings as "businesslike," but with a noticeable air of optimism. The talks mainly focused on arms control at Gorbachev's insistence. According to knowledgeable observers, Gorbachev initially appeared to be better prepared for detailed discussions than did Reagan. At the close of the first day, it became apparent that the Reykjavik talks had become a full-scale summit. The major news from the first day was a joint announcement of the creation of two U.S.-Soviet working groups of aides for detailed discussions on key issues. One working group dealt with arms-control issues, the other dealt with human rights, regional conflicts and bilateral matters. The idea for working groups had apparently been pushed by Gorbachev, consistent with his notion that concrete agreements should result from Reykjavik. According to sources close to the talks the arms-control working group made substantial progress on reductions in strategic nuclear weapons but bogged down in the area of medium-range weapons.

From the start of the Oct. 12 sessions, Gorbachev signaled his willingness to make important concessions on arms. But he also indicated that disarmament had to come in a unified package, rather than separate accords. The two sides reached understandings on a series of arms issues. These included an interim proposal of a limit of 100 warheads each on medium-range missiles, with the elimination of such weapons in Europe, and a 50% reduction in strategic weapons, with mutual ceilings of 6,000 warheads and 1,600 missiles and bombers. In a major concession, Gorbachev agreed that sea-launched cruise missiles would not be counted in the 6,000 ceiling on warheads. The Soviet leader accepted Reagan's proposal that the two sides work toward a goal of eliminating all ballistic missiles within 10 years.

About three hours into the morning session, the talks snagged on SDI. Gorbachev raised a strong objection to SDI, which he characterized as destabilizing. In his view, a workable "Star Wars" system would allow the U.S. to launch a nuclear first strike while protecting America against Soviet retaliation. The argument turned on a proposal by Gorbachev that Washington honor the 1972 ABM (antiballistic missile) treaty for 10 years and adhere to an interpretation of the treaty that would prohibit any work on SDI outside of basic laboratory research. Gorbachev made it clear that his "Star Wars" proposal was an integral part of the understandings reached on reducing offensive nuclear weapons. Reagan adamantly refused to place a limit on "Star Wars" testing. The heated debate over SDI lasted through the afternoon session, pushing all other matters into the background. The two sides were unable even to agree on a date for a Gorbachev visit to the U.S. At about 7 p.m. Oct. 12, a visibly angry Reagan terminated the Reykjavik talks. The two leaders and their aides emerged grim faced from Hofdi House. "There will not be another summit in the near future that I can see at this time," White House Chief of Staff Donald Regan told reporters. "The Soviets are the ones who refused to make the deal. It shows them up for what they are. The Soviets finally showed their hand."

The Courier-Journal & TIMES

Louisville, KY, October 14, 1986

FOR what was billed as only a "pre-summit" chat, the Reagan-Gorbachev meeting in Iceland generated big expectations — and delivered a big disappointment. It also raised again, more urgently than ever, the question of whether President Reagan's Strategic Defense Initiative is a bargaining chip or, as he continues to insist, a firm goal that he will not trade away.

Many scientists believe that a reliable "Star Wars" missile shield over America will never become a reality. But Mikhail Gorbachev and his generals aren't so sure. That's why the Soviet leader, at Reykjavik, was willing to pay dearly for a U.S. pledge to confine SDI research to the laboratory and abide by a strict interpretation of the 1972 Anti-Ballistic Missile Treaty.

Mr. Gorbachev offered deep cuts in strategic offensive missiles and elimination of medium-range missiles in Europe. President Reagan, for his part, was also willing to see a sharp reduction in both sides' offensive forces. But he balked at the proposed restrictions on SDI.

Ironically, many scientists and military analysts believe that Star Wars, if it could work at all, would only do so if the number of targets — Soviet missiles — is dramatically reduced. But as matters now stand, Moscow won't agree to such reductions unless the United States in effect abandons Star Wars.

So progress on arms control is blocked by this Catch 22. But time appears to be on the side of the Russians. They may decide to wait until Mr. Reagan leaves the White House, 27 months from now, before resuming serious negotiations in Geneva. Meantime, America's NATO allies likely would grow increasingly uneasy at the stalemate, and Congress, which already has cut requested funding for SDI, could chop even more.

By spurning a deal in Iceland, Mr. Reagan thus may find that he has weakened the Atlantic Alliance and lost some of his leverage with Congress. SDI, barring a major technological breakthrough, would then be worth less to him, or to his successor, as a bargaining chip.

This approach may be fine with those Americans who want Star Wars deployed at any political and economic cost. But we find it a deeply disturbing approach to the serious business of arms control.

THE DENVER POST

Denver, CO, OCtober 14, 1986

RESULTS of the meeting between President Reagan and Soviet leader Mikhail Gorbachev are best judged by the realistic record of arms-control negotiations over the past half century — not by the frantic hype of the past week.

By that long-term standard, the two-day meeting in Iceland may yet mark a step forward in mankind's struggle to control the deadly cycle of the nuclear arms race.

Some politicians raced to the cameras Sunday to claim that the meeting marked a "breakdown" in U.S.-Soviet relations. But, in fact, nothing "broke." Instead, the two leaders emerged with positions more closely aligned to each other's goals than most experts would have predicted a mere week ago.

It's true that Reagan and Gorbachev failed to reach any final agreements on new treaties limiting nuclear arms, missile defenses, nuclear testing and other disputes. But such dramatic breakthroughs couldn't have been reasonably expected from the Iceland meeting, which, after all, was never planned or billed as a true "summit" that would produce historic agreements on crucial issues. As Reagan had said, it was intended to be more like a "base camp" at which the two leaders would build the foundation for a future summit.

The problem is that this quickly called meeting with its modest aims rapidly snowballed into a case of unrealistic expectations. The Soviets made proposals that were, on their face, much more far-reaching than the Reagan administration had expected. Whatever their motives — propaganda, genuine fear of the U.S. Strategic Defense Initiative, or both — the diplomatic groundwork had not been laid for final agreement on such sweeping proposals.

Even so, both sides demonstrated flexibility at Reykjavik, and both surrendered some ground. The Soviets made bolder than expected proposals for balanced reductions in ballistic missile arsenals. The president, for his part, showed more flexibility than expected on his beloved "star wars" proposal — though far from enough to satisfy Gorbachev. The "failure" to agree merely reflected the fact that both leaders reached for a more significant achievement than ought to have been expected in a meeting called on such short notice without preliminary staff work.

Before the talks reached an impasse, the U.S. and Soviet negotiators had tied together a tentative package that would, among other accomplishments, have cut long-range strategic missiles in half over five years, with provisions aiming to abolish them entirely over a decade. The package also included agreements to eliminate nuclear missiles from Europe and to reduce, and in time end, nuclear weapons testing. There was even discussion of a 10-year extension of the antiballistic missile treaty and a proposed elimination of all nuclear ballistic missiles over a 10-year period — both initiatives described as offered by President Reagan and tentatively accepted by Gorbachev, until the two deadlocked over treatment of the U.S. "star wars" plan.

But if two days can produce so much general progress, then surely more months of patient negotiations by the experts already meeting in Geneva can work out the details. In a few weeks, after all, the U.S. elections will have passed and the fiendishly complicated question of arms control can be viewed with the coolness and objectivity it deserves.

Toward that end, we urge politicians of both parties to avoid making political capital out of the supposed "failure" at Reykjavik. In that regard, Reagan's rhetoric in an address to the nation last night was commendably absent of vitriol. As the president hopscotches around the nation stumping for congressional candidates during the next three weeks, he ought to avoid temptation to resurrect his "evil empire" references to the USSR.

The arms control process is too important — and coming too close to a major success — to be sacrificed in the interest of partisan bombast. At Reykjavik, both Reagan and Gorbachev demonstrated their common interest in controlling the arms race. The world's interest lies in seeing that the many details which stand in the way of their common objective are swept away as thoroughly and as swiftly as possible.

BUFFALO EVENING NEWS

Buffalo, NY, October 14, 1986

THE COLLAPSE of the American-Soviet summit in Reykjavik is a serious disappointment, the more so because progress had for a time seemed to be pushing both sides toward what Secretary of State George Shultz described as "extremely important potential agreements" to limit the nuclear arms race.

The two days of summitry at work brought increasingly high hopes suddenly dashed — a final sinking sense of being so close, and yet so far.

If the failure to consummate far-reaching agreements strikes a body blow to arms talks, however, it must not be allowed to become a fatal blow to this process. The needs and the political dynamics that pushed the Soviets and the United States toward this rare common ground persist as urgently today as they did a week or a month ago.

Much diplomatic terrain was covered by President Reagan and Soviet leader Mikhail Gorbachev and their aides during the grueling two days of talks. Both sides tentatively agreed to reduce their long-range missile and bomber arsenals by half over the next five years, and completely by 1996. They also tentatively agreed to eliminate all but 100 medium-range missiles on each side in five years, and the rest by 1996.

Those conceptual outlines of an arms deal leave a myriad of questions on verification and other complicated details unanswered. But they also reflect enlightened interest on the part of both superpowers not just to freeze their nuclear arsenals at present levels but to drastically shrink them. Such bold accords would venture far beyond the pre-summit expectations of possible progress on medium-range missiles and the scheduling of another summit soon in the United States.

In the end, however, the Soviets doggedly linked any such reductions in offensive weapons to concessions the president was unwilling to make on his Strategic Defense Initiative, the high-tech defensive system he had proposed in 1983. That price, he felt, was too high.

Final judgments about the responsibility for the unyielding differences on SDI by each side must await closer examination of all that went on in Iceland. But we would be reluctant to second-guess Reagan for refusing Soviet demands for significant changes in the 1972 Anti-Ballistic Missile treaty that would confine Star Wars research to the laboratory. Shultz described this proposal as a Soviet effort "effectively to kill off the SDI program" — something the president later said "we could not and will not do."

SDI is a defensive, not an offensive, system. The 1972 ABM treaty has contributed to arms stability for 14 years. Reagan has vowed to keep SDI research, testing and development within the confines of that existing treaty. Why change it now, especially in a future 10-year period of dramatically reduced offensive weapons? Significantly, too, Reagan had agreed to delay any deployment of SDI for 10 years, a concession that provoked criticism from some conservatives.

If the Soviets made concessions, then, so did Reagan. Moreover, the Soviet interest in nuclear arms reductions has intensified markedly since the president proposed the Star Wars program three years ago. There might have been no Reykjavik without SDI. To kill any effective U.S. testing of that system, even within existing treaty regulations, could erode future Soviet interest in arms reduction.

At the very least, what Moscow and Washington must now salvage from the collapse of these talks is a mutual resolve not to allow this failure to embitter future ones. If Reykjavik failed to produce agreements, it still need not have been wasted dialogue. The limits of compromise were tested, but substantial opportunities for willing compromise also emerged.

There are signs that both sides, though disappointed, are avoiding despair. "Never have our positions been so close," said Gorbachev at the end. Larry Speakes, the White House spokesman, said: "We went 99 yards. Both sides did. We didn't cross the goal line. We made very important progress."

With patient determination, after what may prove to be a period of cooling off and reassessment, Washington and Moscow ought to strive to recapture the constructive momentum leading up to this summit, to move closer still and finally to cross that goal of arms control glimpsed, however briefly, at Reykjavik.

The Boston Globe

Boston, MA, October 11, 1986

Like the island itself, set in the autumn North Atlantic, a way station for fliers and mariners on an ocean feared for winter storms, so the Iceland mini-summit represents a harbor of refuge from a cold war that is growing perceptibly colder.

The issue of nuclear weaponry is paramount. The economic drain of the arms race has become staggering, and the cost of miscalculations is now too high for any sane person to accept. Sensible leaders everywhere – including those in the United States and the Soviet Union – want arms-control progress this time, not just a summit charade.

The Iceland meeting could be a turning point, and all Americans will applaud President Reagan if it is. Yet the developments in US arms policy that have unfolded, especially in recent months, suggest that this may turn out to be another opportunity missed.

The traditional view of the Soviet Union taken by US hard-liners is that the Soviets always seek military advantage by piling up superior military forces and that they negotiate only to "codify" what they have built. Judging from the behavior of the two sides over the past few years, that caricature seems to apply to the White House, not the Kremlin.

The president has gone to Iceland having made a series of policy shifts that almost all proponents of arms control believe are steps in the wrong direction.

He has rejected the goal of six predecessors to end nuclear-weapons tests. He has announced plans to conspicuously violate the SALT treaty in one month. He has unilaterally redefined the Anti-Ballistic Missile Treaty in a way that makes it meaningless and, if taken at face value, will make an offensive and defensive arms race inevitable.

Having accomplished all that, Reagan, before departing for Iceland, assured his hard-line supporters that he plans to hold firm.

What is distressing is that a historic success in arms control, for which President Reagan could take full credit, should be there in Reykjavik for the taking.

Over the same period, extending back for more than a year, Soviet leader Mikhail Gorbachev has made concessions that indicate a clear readiness to deal. He has maintained a unilateral moratorium on nuclear-test explosions and the testing of antisatellite weapons, abandoned opposition to on-site treaty verification, and made major concessions on offensive missiles.

Both sides have been hinting that there may be an agreement in Iceland to reduce medium-range missiles in Europe and Asia. Like any sign of progress, this agreement would be welcome, but in the context of what could happen, given the Soviet willingness to meet the United States halfway in a range of more important areas, it would be thin gruel.

Even if there is an agreement in principle on a cut in medium-range missiles, numerous details of implementation will remain. The hawks around President Reagan are still in a position to obstruct for months and years an agreement on matters such as verification of a treaty. If Iceland distracts attention from the important arms issues – space weapons and strategic arsenals – it will represent a net loss.

Gorbachev presumably has to deal with his own hawks and weaponeers. He has held them in check over the past year, extending the moratorium on nuclear tests, and making a series of conciliatory moves on verification and negotiating objectives, despite continuing US rebuffs. But it is unlikely he can hold them in check indefinitely. When he can do so no longer, the tit-for-tat arms race will resume.

Striving to give the president negotiating room, House and Senate leaders yesterday struck a compromise that eased several arms-control provisions that the House had attached to the 1987 spending bill. Although that is a loss for arms control, the provisions that remain make it clear that Congress disapproves of the path the administration seems to be on because it threatens the ABM treaty and SALT. The compromise was an unmistakable, though diplomatically phrased, expression of concern.

It is a shame that President Reagan will not sit down today with a vote of confidence ringing in his ears. At least he carries the hopes and best wishes of millions.

The Philadelphia Inquirer

Philadelphia, PA, October 14, 1986

For months one urgent question has underlain all talk of arms control: Would President Reagan swap a slowdown in his program to build missile defenses in space for Soviet agreement to slash back heavy nuclear missiles?

Now, after the collapse of the mini-summit in Iceland, Americans know the answer. It is negative. At Reykjavik Mikhail Gorbachev challenged the President to play his "Star Wars" bargaining chip. Mr. Reagan refused. The Soviet leader offered to cut long-range missiles and weapons in Europe in exchange for agreement to limit "Star Wars" to laboratory research for the next 10 years. But the President chose full commitment to the unlikely dream of a total shield against nuclear attack over the tangible achievement of cutting strategic weapons in half or even lower.

To understand the historic opportunity that was missed at Reykjavik one must recall why "Star Wars", also known as the Strategic Defense Initiative or SDI, was conceived in the first place. The concept emerged because the administration worried that America was vulnerable to a first strike by Moscow, which possessed larger numbers of multi-headed missiles. At the time, U.S.-Soviet relations were strained, and arms cuts looked unlikely. But as the Star Wars program has come under scrutiny, scientists have challenged its potential to protect civilians, and legislators have questioned its stupendous cost. Mr. Reagan told the nation last night that he could not give up SDI because it is the "key to a world without nuclear weapons." But the Soviet reaction to SDI deployment would be to produce more weapons to penetrate U.S. defenses.

In fact, Star Wars' most useful function has been to drive the Soviets back to the bargaining table. The administration says this proves the Kremlin believes the system will work. But, all signs are that Moscow is worried less about the success of Mr. Reagan's particular vision and more about a costly high-tech weapons race that will hurt its weak economy and show up its technology lag.

Those worries brought dramatic Soviet proposals at Reykjavik: a 50 per-cent cut in strategic missiles and removal of all medium-range missiles from Europe. There was, of course, a quid pro quo. The Soviets wanted agreement that neither side would withdraw before 10 years from the antiballistic missile (ABM) treaty that forbids deployment of a Star Wars system, and that Washington would limit SDI to laboratory research.

On that demand the whole deal fell apart. President Reagan claimed that Mr. Gorbachev had demanded that he scuttle Star Wars by giving up the right to develop and test. But in fact, the ABM treaty expressly *rules out* the development or testing of anti-missile systems in space. If Mr. Gorbachev was trying to tighten the treaty, he was only reacting to on-going administration attempts to unilaterally loosen the treaty to include Star Wars tests.

The issue in Iceland was not "scuttling" SDI, since SDI does not exist. Limiting research to the lab for 10 years wouldn't set the program back much scientifically; what the President apparently feared is that an agreement would cut lagging political and financial support for the program.

But that support will lag further now that the President has made clear he won't trade Star Wars, even for the purpose that many believe it best serves. Surely the exchange of a blueprint — on which research could continue — for the destruction of half the superpower arsenals is a price well worth paying. Yet at the critical moment Mr. Reagan chose to stick with his dream. When offered a better deal than most experts hoped for, Mr. Reagan refused to bargain.

It was because the President remains committed to an impractical vision that he was caught unprepared by Mr. Gorbachev in Iceland, in a major political setback that could cost America dearly among its European allies. Both leaders say their proposals are still on the table. But, ultimately, if the superpowers are to move towards a safer world, the compromise outlined at Reykjavik still holds the best beginning. But it won't happen in the present administration unless President Reagan moves from dreams to reality.

Los Angeles Times

Los Angeles, CA, October 14, 1986

As they left Iceland, President Reagan's advisers, many near exhaustion after sleepless hours that promised a breakthrough in arms control but left both sides empty-handed, brooded over what might have been. But the weekend summit meeting has come and gone and cannot be recalled or rewritten. What remains is what might yet be.

What might have been was an agreement, one half of which would have cut both American and Soviet nuclear forces by 50% in the next five years and wiped them out altogether at the end of a decade. The other half would have kept the President's Star Wars genie bottled up inside laboratories during that time.

Most American scientists would have jumped at the deal, as would we and as the President should have done. Most scientists doubt that the Star Wars defense system which the President insists will some day throw a missile-proof shield over this country can be built in less than a generation, if ever, and then only at a cost in the hundreds of billions of dollars. By eliminating the Soviet arsenal of strategic weapons, the bargain, if it worked, would have destroyed more Soviet missiles than Star Wars could ever hope to do, and at a fraction of the cost, in dollars and in wear and tear on the nerves of American allies.

What might yet be is careful negotiations at Geneva, with days and weeks to ponder proposals and counterproposals rather than the minutes and hours available at Reykjavik. They would start at the point that Reagan and Soviet General Secretary Mikhail S. Gorbachev had reached before they said goodby. Both leaders indicated that talks will go on. "I'm still optimistic," Reagan said in a television address Monday evening. "Let us not panic," said Gorbachev as he headed back to Moscow.

But future negotiations will be no picnic. As nearly as anyone on the outside can tell, Reagan and Gorbachev went to Iceland to sound out the limits to which each was willing to go in order to give some guidance to their negotiators.

What happened there may well reflect the outer limits for both men, outer limits that were not new and should not have come as a surprise. The last time he addressed Star Wars publicly before the Iceland meeting, Reagan said, "Our response to demands that we cut off or delay research and testing is: No way." His language may have been different at Reykjavik, but his message was the same.

As for Gorbachev, his parting remarks were in line with what he told Reagan during last November's summit meeting: Star Wars would not necessarily be a defensive weapon but could as easily be an offensive weapon, attacking Soviet targets from space.

The Soviet leader said that as they were saying goodby, Reagan asked him, "Why are you so intransigent" over one word, testing? It was not just one word, Gorbachev said he replied, it is "what the U.S. Administration really intends."

Still, the Reykjavik weekend provides a place to start thinking of ways to break the impasse. Gorbachev says the Soviet offer to cut its long-range missile force in half over the next five years will remain on the bargaining table in Geneva. A vastly reduced offensive threat should mean a vastly reduced defensive system. A reduced defensive system, more precisely described, might well be easier to negotiate than the grandiose, but imprecise, schemes that the Administration has promoted in public as part of an effort to keep the Star Wars budget growing.

The fact that the Administration has had so much trouble explaining just what Star Wars would look like is a major source of Congress' skepticism about the project. That skepticism itself will be a factor in determining the future course of arms control negotiations. Even before Reykjavik, Congress was moving to hold down the budget for Star Wars research to help reduce the federal deficit. A number of members were talking Monday about giving the program even closer scrutiny in the future.

The failure to come away from Reykjavik with something in writing that would slow down the arms race is a setback, but it is not the end of arms control negotiations. As long as the President and the General Secretary do not lose each other's telephone numbers, there is reason to hope that they will find a way to talk themselves out of the corners they talked themselves into at Reykjavik.

THAT'S RIGHT, MISSION CONTROL! HE WANTS US TO TAKE HIM TO JUPITER!
NASA
©1985 MIAMI NEWS

Part IV: Interplanetary Probes and Space Stations

The first interplanetary probes—the Soviet Lunas and the American Pioneers—were sent into space to find out what was there. Carrying a few measuring devices and a radio, they were simple craft. Scientists knew so little about the nature of space in the early days of space flight, that even such minimal instrumentation could ensure a vast advance in their understanding. Since those first pioneering launches, the United States and the Soviet Union have worked through several programs of probes ranging from the exploration of the Earth's nearest planetary neighbors, to the awesome voyage of Pioneer 10, the first spacecraft to leave the solar system and enter interstellar space. The probes themselves have become as increasingly technologically advanced as the objectives of their flights have become ambitious. Their designers have to take account of the fact that their vehicles will generally spend much longer operational periods in space and may travel many millions of miles from the Earth. In addition to the protection required against the dangers of meteors or radiation, the communications subsystem has to be powerful enough to receive and send signals over immense distances. Panels of solar cells usually provide the power, converting the sun's rays into electricity.

In the 1960s, the U.S. launched their Mariner probes that investigated Mars and Venus. While their findings were considered dramatic and invaluable at the time, they paled by comparison to the Pioneer, Viking, and Voyager missions of the 1970s and 1980s. Two American landers in the Viking series touched down on Mars in 1976. Wonderfully detailed panoramic views of the Martian landscape were transmitted back to Earth while complex remote-controlled experiments were performed to sample the surrounding soil and test for signs of life. While the results were inconclusive, some scientists believe the Viking results proved the existence of primitive Martian organisms. Most scientists, however, prefer to await the results of later missions. Pioneers 10 and 11 provided some of the most exciting missions. Designed for extremely long journeys, they were able to function for over seven years at up to 1500 million miles from the Sun. But perhaps it is the missions of Voyagers 1 and 2 that have provided the scientific community with the most stunning data collected on the mysteries of the solar system. Their investigations of Jupiter, Saturn and Uranus provided astronomers and astrophysicists with once-in-a-lifetime discoveries that many hope will inspire similar explorations.

After the success of the Apollo lunar expeditions, NASA began serious development of the first manned space station to be placed in Earth's orbit. Skylab, an unmanned 77-ton U.S. space station, was launched in 1973 carried eight telescopes and various sensors, which were invaluable aids to astronomers. Originally, it had been thought that the craft would remain in orbit long enough for the space shuttle to ferry rockets up to it in an attempt to prolong its orbit. But these plans never materialized as NASA underestimated the level of solar activity which exerted a cumulative drag on Skylab. When Skylab's orbits deteriorated, there was a lacking of funds, sapped by the shuttle program, to save it. Eventually it entered the atmosphere in a shower of pieces over the Indian Ocean and Australia. Since then, there have been many calls for the development and launching of a new space station. But with the success of the shuttle and the lack of a coherent space station strategy, these plans have yet to fruitfully materialize.

Voyager I Flies by Saturn

The unmanned United States spacecraft Voyager 1 flew by Saturn November 12, 1980, gathering information about the giant planet, its complicated ring structure and its varied collection of moons. Another American spacecraft — Pioneer 11— had flown by Saturn in 1979, but the one-ton Voyager 1, with its more sophisticated equipment, amassed far more data. Voyager 2, a twin to the Voyager 1, passed by Saturn in the summer of 1981.

At its closest approach, Voyager 1 came within 78,000 miles of Saturn's cloud tops. It came much closer to Titan, Saturn's largest moon and the only planetary satellite in the solar system known to have a substantial atmosphere. The spacecraft skimmed within about 2,800 miles of Titan. Titan's clouds prevented Voyager from observing the moon's surface. But spectrometer and radio-wave analysis afforded much new information. The atmosphere turned out to be mainly molecular nitrogen, together with atomic and ionized nitrogen. Methane, which had been observed from Earth, apparently made up less than 1% of the atmosphere. Atmospheric pressure was much higher than expected: 1.5 times that of Earth at one level in the atmosphere, and possibly twice or three times that of Earth at the moon's surface. Titan's atmosphere appeared similar to that of the inner planets--Earth, Venus and Mercury--at an early stage in the evolution. The cold temperatures on the moon had blocked the development that occurred on Earth and presumably made it possible for life to evolve. Voyager 1 discovered three new moons as it approached Saturn, raising the known number of satellites circling the planet to 15. Another discovery made as the space craft approached Saturn was a vast cloud of hydrogen gas circling the planet. The tenuous cloud was distributed in Saturn's equatorial plane and apparently ranged from 300,000 to 940,000 miles out from the planet. Scientists speculated that the hydrogen came from methane in Titan's atmosphere that split up under solar radiation.

On Saturn itself, the Voyager 1 instruments found a brand of equatorial winds that blew at speeds up to 1,000 per hour. This was four times the speed of winds on Jupiter. It came as a surprise to scientists, since Saturn was both smaller and cooler than Jupiter. One theory advanced to explain the strong winds was that they were powered by cyclic rising and falling of cold clouds of ammonia, which exerted a pumping action.

The most surprising and apparently inexplicable findings of Voyager 1 dealt with the rings that circled Saturn. Previous observations had indicated that there were six rings, or possibly a few more. As Voyager produced more and more detailed photographs, scientists increased their estimates of the number of rings, at one point counting 95 and later figuring that there were several hundred. This was not the result of observing new material circling Saturn. Rather, the previously observed broad rings, under closer examination, proved to be made up of numerous finer rings. Scientists compared the ring system to a phonograph record, with its numerous close-set ridges and grooves. It was not just the number of rings that was surprising. Previously, it had been thought that Saturn's rings were held in place by a pattern of gravitational resonances from the moons. While the gravitational force of the moons obviously played a role in the shaping of the ring system, the structure of the rings appeared to be too complicated for that to be the only effect responsible. One of the features noted by Voyager was the presence of dark spokes in the rings, directed radially outward from the planet. These spokes appeared to be dark gaps when Voyager viewed them from above the plane of the rings, but when the spacecraft dipped below the ring plane they were seen as bright. This meant, scientists said, that the spokes held particles that scattered sunlight forward instead of reflecting it back. This in turn meant they were small in size--in the range of the wavelength of light. As Voyager came nearer, it discovered that several rings were eccentric--that they were not circular, as were the other rings around Saturn. Then it was discovered that two rings appeared to be braided: they crossed over each other like threads in a rope. Mission scientists said that other forces besides gravity might have to be invoked to explain these effects.

WORCESTER TELEGRAM.

Worcester, MA,
November 6, 1980

One way to get a perspective on things is to think about the Voyager I spacecraft, now almost a billion miles from earth and heading for Saturn. One week from today, the sophisticated robot will be transmitting photographs back to earth from a range of 2,500 miles from that remote planet. It will have gone through the famous rings, identifying new moons and other objects, and will give man his best view yet of the strange body first spotted by Galileo in 1610.

The immensity of space daunts human imagination. More than 300 years ago, Blaise Pascal made his famous comment: "The eternal silence of those infinite spaces terrifies me." Anyone who thinks about the strange, distant realm that Voyager I is passing through will understand what Pascal meant. It is awe-inspiring to think of those infinities beyond infinities where the galaxies pass in parade, separated by light years of space beyond reckoning.

Saturn seems almost at the utmost bound of human thought, yet it is a close neighbor, relatively speaking. It is part of our own solar system and may hold clues to earth and the beginnings of life. It is bathed in methane and possibly acetylene, ethane and ethylene, as well as other hydrocarbon compounds that may be the building blocks of living matter. Saturn even has a moon, Titan, larger than our own and holding its own atmosphere. Is there life on Titan? On Saturn? Probably not, but the possibility grips the imagination and spurs the scientists on.

Beyond Saturn lies Uranus, then Neptune, and finally, at the edge of the sun's gravitational pull, Pluto, 3.6 billion miles away. And beyond Pluto lies those incredible empty reaches of space and time that so terrified Pascal.

At a time when we are caught up in the political hubbub and international turmoil here on this fleck of dust called earth, a glance outward at the universe produces reflection and perspective on the great scheme of things.

ARGUS-LEADER

Sioux Falls, SD, November 14, 1980

There is pride in the scientific achievement of Voyager I soaring past Saturn and transmitting hundreds of pictures and mountains of data 947 million miles to the Jet Propulsion Laboratory in Pasadena, Calif.

Scientists are acclaiming pictures of Saturn's complex rings and "braided" ringlets as "astounding" results, giving earth-bound mortals a view of things in space so remote that there were no suggestions about their existence.

President Jimmy Carter telephoned his congratulations to the Voyager team in Pasadena for "a superb scientific achievement." It is all of that — and new evidence of American knowhow. The voyage of the robot spaceship around Saturn came in the 38th month of its voyage of 1.24 billion miles.

It's also encouraging to note that both Voyager I and Voyager II each carry 12-inch copper records with greetings from earth in 60 languages and a sample of 20th century artifacts.

If beings from another galaxy intercept either spacecraft millions of years hence, they'll have an opportunity to learn something about planet earth in the 20th century. Meantime, thanks to Voyager, earth's people have a much better idea of Saturn and its icy moons.

Chicago Tribune

Chicago, IL, November 18, 1980

Voyager I's close look at Saturn is a thoroughly satisfying scientific adventure—astronomy for artists and poets as well as mathematicians and physicists. How fortunate we are that the computerized data streaming hundreds of millions of miles back to earth at the speed of light is giving us more mysteries than solutions, more enigmas than answers. Like the ancients who watched the planets wander among the stars in the nighttime skies, we are still baffled, astonished, awed, and humbled.

And what a Saturnalia of phenomena we now have to feed our imaginations: Fifteen moons of marvelous diversity instead of five or six. As many as 500 separate rings, some of them lopsided or braided in majestic defiance of any known laws of orbital physics. A vast trench on the icy surface of a Saturn moon, a bright orange-yellow fog wrapping another. Cobwebby markings circling Rhea and Dione. Methane skies that may drip hydrocarbon rain into powdery plastic terrain. Lakes of liquid nitrogen as cold as death. Jules Verne couldn't have done it better.

"Mind-boggling," say the scientists. "Absolutely astounding." "Unimaginable." "We are seeing new things so remote from our experience that we're not able to come up with even a hint of a suggestion as to what some of them are."

What better argument could there be for making sure this country shares the next great celestial opportunity—the chance to study Halley's comet when it skims past Earth in 1986? Mind-boggle is rare these days and worth pursuing, even across a universe.

THE DAILY OKLAHOMAN

Oklahoma City, OK, November 18, 1980

LAST Wednesday afternoon, the remarkable flight of our Voyager I unmanned spacecraft took the vehicle closer to the planet Saturn than any prior exploration. Scientists will be evaluating the discoveries made during this probe of the ringed planet for decades.

This successful mission will be followed next year by Voyager II, and the trajectory and missions of that vehicle will be adjusted on the basis of what has been learned from Voyager I.

Manned flights to Earth's moon are more glamorous, but nothing so far attempted is as significant to astronomy and astrophysics as these deep planetary probes. We have no idea, yet, how the knowledge gained from these first visits will affect our lives. But we can be certain that it will.

Meanwhile, all Americans are indebted to the unsung scientists and technicians of the National Aeronautics and Space Administration whose pioneering work is keeping the United States ahead of the world in deep space exploration.

RENO EVENING GAZETTE

Reno, NV, November 29, 1980

Voyager 1's highly successful Saturn mission proves once again the ability of American science and technology. American skills might be buried under blindness and a backward-looking view, as in the auto industry, or hampered by bureaucratic fumbling, as in Washington, but when they are freed to do their task, they can still perform superbly.

Voyager 1 performed flawlessly, sending spectacular photographs of Saturn millions of miles, along with a multitude of data that will take years to decipher and understand. Voyager discovered new moons, learned that the moon Titan has a heavy atmosphere perhaps similar to that on earth eons ago, and disclosed scores of rings around the planet where we thought there were only a few.

Altogether, the mission has added immeasureably to our understanding of Saturn, its rings and its moons. Eventually, this information will add to our understanding of our own plant — how it formed, how its systems work. New theories might be needed on the gravitational interplay of bodies in space; we might be able to discover more about how life evolved on earth; we will be able to piece together more information about how solar systems form.

The addition of knowledge simply by itself is tremendous. Ever since mankind learned that the planets are more than lights in the sky, and that like earth they revolve around a star, there has been a great hunger to learn more about them. And the United States space program, along with that of the Soviet Union, has told us much. The United States program, in particular, has brought us a wealth of information about Venus, Mars, Mercury, Jupiter, Saturn and our own moon. For scientists and much of the general public, chained to one planet, the space probes have been a glorious adventure.

For this reason alone space research is worth continuing. But there is more to it than this. The space program can also achieve goals of immediate practical importance. Already it has brought significant advances in computer science and electronics. We know a great deal more about the weather and with future projects could learn even more, both through the study of earth's weather from satellites and through the study of weather systems on other planets. We have increased our knowledge about geology, which is aiding our search for minerals.

There is no telling what breakthroughs might come from continued space research, particularly from probes of other bodies in our solar system. Knowledge has a way of building up randomly at times, then, all of a sudden, coming together to provide amazing new insights into ourselves and our environment. We cannot predict precisely how space research will augment our own lives, any more than we can predict what we will find in space. But we do know that something important should come from all this.

So it is greatly disappointing that the United States space program is withering away. Voyager 2, now in space, will rendevous with Saturn next year and could then be instructed to bypass Uranus in 1986 and Neptune in 1989. The space shuttle program is limping along. But beyond that — nothing. Even a space meeting with Haley's Comet in 1986 has not been approved, although the Soviet Union, Japan, and the European Space Agency plan probes of lesser extent than that suggested for the United States. Time is short; if the comet probe is not included in the 1982 budget, the project will have to be scrapped.

The mood of the nation is for financial conservatism. This will be the mood of the new Congress and presidential administration when they take the reigns of government next January. But most proposals for space probes do not require extraordinary amounts of money or creation of new technology; surely funds can be found to sponsor some of these projects and keep the brilliant NASA research team together. The eventual benefits will be well worth the relatively small amounts of money spent. We are only on the threshhold. Let us step forward, not back.

THE MILWAUKEE JOURNAL

Milwaukee, WI, November 18, 1980

It is good for one's mental health to let the imagination soar occasionally. Voyager 1's sweep past Saturn offered that opportunity.

Voyager 1 provided Earth with its first detailed look at the second largest planet in the solar system. And it was a spectacular view. No less striking than the yellow hues of the planet were the scientific discoveries — some defying explanation. For instance, Saturn's legendary rings turn out to number not 6 but closer to 120. And one of the rings is "braided," a phenomenon that appears to contradict the laws of physics.

Down-to-Earth realists may complain that there is no "return" on the large investments required to launch programs such as Voyager. Agreed, Voyager is not producing Teflon frying pans in the kitchen, but it is providing tremendous amounts of new knowledge. To list some of the satellite's accomplishments: Voyager, in its journey so far, has discovered two new moons of Jupiter and three new moons associated with Saturn. The satellite's high-resolution cameras have sent back images of volcanoes on the Jovian moon Io that are larger than anything on Earth. A ring around Jupiter also has been discovered.

Voyager 1 represents only the beginning. Voyager 2 already is on its way to examine Saturn again. Scientists hope to launch a Galileo satellite in 1983 to return to Jupiter and probe deeply into its atmosphere. There are plans to send another spacecraft to Saturn in 1986 and to land a spacecraft on Mars that could return to Earth with Martian soil.

These wondrous close looks at the solar system underscore how little we really know, and whet the appetite for learning even more. In probing the universe, we on Earth are discovering our own heritage, our celestial "roots," so to speak. That gives worth to the investment already made in an American space program — and provides justification for future funding at a substantial level.

ST. LOUIS POST-DISPATCH

St. Louis, MO, November 4, 1980

Despite all the media attention riveted on the U.S. presidential election, the impending rendezvous of the spacecraft Voyager 1 with the golden planet Saturn and its shimmering rings may be no less a significant event. Like the September 1979 visit to Saturn by the Pioneer 11 space vehicle, which gave man his first closeup view of the ringed planet, Voyager's three-year trip of almost 1 billion miles represents a vast extension of man's reach and an awesome triumph of the human intellect.

While en route to Saturn last year, Voyager's television cameras spectacularly lifted the curtain on some of the mysteries of the giant planet Jupiter. No one can predict what surprising information Voyager will transmit back from Saturn across the vast void of space. The signals, travelling at the speed of light, will take an hour and a half to reach observers on Earth. Scientists at the Jet Propulsion Laboratory predict that, when the spacecraft's encounter with Saturn is over, "all the textbooks in the world dealing with the planetary system will have to be rewritten." A week after election day here Voyager will pass within 2,500 miles of one of Saturn's moons, Titan, which is larger than the planet Mercury; the television pictures ought to be full of fascinating detail. After that, the $500 million Voyager mission will continue toward the edge of the solar system and then disappear into deep space.

One more probe of Saturn is in the works. Voyager 2 will pass close to the planet about a year from now;, and for a time at least, that will be that. But human curiosity is insatiable. Man has walked the moon, operated a space vehicle on the surface of Mars and pushed his vision to the edge of the solar system. Saturn has not seen the last of man, and vice versa.

THE BLADE

Toledo, OH, November 21, 1980

VOYAGER I has performed magnificently, resolving some puzzles about the sixth planet from the sun, Saturn — the most remote of the planets known to the ancients — while raising other questions.

Among the discoveries are the amazing complexity of Saturn's ring system, its superhurricanes with wind velocities of as much as 1,000 mph on a surface once thought to be relatively placid in comparison with its giant neighbor Jupiter, the existence of at least six major ring systems instead of the traditionally accepted three, the possibility that the rings themselves may have an atmosphere, and the fact that there are at least 15 moons where previously the commonly accepted estimate was 11 or 12.

The largest Saturnian moon, Titan, was something of a celestial flop. It was found to possess a largely opaque, "smoggy" atmosphere, apparently composed in large part of nitrogen which obscured most surface features but did not disguise the fact that the satellite would be an inviting place neither to live nor to visit. And there was the braided strands of the newly discovered "F" ring, the existence of which, in light of commonly accepted theories of orbital mechanics, "boggles the mind," according to one of the astronomers at the Jet Propulsion Laboratory in Pasadena, Calif.

But scientists today are not the first to be baffled by the ring system of Saturn. When Galileo first observed the rings through his crude telescope in 1610, he was moved to conclude that "Saturn is not one alone, but is composed of three, which almost touch one another." Two years later, the rings, which are at most about 10 miles wide, presented an edge-on appearance to the earth and were invisible, causing the renowned Italian astronomer to wonder if Saturn had somehow devoured its children. The rings were again visible in 1613, but Galileo was unable to account for the phenomenon. It was a Dutch astronomer, Christiaan Huygens, who wrote in 1659 that "Saturn is surrounded by a thin, flat ring which nowhere touches the body" of the planet.

This "star of wonderment" is matched, in another way, by the insouciant fashion in which Voyager has flawlessly photographed and transmitted thousands of pictures to the earth even as it continues on its lonely course toward the stars, already more than two million miles from Saturn. The satellite, whizzing past lonely moons, giant planets, and spectacular rings, is a fitting monument as well to the unbounded curiosity of those who designed and sent the spacecraft on its way and even now monitor its wealth of information.

There are those with tunnel vision who are quite willing to wind down such epics of 20th-century space exploration, but in the long run it is inconceivable that man should not desire to know more about what is out there in the remote reaches of our solar system.

The Virginian-Pilot

Norfolk, VA, November 13, 1980

Take a good look at the photographs of Saturn being transmitted by Voyager 1, as well as those expected to be sent back to earth next August by Voyager 2. For not until 1986, when Voyager 2 whizzes past Uranus, is the National Aeronautics and Space Administration scheduled to stage another planetary picture-taking spectacular.

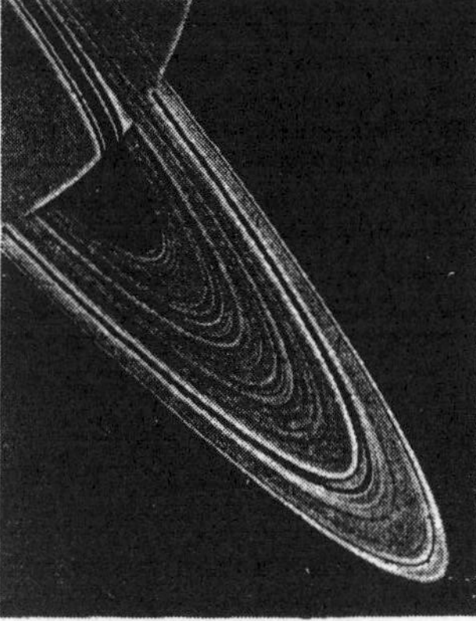

The Rings of Saturn

Spirits were lifted by the Apollo missions to the moon. But earthly woes—soaring energy costs, inflation, joblessness, hunger, nuclear proliferation—soon grabbed the nation's attention. Saturn is a technicolor diversion, a transitory stimulant to the imagination. But it would be remarkable if the stir it causes prompts Congress to boost space-agency funding.

Americans can now take or leave the space program. U.S. prestige has sunk. Defenses must be shored up. Much of the industrial plant is obsolescent. Families struggle to make ends meet. The past seems to be brighter than the future. The forthcoming Reagan administration, moreover, favors a slowdown in federal spending.

So what if Voyager 1 has spotted three more moons—that makes 15—in Saturn's orbit. So what if Saturn's rings are pale yellow, golden brown, and reddish brown. Gasoline's more than $1.11 a gallon.

Pioneer 11 flew by Saturn 14 months ago. But it provided relatively few measurements and photographs to waiting scientists. The two Voyagers, launched in 1977, are more sophisticated observers, as they proved with the information collected last year on their swings past Jupiter and its moons.

As for those puzzling Saturn rings, which continue to baffle despite Voyager 1's close-up view of them, they fluttered pulses at Pasadena's Jet Propulsion Laboratory, which controls the Voyagers' missions. The rings appear pale yellow through telescopes; that there were other colors was a revelation. Also startling were clumps of concentrated particles in the rings; their meaning has yet to be divined.

All of which has nothing to do with the price of peas. Unless Congress increases space funding, NASA will drift toward the interplanetary age, dismantling space teams as it goes.

That's regrettable. Innumerable benefits have flowed from the space program—the nation's technological lead was lengthened by it, a plus for exports. We know more about Earth because we sprang toward the stars. In unlocking the secrets of faraway worlds with strange-sounding names, we unlock secrets that extend our mastery in the world in which we live and move and have our being. But that no longer plays well in Peoria.

The Boston Herald American

Boston, MA, November 18, 1980

PROBLEM: Galileo boasted of discovering the new planet, Saturn, in 1610. But he couldn't explain the "cup handles" he found attached to it. And he was especially perplexed when he took another look and the "cup handles" had disappeared.

SOLUTION: After 45 years of puzzlement, Dutch astronomer Christian Huygens published his explanation: "aaaaaaa cccc d eeeee g h iiiiiii llll mm nnnnnnnnn oooo pp q rr s ttttt uuuuu." In its proper sequence it says: "Annulo cingitur, tenui, plano, nusquam cohaerente, ad ecliptican inclinato (It is girdled by a thin ring, nowhere touching, inclined to be ecliptic.)"

Suddenly it became clear, and astronomers have been unraveling the mysteries of Saturn and the rest of the planets or "wandering stars" ever since.

Their machinery has changed dramatically, from Galileo's inch telescope to something as complex as Voyager I and its network of magnetometer, ultra-violet and infra-red spectrometers, photo polarimeter, and cosmic rays. But the driving force in this quest has been the same.

Curiosity.

Just as it has been with much of the adventure and progress achieved by the human race, curiosity drove Galileo. It drove Huygens and astronomers and scientists in the centuries since, in their exploration of the moon, of Mars and Venus, of Jupiter, and now of Saturn with its braided rings, its tiny moons, its giant moon Titan, and its mysterious atmosphere.

But now what?

Now in the name of a balanced budget and economy, the President and Congress have put the brakes on what little exploration of space the U.S. does undertake. The space shuttle, already long overdue and way over-budget, won't be launched until next year, and even then its first priority is leaving a trail of Defense Dept. and intelligence satellites.

The only space exploration now on the books is a new assignment for Voyager II which will proceed to Uranus after it passes Saturn next summer, and Galileo, a space station that will orbit Jupiter some time after 1986.

But a joint U.S.-European venture to explore the polls of the sun has been delayed at least three years. Plans to orbit the moon, to study Saturn, to map Venus and to rendezvous with Halley's Comet in 1986 have been scrapped, and a plan to gather soil samples from Mars has been deferred to the 1990s.

No one rails against waste and excess government spending more than we do. But we now spend less than one percent of a $500 billion-plus federal budget on space research and exploration. Somewhere in sorting out our national priorities we ought to be able to find a place for curiosity. There's something a little frightening and perhaps self-destructive about putting a clamp on it.

A flotilla of space ships from the Soviet Union, Japan and Western Europe are already planning to intercept Halley's Comet, with its mysterious nucleus and vaporous tail, when it comes into view in 1986. As of now, the U.S. has scrubbed its plans to launch what would have been a fascinating combination of technical wizardry that would have "sailed" through space powered by a phenomenom known as solar wind. What we're left with instead is little more in the way of research tools than we had on the comet's last visit in 1910. That was when hawkers used to sell "comet pills" to ward off any influences that "the dreaded star may import."

PROBLEM: The federal government overspent its budget by more than $50 billion. Inflation is soaring, and taxpayers need a break. Justify sending a space ship to Halley's Comet.

SOLUTION: b d ee ii ss t.

Or, put another way: Sed ibi est (because it's there).

The Pittsburgh Press

Pittsburgh, PA, November 20, 1980

One could spill superlatives all over the page trying to do justice to the accomplishments of the Voyager space probes.

But the assessment of a spokesman for the Jet Propulsion Laboratory, which is managing the project, should suffice:

Thanks to Voyager I, more has been learned about the rings and moons and planet of Saturn in one week than in all the billions of years that transpired before this.

★ ★ ★

The encounters of Voyager I and II with Jupiter and its system of moons last year was similarly successful. And if all goes well, there will be more spectaculars when Voyager II reaches the planet Uranus in January 1986 and Neptune in August 1989.

Talk about getting the most bangs from your bucks! (In the Voyagers' case, about 460 million of them.)

Despite all this, the U.S. space program is losing its zip. For it is really being sustained only by past commitments.

In real-dollar terms, the budget of the National Aeronautics and Space Agency is only about half of what it was in 1966, in the heyday of the Apollo moon-landing program.

★ ★ ★

Granted, Americans will once again go into space when the shuttle finally begins its earth-orbital flights sometime in this decade.

But the trouble with the shuttle, besides its myriad technical problems, is that it is not part of some longer-range concept.

For example, it could be used to assemble a permanent manned earth station. Or to construct orbiting solar collectors to beam energy to earth. Or to plant a scientific-industrial colony on the moon. Or to launch a manned mission to Mars.

Of course, the United States cannot afford to do all of those things, but we should have some goal.

Without a larger vision of what the future of mankind should be in space, we are really indulging in little more than stunts.

THE INDIANAPOLIS STAR

Indianapolis, IN, November 15, 1980

Voyager 1's success, traveling 1.24 billion miles toward the outer reaches of the solar system in 38 months and sending back color pictures of the fierce swirling beauty and messages that begin to unravel the long-locked mysteries of Saturn, speaks for itself — and breaks into new realms of the human adventure.

Saturn with its rings and clustered moons is the most exotic planet of them all, even to the hilltop amateur stargazer scanning the skies with a small telescope.

But the 38,000 mph two-ton spaceship with its one-ton laboratory will not unlock all of the enigmas of the golden planet with its violent storms and churning yellow clouds, or even all of the puzzles of its moons or rings.

In fact, the mysteries of the rings keep getting deeper as the data, having traveled nearly a billion miles, accumulates, one scientist said.

And Voyager's closest encounter with Saturn was 77,000 miles. A visitor that far from Earth would miss seeing New York, London and Tokyo, to say nothing of the Empire State Building, the Sears Tower, Las Vegas, the Fighting Irish and Bo Derek, among other much-talked-about attractions.

But slowly and certainly many riddles of Saturn and its rings and satellites will yield answers as the scientists pore over the amassed information that spent an hour and 25 minutes traveling through space by radio at the speed of light.

One Voyager scientist called the results "astounding." Even more fantastic things are likely to emerge as science does its wizardly work.

There's little doubt no life-forms that would be recognizable to humankind exist on huge gaseous Saturn or its rings or icy moons or the giant of these, methane-covered Titan.

In fact, one conclusion that most consider certain after all planetary probes and studies to date is that mankind is alone in the solar system aboard "the good ship Earth." As yet there is no proof that he is not alone in the universe, although the bets are still out.

Powerful waves of thought as violent in their consequences for man as celestial storms are for distant planets, stars and galaxies sprang from earlier discoveries — the roundness of the earth, its place in the solar system, gravity — and the theories of evolution and relativity.

Out of these seeds of knowledge and thought grew optimism perhaps exceeding the vanity of vanities, and prophecies of an imminent golden age. Out of them also came the deepest of pessimism, reinforced by inhumanities beyond measure.

The rendezvous of Voyager 1 with Saturn was like many great human adventures and Odysseys — astounding, brilliant, amazing, colorful, spectacular and a wellspring of new knowledge, new legends, profundities and questions.

Like all human adventures, the rendezvous has ended and Voyager has begun a lonely journey through the cold infinity of outer space, the dark and boundless prairies of the universe.

Was the amazing voyage proof that, in the last analysis, the human being must search within his mind for the meaning to the riddle of the cosmos and his place within it, that the Creator has made it possible for him to be as big inside as the whole universe is outside?

THE DENVER POST

Denver, CO, November 14, 1980

As VOYAGER I skimmed away from Saturn Friday, climaxing its three-year voyage of discovery, it began what one news story called "an endless journey among the stars."

The phrase is prophetic — not because of the trajectory of this small, clever machine — but because it challenges man's intellect, also, to make the ultimate journey beyond its earthly moorings.

The challenge is difficult. So much of what we receive from space is alien to our human reference points. Even that word, "endless," suddenly becomes breathtaking. Until recent decades, we used it when things were older than a few lifetimes or, reaching a bit further, to speculate in the realms of Biblical or geologic time.

Now Voyager I, rising from a decade of remarkable space and laboratory discoveries, is challenging us to search for the endless, the infinite. It may help us to find what the universe is made of, how it formed, how life evolved.

Already we are thrusting into far corners. Physicists are seeking divisible particles in the proton and the neutron. We see patterns by which the heavy elements in our world — and in our bodies — trace their origins to ancient supernova explosions so vast as to dwarf comprehension.

What happened next? How was the debris collected into swirling bodies of material which became planets? Why is Earth so special — so rich in atmosphere and the heavy elements needed by our living structures? Scientists are at work applying Voyager's lessons to such puzzles. A wealth of data has flown the billion miles from Voyager to Earth. Included are not only pictures but subtle measurements of infrared, ultraviolet and radio waves that may be more factual than what we can see.

The spaceship, for example, has found intriguing mixtures of ammonia and methane in the vicinity of Saturn and its big moon, Titan. In the laboratory, a spark applied to such mixtures will produce amino acids — the building blocks of proteins and possibly life. Is Voyager showing us how our own world looked a few billion years ago? The new pictures of Saturn's rings may teach science new lessons on the dynamics of planets.

So we envy the 1,000 Littleton employes of Martin-Marietta who helped build the Titan IIIE/Centaur launch vehicle which propelled Voyager to its splendid destiny. They helped extend man's reach into the solar system and beyond.

In a way, this event was inevitable — a way station in our insatiable curiosity to find what things are and how they work.

Partly, we are fulfilling Galileo's quest.

When he invented the telescope nearly 400 years ago — and observed the rings of Saturn — this great Italian scientist expressed hope that future men would extend his observations.

We have built remarkably on his genius, adding our own hopes and asking the next generations to carry the search deeper into space. And, like Galileo, we are still probing the scope and substance of what previous generations have always regarded as the province of the gods. Maybe that helps explain why the search is endless.

The Charlotte Observer

Charlotte, NC, November 17, 1980

Once again we're reminded, thanks to the billion-mile trip of Voyager 1, that the more we learn, the less we know.

Consider:

- Photos from the 10,000-pound spacecraft revealed that the atmosphere around Jupiter, the largest planet in the solar system, is full of violent, erratic clouds. Scientists had thought the clouds circled the planet uniformly and quietly.
- Photos last week showed that Saturn, second in size, may have as many as 1,000 rings, instead of 6 or 7. Many of those rings were found in what scientists had thought were empty gaps of space. Scientists were puzzled also by Saturn's outermost ring of moon-like particles which were intertwined, like braids. Laws of orbital mechanics say those particles should either cruise out into space or into Saturn itself. But that hasn't happened and for now, scientists can't say why.
- Instrument readings and photos of Titan, Saturn's largest moon, hinted at atmospheric conditions close to Earth's. Titan's atmosphere is at least as thick as Earth's and composed mainly of nitrogen, the main gas in the Earth's atmosphere. Scientists had previously thought the atmosphere was mostly methane.

Scientists have just begun scrambling to piece these tidbits together, like eight-year-olds trying to work an adults-only jigsaw puzzle. Their laws, which explain a simple three-body system like the earth, sun and moon, don't explain more complicated systems.

Why should scientists bother? One reason has to do with expanding knowledge about life on Earth. The discovery of the carbon dioxide blanket around Venus, for example, — which entraps heat from the sun, keeping temperatures on that planet well above 900 degrees Fahrenheit — tells human beings something about what could happen if that same gas builds up in the Earth's atmosphere.

If even the slightest trace of organic life is found on Titan, or elsewhere in the solar system, that opens up all kinds of possibilities about life elsewhere. If no trace is found, that says something else about the Earth's most precious resource.

It is human nature to explore and to explain. Exploring is what keeps us alive and gives us hope; explaining is what gives us purpose. And though that process may lead us to the errors of our earlier thinking, it also keeps us humble.

THE CHRISTIAN SCIENCE MONITOR

Boston, MA, October 14, 1980

The United States is approaching a critical juncture in outer space. What happens over the next six months will provide important clues as to whether Americans, after dramatic early achievements highlighted by putting man on the Moon, have now lost sight of the stars – and the challenges and opportunities space holds for this and future generations.

Next month's scheduled Voyager 1 flyby of the planet Saturn will signal the beginning of the end of America's long-running program of planetary probes which in recent years have measurably increased what we earthlings know about such heretofore little-known neighbors as Mars, Venus, and Jupiter. Beyond the companion Voyager 2 flyby of Saturn in August of next year there are no major planetary probes planned.

The Russians, meanwhile, continue to push ahead with their manned space program. Although the US retains a technology lead, the record 185 days in orbit set by cosmonauts Leonid Popov and Valery Ryumin over the weekend provides a dramatic reminder that the Soviets are far out front of the US (with its 84-day endurance record set in 1974) in the number of man-hours devoted to conducting experiments and learning to adapt to long stays in space. In short, the Soviets are making steady progress toward establishing a permanent manned space station in orbit. It would be a pity if it took such an event – another Sputnik-like shock – to reawaken Americans to the need to revitalize their own space program.

Poised with considerable new scientific knowledge already in hand and the promise of still greater breakthroughs within reach, the US planetary program is in danger of being left to drift in a sea of official indifference. Confronted with a national economy that demands tough budgetary decisions, recent administrations understandably have been reluctant to expand spending on space. The benefits of space research, although enormous, tend to be long-term and thus harder to justify to constituents in the short run. In the face of such choices, it will take new vision and strong leadership in Washington to keep the US in the forefront of exploring mankind's last great frontier.

Officials at the National Aeronautics and Space Administration are hanging their hopes on next March's oft-delayed launching of the reusable space shuttle, which is heralded as opening a new era in space travel. They already see hints of renewed public interest in space. The next logical evolutionary step in manned space exploration, the shuttle is seen as advancing space travel beyond the experimental stage and for the first time giving humans the capability of working and building in space. The hold of the shuttle will carry virtually any cargo a business firm or a government agency wants to send aloft. Already booked far in advance, the shuttle will provide a basic transportation system for the hoped-for eventual construction of permanent orbiting space stations, solar energy generating plants, and gravity-free manufacturing facilities.

Even if, as NASA hopes, the shuttle succeeds in reviving public interest in space – and reassuring the world that the US still has the technological know-how it demonstrated in the glory days of Apollo – it is far from clear how large a role NASA itself will play in future space exploration. The Defense Department is assuming an ever larger role; the Air Force already has taken one-third of the shuttle's advanced bookings. It would be a tragic turn of events if military considerations were to become the primary driving force behind future space ventures. There can be little doubt that space technology will be needed to help provide solutions to world energy, environmental, and food problems. But after the shuttle, will NASA be relegated to serving as little more than a transportation or shipping agency for private communications firms and the military? Or will the R & D expertise of NASA be used, for instance, to tap the potentially unlimited reservoir of solar energy that exists outside earth's atmosphere to meet the growing demand for energy?

Moreover, it should be kept in mind that the prospective laying-off of skilled NASA employees once the current projects are completed will leave the agency ill prepared to assume a larger role later on. Once its experienced technicians and engineers find new positions outside the space agency, reassembling a team with comparable experience will be much more difficult.

The lack of enthusiasm in the White House and on Capitol Hill for spending more on the civilian space program is reflected in the $5.2 billion NASA budget for 1981, virtually unchanged from its 1980 budget or, for that matter, not that much larger than the $4 billion NASA spent in 1969, the year Apollo 11 landed on the Moon. With inflation and the decreased value of the dollar taken into account, that amounts to a sharp reduction since 1969.

The spinoffs of space exploration already are paying huge dividends in advances in computer technology, electronics, and countless other fields that touch the lives of all Americans. Scientific breakthroughs on the horizon hold out the promise of further increasing our understanding of our own planet – its weather, environment, geology, and oceans and how they evolved – as well as shedding valuable new light on how we fit into the universe around us. The possibilities seem limitless – certainly too vast to allow the space program to continue its current drift.

Topeka Capital-Journal

Topeka, KS, November 16, 1980

Mankind has learned more about Saturn in the last few days than in all of previous recorded history, in the view of one scientist at the Jet Propulsion Laboratory.

The journey of Voyager I has provided hundreds of exciting pictures of the giant ringed planet and its 15 moons. Scientists will spend years analyzing the data, but already have learned some new things about Saturn.

The fact that it will take years to draw conclusions from the photos is testimony only to the massive volume of information the space probe is providing.

Everyone's first question already has been answered: No, Saturn and its moons do not appear able to support life forms as we know them. This was expected. But the research that will be done on Voyager's photos may unlock other puzzles about the universe.

Manned and unmanned space missions have explored Earth's moon, Venus, Mars, Jupiter and now Saturn. Our horizons have been expanded by a billion miles, just 77 years after man's first airplane flight.

Not everyone shares the scientists' enthusiasm and quest for pure knowledge, but critics of the space exploration program should remember that there are many practical benefits that accrue eventually to nearly all of us as a result of this pure knowledge.

Pocket calculators, digital watches, microelectronics used in automobiles and other products all are direct results of the technology in which the space program was the catalyst.

The Voyager mission is a technological triumph which deserves all the attention it has received.

Minneapolis Star and Tribune

Minneapolis, MN, November 19, 1980

Last week Saturn stopped the chatter at our morning bus stop. Well, not stopped, exactly, but steered it down an unexpected track. Those pictures sent back by Voyager I had loosened the tongues of astronomers everywhere. The astronomers' astonishments put commuters' minds in orbit, too.

Reminiscing, we broke the ice. When Saturn first caught our adolescent eye, its message was of symmetry, of order on high, cloaked in a surprise. A junior-high teacher took us to the football field at night. With his hobby-kit telescope set on a tripod, look, he said, at that yellow star. We looked. In the dancing winter darkness, through a bit of ground glass, the star stood forth as golden Saturn. See the rings? asked our teacher. We could see little else. Slantwise that year, upper left to lower right, they banded the distant planet with a perfect grace. Circles of quiet light surrounding a silent sphere. Measured, serene, passionless, lucid. A bright sensuous jewel of celestial right reason in the velvet sky. Such was Saturn to us at feverish 14.

Well, your teacher should have taught you better, said a literal lady who is a teacher herself. Didn't he point out that mighty Saturn is flattened at the poles and fat in the middle? Comes from spinning too fast. And that one of its many moons goes around backwards? Some symmetry! Saturn should be a lesson, even for kids, that nature, too, may be out of control. No wonder Voyager finds braids in the rings and kinks in the braids and shifting spokes that no one can explain. Like the planet you know best, Saturn close up seems rather a mess.

Wrong approach, interrupted a closet classicist. As the ancients knew, these planets afar are gods, not satellites. Saturn was a favorite. He taught people early the arts of farming and civilized life. In Rome, they liked him so well his annual holidays lasted a week. Everyone gave gifts. The courts were closed. Slaves and masters supped together. So why quarrel now with Saturn the planet as a symbol of goodness? Voyager's vision may yet pave the way for space-age Saturnalia.

But, reminded a philosopher among us, don't forget the clouds with dark halos. Dark halos? Yes, the spacecraft took their picture. Poetic science sees something sombre in Saturn's saintliness. Surely the god's celebrants in old Rome saw it too. You can't close the courts unless crime and contention have brought courts into being. Slaves and masters may not consort unless first enslavement is practiced and approved. Perfect planetary rings harbor twists and smudges that are riddles for the rational. Halos are dark, even a billion miles away.

Is that what the news from Saturn means? Does the good exist everywhere only against a backdrop of equal evil? Boarding the bus, we sat very quiet all the way downtown.

Voyager 2 Flies by Saturn

The U.S. unmanned spacecraft Voyager 2 made its closest approach to Saturn August 25, 1981, coming within 63,000 miles of the giant planet's cloudtops. In the course of the encounter, Voyager 2 took photographs and gathered other information about Saturn itself, its moons and its complex system of rings. In late 1980 Saturn had come under the surveillance of the twin Voyager 1 spacecraft. In light of the Voyager 1 findings, the sequence of observations to be made by Voyager 2 was reprogrammed to focus attention on aspects of the Saturn system that had revealed themselves to be particularly interesting and baffling--in particular, the ring system. But the Voyager 2 did not merely refine the data obtained from the previous mission: it also made new findings of its own, passing closer to a number of Saturn's moons than had its predecessor. Saturn was known to have at least 17 moons, several having been discovered by Voyager 1. The 1,800 pound Voyager 2 functioned perfectly for much of the Saturn mission, beaming back to Earth photographs superior to those obtained by Voyager 1. NASA administrator, James Beggs, hailed the Voyager mission as "one of the great scientific achievements of our age."

While Voyager 1 had stunned scientists by showing that Saturn's rings, thought to number six or perhaps a few more, actually were divided into at least several hundred ring formations, Voyager 2 with its improved cameras revealed that even that number was too low. "We're that much closer this time, and we see literally thousands of rings around the planet," said scientist Bradford Smith. He added, "Once again, we underestimated the scales we see when we fly closer to the outer planets."

Some of the most unexpected findings of Voyager 1 remained mysterious under the scrutiny of Voyager 2 such as the "braiding" of some rings and the presence of "spokes"--dark sections directed radially outwards from the rings. However, the colors reflected by different rings gave a clue to what they were composed of, said scientist Eugene Shoemaker. Small ice particles apparently predominated in the inner rings, while larger pieces of rock apparently made up the outer rings.

Since Saturn takes nearly 30 years to circle the sun, the nine and a half months between the two Voyager fly-bys were equivalent to about a week in the progression of seasons on the giant planet. Scientists found substantial changes in Saturn's weather, but they also found that they were able to identify storm systems that were in existence when Voyager 1 passed Saturn.

Voyager 2 investigated several of Saturn's moons, including Titan, Iapetus, Hyperion, Enceladus, and Tethys. While Voyager 2 did not pass as close to Titan as had Voyager 1, the later spacecraft's photographs of the moon revealed changes in the atmosphere since the Voyager 1 fly-by.

The Washington Post

Washington, DC, August 26, 1981

THE VISIT VOYAGER 2 is now making to Saturn completes two decades of exploration of the planetary system. No more information about those fascinating objects in the sky will pour in from American spacecraft for at least four years, once Voyager's equipment is powered down. And the next data from outer space may be from Voyager itself if it and its equipment survive the journey to a planned rendevous with Uranus early in 1986.

In one sense, this halt in the acquisition of knowledge will be useful. More has been learned about the other planets that circle our sun in the years since Mariner 2 flew past Venus in 1962 than had been learned in all prior human history. That knowledge—some of it illuminating, some of it mystifying—does not all fit together neatly, and a pause for reflection, theory-building and planning can help define the questions to which a new generation of spacecraft can seek answers.

The flights of the unmanned spacecraft—from the rudimentary Mariners through the Vikings, Pioneers and now Voyagers—have been technological marvels. The idea that human beings could design, build and send through millions of miles of space instruments that could send home pictures, temperature and radiation readings and much other scientific data was still a dream whem most of the Americans alive today were born.

The people of ancient times knew about Mars, Venus, Mercury, Jupiter and Saturn long before the telescope was invented. In their imagination, such wandering stars became the homes of other civilizations. We now know civilization does not exist upon them; they and their moons are home to such strange things as oceans of methane and rains of sulphuric acid as well as ice and rock. There is no other Earth out there, and each planet is unique and strikingly different from its neighbors.

The first era of planetary exploration has produced the data for a revolution in the way scientists think about the solar system and the Earth's role in it. The next era—which must not be too long delayed by the budgetary pressures that have brought about this pause—may change it all again. The planetary system and the universe are as grand as any ancient observer ever dreamed they were, and craft like the Voyagers are just beginning to provide a glimpse of what they are really like.

The Pittsburgh PRESS

Pittsburgh, PA, August 30, 1981

Many Americans may not have been aware of it, but the past two decades have been a veritable "golden age" in the exploration of the solar system.

It will be a long time before there is another one.

Beginning with the first flyby of Venus in December 1962, the United States has sent 19 unmanned probes to five planets — five to Venus, six to Mars, one to Mercury, four to Jupiter and three to Saturn.

Actually, the golden age encompassed only 15 years. Mariner 2, the first of the missions, was launched in 1962; Voyager 2, the last, was launched in 1977.

Now receding past Saturn, Voyager 2 is scheduled to encounter Uranus in 1986 and Neptune in 1989. If, against all odds, its instruments are still working then, it will send back the first close-up pictures of those unimaginably distant bodies.

Otherwise, the burst of planetary exploration is over — the victim of budget cuts and a shift in NASA's priorities to Earth-orbiting space shuttles.

The launch of a spacecraft from a shuttle in 1985, designed to go into orbit around Jupiter, is the only interplanetary mission firmly on NASA's schedule in this decade.

But what a productive era it has been!

In 18 years we have learned more about the planets — and encountered more new mysteries — than in all previous human history. Analyzing the data from Voyager 2 alone will keep scientists busy for years.

Each planet, as well as numerous moons, has been found to be unique. From the deep rifts and ice caps of Mars to the inexplicable complexity of Saturn's rings. From the perpetual cloud cover of Venus to the extraordinary volcanic activity on Io, one of Jupiter's moons.

★ ★ ★

More questions than answers about the origin and history of the solar system have been provided by these 19 missions. But that is as it should be, for the human spirit thrives on challenge.

If nothing else, the Mariners, Vikings, Pioneers and Voyagers have given us a new appreciation of how special and fragile is the Earth — the only planet we know of harboring life, and the only life we know of impelled to search out the worlds beyond it.

The Times-Picayune
The States-Item

New Orleans, LA, August 28, 1981

Saturn, the jewel of the solar system, has been revealed by Voyager 1 and now spectacularly by Voyager 2 to be a veritable jewelry store stocked with an assortment of glittering moons and thousands of rings spinning like Mardi Gras beads around the central gem.

After travelling almost 1¼ billion miles over four years, Voyager 2 came within 30 miles and 3.1 seconds of its scheduled arrival point, and skimmed by Saturn at 54,000 mph and an altitude of 63,000 miles, swivelling its camera and sensors like a kid at his first parade. "This is exploration at its finest," said Andrew Stofan, acting space sciences chief of the National Aeronautics and Space Administration.

He meant, clearly, all aspects of achievement — human, technological and scientific. "The theories are falling by the board," commented one scientist. Fly-by day, said another, was "the day of challenge, and the rest of the week you'll see how we meet the challenge." That week will stretch into years, of course, as scientists try to solve the puzzles Saturn now poses and to relate the results to earth studies. "Everything we're seeing," said yet another scientist, "is new."

The single hitch in this remarkable mission — the jamming of the camera mount on the other side of Saturn — seems to have been fixed, so the truly pioneering leg of Voyager 2's endless flight should continue on schedule. It is now on a heading to Uranus, an odd gassy planet that rotates on its side and has rings as yet unphotographed. It will take Voyager five more years to get there, and after that it will spend four more years getting to Neptune, which is so far out that it has not completed a full revolution around the sun since its was discovered in 1846.

But until Voyager reaches Uranus no American ship will visit another planet. A probe of Jupiter dubbed Galileo is set for a 1985 launch and 1987 arrival, but budget cuts and concentration on the space shuttle have squeezed planetary exploration — even a bargain $350 million scout to check out Halley's Comet, due to pay the earth another visit in 1986. The Soviets and the Japanese are sending craft up to meet it, and we could miss an opportunity to be in on fundamental discoveries, for comets are thought to preserve original matter from the solar system's birth.

White House Counsellor Edwin C. Meese told the jubilant Voyager team that space exploration would find "a very receptive administration — obviously within budgetary restraints." But such exploration is something only the federal government can do, and decisions should be made on the basis of investment in the nation's position as leader in space as well as on that of of simple disbursement of funds.

The Boston Globe

Boston, MA, August 27, 1981

If the Red Sox broke your heart in Anaheim, there was consolation down the road in Pasadena, where the Jet Propulsion Laboratory was relaying the latest pictures of Saturn, taken by the spectacularly successful Voyager 2 satellite.

Launched four years ago, Voyager's achievement is hard to overestimate. It has added enormously to mankind's knowledge of the solar system and has added tantalizing new questions for scientists today and future probes in the years to come.

The last-minute disruption of the camera platform, just after Voyager had passed through Saturn's complex ring system, takes nothing away from the achievements of the flight. The failure of the camera aiming device, possibly caused by collision with a particle in space, was just a reminder of the hazards which the spacecraft had to overcome to succeed.

The list of accomplishments is long. Fresh pictures of the ring system and of weather patterns on the surface of the giant planet will undoubtedly provide planetary scientists with a better understanding of the dynamics of satellite formation. Startling pictures of the misshapen Saturnian moon Hyperion will be pored over for clues to the forces that left it looking like "a thick hamburger patty," in the metaphor of Dr. Bradford A. Smith, NASA photo-interpreter.

Nothing is perhaps so impressive as the elegant accuracy of the shot itself. In four years of travel since its launching, Voyager had traveled more than 1.2 billion miles, passing Jupiter and its moon system before arriving at Saturn within three seconds of the scheduled time and passing about 30 miles from its theoretical aim point.

It is hard to find an earth-bound analogy, but the shot is rather like a golfer teeing up in Boston, hitting a drive that soars across the country. It hits exactly the right freight car in a speeding train in North Dakota (on the left side, please, not the right) and bounces southwest across the rest of the country and out over the Pacific. Near Hawaii, the ball passes as planned through a predetermined porthole in a tanker steaming south, and then bounces off the deck and off to the beyond. Now that is American technology at its exciting best.

Voyager, if all goes well, will continue for another five years to the frigid planet Uranus and then on for four more years to Neptune. Saturn, alas, will lapse into the loneliness that has marked its past because no further flybys are scheduled by NASA or any other country. It is a shame. Mankind has reached a long way to examine the solar system. A deep-seated drive to know and understand the world beyond ours ought to have a more satisfactory sequel.

That drive has somehow escaped a few Americans. Among them is the senior senator from Wisconsin, William Proxmire, who, in the glow of Voyager's triumph, criticized funding for additional Saturn probes. "The planets will still be there for years to come," he said on ABC's "Nightline," as he called for a balanced budget.

Had Proxmire been fiscal adviser to Queen Isabella, he would no doubt have urged her to save money by giving none of her jewels to that bumptious sailor from Genoa.

Proxmire could make amends by supporting funds, while there is still time, for a launching to greet Halley's Comet, due in 1986. Voyager may yet lend him and others the necessary vision. Without such vision and confidence, we'd still be waiting for Columbus.

The Evening Gazette

Worcester, MA, August 29, 1981

The reports on the climax of the four-year journey of Voyager 2 to Saturn and beyond continue to be mind-boggling. Not only is there the spectacular recording of the sights and sounds of Saturn, but there's also the incredible freeing of the stuck camera platform across billions of space miles.

Anyone who has ever tried to get a balky radio, camera or latch to work properly can imagine the glee that greeted the loosening of the platform frozen in place while Voyager 2 was on the far side of the planet Tuesday. The jold from an apparently incorrect computer command freed the equipment which had been stuck an entire day.

Although Voyager 2 is millions of miles away and speeding to an even more remote rendezvous, the spacecraft is still under command from Earth. Man has attempted the impossible and prevailed in this link with the other planets in the galaxy. Voyager 2 is scheduled to visit Uranus in 1986, and then take a peek at Neptune in 1989.

Meanwhile it has delivered to scientists at the Jet Propulsion Laboratory in Pasadena, Calif. and to kibitzers at space observation towers around the world, an incredible amount of information — sonic readings, photographs in black and white and color, temperatures and data of unknown elements such as the "kinky ring" that looks like twisted wire, discovered just before the camera platform stuck. The studies of all this information will take as long or longer than Voyager 2's continuing trip. As one of the scientists, physicist Dr. Edward C. Stone, remarked: "We have an overwhelming amount of new data."

That new information includes detailed views of Saturn's 11 moons, each of which appears to be different in shape, size, composition and distance from the planet. The two seen most recently are Hyperion, which is flat, hamburger-shaped, and pockmarked, and Enceladus, which may be a huge ice ball with a smooth surface as the result of melting and refreezing.

While the scientists and explorers may be the ones concentrating on the information Voyager turns up, everyone can take pride in the accomplishments of science as shown by the spacecraft's incredible journey.

The Chattanooga Times

Chattanooga, TN, August 29, 1981

By any measure, the celestial expedition of Voyager 2 and the spacecraft's voracious collection of data on the planet Saturn, beamed back to eager scientists at California's Jet Propulsion Laboratory, is a stunning triumph of technology. It must not be forgotten, however, that the technology would have been useless without the scientific vision that resulted in Voyager 2's launch four years ago on its billion-mile journey.

We have learned several things about Saturn this week. The shape of one of its moons, Hyperion, for example, has been variously likened to a brick, a bent beer can and a hockey puck. Another moon, Enceladus, is described as "wrinkled," suggesting the possibility of recent surface activity, such as a water flow that later froze. And scientists are still attempting to digest the avalanche of information on Saturn's mysterious rings.

Just as intriguing are the mind-boggling distances that characterize Voyager 2's still-uncompleted trip. It has traveled more than one billion miles from Earth. Its data transmissions travel at the speed of light — 186,000 miles per second — but still take about 90 minutes to reach JPL scientists and their decoding computers. But the wealth of information makes the wait worthwhile.

We have learned, further, that the distance from one edge of the Saturnian rings to the other is about 265,000 miles. That means, in effect, that the planet occupies about the same space as that between Earth and its moon. And Saturn itself appears to be, in a word, unfriendly — at least to life as we know it. It features a hostile terrain, an atmosphere of uncertain makeup and wild electromagnetic storms, including a 1,100 mile-per-hour wind that lashes the planet's equator.

There is some concern among space scientists that Voyager 2 represents this country's last major project to explore the planets. However, that doesn't necessarily mean the U.S. space program is being crippled; despite costly delays, the Space Shuttle is due for another launch next month, during which it will conduct numerous experiments that will be of more immediate benefit.

Similarly, there are concerns among some that the cost of Voyager 1 and 2 — about $500 million — is not money well-spent when other government programs are being cut back. The argument, essentially, is that the taxpayer is not getting much of a return on his investment. But that is a narrow view. The Voyager series of explorations — upon which, incidentally, the "Star Trek" film was based — is a valuable expression of man's search for knowledge. And that search will continue, even as Saturn's gravitational field "slings" Voyager 2 toward its next stop, the planet Uranus. Even traveling 50,000 mph, the one-ton, nuclear-powered robot ship won't arrive until 1986. Maybe by then, we'll have absorbed the information it has transmitted so far. Maybe.

THE BLADE

Toledo, OH, August 29, 1981

THE more man learns about the giant planets of the solar system through the amazing travels of Voyagers I and II, the more puzzling they and their extensive satellite families become.

Voyager II made its closest pass at Saturn, the sixth planet, this week and now, aided by the slingshot force of that planet's gravity, is presumably on its way to lonely encounters with Uranus in 1986 and Neptune in 1989. Then, like its predecessor, it will disappear into the vast reaches of space that so far as mankind is concerned have been traversed only in the imaginations of earthbound science-fiction writers and movie producers.

Saturn's ring system apparently is even more complex than the data transmitted by the first Voyager had led scientists to believe. Its entourage of moons, at least 17 of them, is more varied in range, size, and characteristics than had been earlier assumed. One moon, Iapetus, appears to be virtually a snowfield on one hemisphere and a coalbin on the other — a contrast in black and white for which there is no easy explanation.

The surface of the planet itself is rather bland compared with the gaseous turbulence of its giant neighbor, Jupiter, which is sometimes referred to as a failed sun. But Saturn does have rapid atmospheric changes and ferocious jet streams. It is cooler and more mysterious, it will not give up its inner secrets readily, and it may be generations before much more is known about what lies beneath its clouds.

Although there was trouble with Voyager's photographic equipment during the Saturn flyby, one can marvel at the precision with which both Voyagers have carried out their exotic missions. The next target, Uranus, lies nearly a billion miles farther out from the sun and Neptune a billion miles beyond that.

These incredible distances tax the imagination and no doubt will impose bounds on explorations for the foreseeable future. But it is in the nature of man for his reach to exceed his grasp, whether it comes to everyday physical objects or knowledge of the universe.

The Voyager missions have been feasible in part because the conjunction of the four giant outer planets has been favorable to their flights. These conditions will not exist again during the rest of this century. Still, it is inconceivable that there will ever be a time when scientists no longer seek to expand their knowledge of the marches of space, unreachable though that goal may seem now.

The Dispatch

Columbus, OH, August 28, 1981

WHY DO WE walk among the stars? Why do we send our machines billions of miles into space to photograph worlds we may never inhabit? What drives us to question the order of things when the answers are so complex that they confound our knowledge?

We send Voyager 2 to Jupiter, then Saturn and hopefully Uranus and Neptune in a bold challenge to the mysteries of the universe, and we are humbled by the realization that our knowledge is finite, our understanding feeble.

We probe in space but really probe ourselves. Centuries of theories about the stars, the planets, their origin, their movements are put to the test in the startlingly short span of a few years.

We add to our knowledge and evaluate each theory. And we evaluate the leap of reason — the mental process of forming theories that takes us from verifiable fact to hypothesis. We evaluate not only our conclusions but the way we reach conclusions, the way we think. Space travel is, utterly, a voyage of the mind.

Scientists term each mission a success when they fidget — long range — with their complicated widgets. They proclaim success not only in joy but in an effort to justify the millions of dollars invested, and in hopes that the remote manipulation of cameras, transmitters and rockets will generate more funds for future missions.

But the real justification is the quest for knowledge and a fuller understanding of who we are. We walk among the stars because they are intimately a part of our existence.

We send mechanical surrogates to distant places because we cannot yet go ourselves and we need to know — now — what awaits us. And we question because the search for knowledge is the ultimate human endeavor: Fulfillment comes as much in the quest for knowledge as in its possession.

The triumphant journey of Voyager 2 is a celebration of human ability, a herald of our presence in the universe. And it is testimony that one creature in the vast reaches of space has the will and courage to probe its mysteries.

The Register

Santa Ana, CA, August 28, 1981

Few would dispute the achievement represented by the recent encounter of *Voyager II* with Saturn. The mission, and the pictures returned to earth have simultaneously added a great deal to our knowledge of the cosmos we inhabit, and impressed upon us just how much remains to be learned. We inhabitants of this small planet may seem insignificant in the perspective of interstellar space, yet the fact that we reach out into the cosmos gives us reason to believe that there is something special about us. An event like the *Voyager* flight seems to do something for our sense of shared humanity; it increases our awareness of the possibilities inherent in life, and stimulates our imaginations.

While the *Voyager* flight itself has served to enhance our image of ourselves, a good deal of the discussion surrounding it tends to bring us back to earth. One might have hoped that an event that demonstrates the magnitude of the possible frontiers that still await us would unleash some genuinely imaginative thinking, some verbal exploration of alternatives, some farsighted musings on what could conceivably come to pass when the imagination of individual people is truly unleashed.

Instead, most of the media gave us discussions of the NASA budget and conventional wisdom on the relationship of the domestic economy and future space exploration. The unstated premise underlying most discussions is that space exploration and exploitation of the possibilities of space are by their very nature government operations.

Are they really? In the atmosphere of possibilities and potentials that surrounds a successful space mission, we invite you to consider that question.

The commercial potential of satellites is already apparent. Indeed the satellite business has become a quasi-corporate, quasi-government undertaking. A small company in Texas, despite recent setbacks, still plans to launch a totally private satellite in the near future. There's no logical reason that satellites couldn't become wholly private endeavors in the near future, though there may be practical and political reasons that make such a development unlikely. But let's think in terms of what *might* be for the moment.

There's bitterness in the space community that the government doesn't seem to want to fund the kind of extensive probing of Halley's Comet that many scientists would like to see. If the government doesn't want to do it, does that mean it can't be done? Perhaps it does, but perhaps we're trapped by conventional thinking. Would it really be impossible to raise funds for such an endeavor privately and voluntarily? Frankly, we doubt it, if the decision were made and the resolve to stay with it existed. The TV rights alone might finance most of the project. Such fund-raising might involve commercialization and a carnival atmosphere, which might offend some serious scientists. But we've always been delighted at the odd blend of the serious, vulgar, exploitative, greedy, adventuresome and philanthropic that seems to characterize human beings as they really are. An enterprise that recognized these characteristics, reveled in them, and combined them for a noble purpose just might make it.

The argument against funding space exploration that is most persuasive revolves around the notion that so long as our economy is in such bad shape, and we're beset by inflation, unemployment and inefficiency, the leap to the stars must be considered a luxury, to be financed, if at all, when we find a way to achieve a surplus of funds consistent with some measure of prosperity here on the third planet from the Sun.

This argument has great merit if space exploration is to be considered the exclusive province of government. But if we open our minds to other possibilities, we may find a different perspective. Human society, though it may be a tiny dot in interstellar space, is remarkably complex and sometimes apparently contradictory. That some are rich while others are poor may be viewed as tragic, or it maybe viewed as a reflection of the diversity inherent in humanity. If we view humanity only as closed system, we miss the surpluses that exist in some sectors — surpluses that may be available to finance dreams and aspirations.

Humankind on this earth has achieved its most notable successes when people had the maximum possible personal freedom to dream their dreams and do something about them. Viewed in this perspective, our problems may not be so much the result of ignorance and greed, but of unwise choices. Solutions to our problems on earth, while perhaps not likely, may just be possible if we understand the importance of liberty and rescind the choices that have led to its constriction. Experience indicates that a full flowering of liberty will lead to innumerable surpluses — of money, talent, restlessness, venturesomeness and dreaming — that could stimulate exploration to the stars and beyond, for those who care to participate in the adventure.

Is that an impossible dream? Perhaps. But *Voyager II* has given us a license to consider the impossible and wonder if it couldn't be done.

The Burlington Free Press

Burlington, VT, August 27, 1981

Sailing gracefully across the uncharted seas of space, Voyager 2 is an electronic version of the explorers of the past who embarked on similar journeys to push back earthly frontiers.

In its incredible voyage of discovery, the spaceship fulfills man's dreams of breaking the bonds of Earth and satisfying his curiosity about the nature and origins of the universe.

As Voyager 2 visits each planetary port, it is gathering and transmitting to Earth the sights and sounds of forbidding worlds which share our solar system. And the information being received not only is changing scientists' perception of those planets but also is providing data which will be analyzed for years to come.

To describe the breathtaking photographs of Saturn and its moons would be to strain the capacities of superlatives. The pictures are at once awesome, magnificent and spectacular. Views of the misshapen Hyperion, icy Enceladus, giant Titan and Saturn's complex rings are stunning. Through the eyes of Voyager, the people of the earth are able to see phenomena that could only be the subject of theory in the past. That the spacecraft could transmit such pictures a billion miles represents a triumph in technology for American science.

For as long as man has gazed up at the heavens and wondered about the mysterious worlds that existed beyond the boundaries of the earth, there has been a profound human yearning to explore the seas of space in search of its secrets. The possibility of life on other planets tantalized scientists and laymen.

Having discovered that neighboring planets are desolate and lifeless has not diluted scientific interest in their character. And many laymen still believe that other forms of life exist somewhere in the universe.

But for now the world's attention is focused on Saturn and the magnificence of a world that is so far away and has been brought so close by man's technological genius.

The Oregonian

Portland, OR, August 31, 1981

The sticking gears of Voyager 2's camera platform are no longer jammed, but all is not well with the spacecraft. It is suffering from a little red-eye as it plunges outward bound on a five-year journey to the distant planet Uranus.

Doctoring a sore-eyed camera at a distance of more than a billion miles on a tiny speck in the solar system is akin to tracking a minnow in the ocean. The crews of the National Aeronautics and Space Administration rate raves for their job in steering the Voyagers to Saturn. A superb job was done in the latest encounter, and despite probable dust particles in the gears, most of the pictures and data sought were obtained.

If Voyager 2's sore eye can be cleared up soon, we will get pictures Sept. 4 of Phoebe, the outermost of Saturn's 17 known moons. Voyager 1 did not come close enough for a useful Phoebe picture, but Voyager 2 will get 7 million miles nearer.

Even if the camera's directional gears remain sticky, it will be possible to so maneuver Voyager 2 that it would get some pictures of Uranus from a fixed camera. But in five years of travel to the most distant and supercold reaches of our solar system, Voyager 2 couldn't be blamed if it came down with a fairly difficult case of the frozen shakes.

THE SACRAMENTO BEE

Sacramento, CA, September 1, 1981

As lay observers of how things tick, we can share only vicariously the excitement of the world of science over Voyager 2's exploration of Saturn, the high point of a journey into space that began over four years ago and has now traversed some 950 million miles. Right at the start we're almost lost in a wonder bordering on unbelief at the thought of anything coursing such a distance and being able to arrive at a point which seems but a dot in the vast expanse of the cosmos.

And what is it that the uncanny instruments aboard Voyager 2 beamed back to the learned physicists, astronomers and chemists on planet Earth? Their fascination, we gather, centers on the composition of the mysterious rings around Saturn, and why they hold their relative position to the giant gaseous sphere, and why at one moment they seem to express the radiating form of spokes and another the ripple of waves, and why the moonlets that may — or may not — be contained in them work gravitationally to affect their character.

A still deeper mystery, exciting the imagination of radiophysicists, is the apparent presence of electrical charges in the ring plane, something detected by the earlier Voyager 1 whose sensitive instruments picked up radio noise characteristic of electrical discharges.

We confess all such speculation on the exotic objects of scientific curiosity find us intrigued, but baffled. The very idea that Saturn's rings somehow relate to the common universe of matter in ways that might further unlock the molecular structure of a human gene is as beyond our ken as is the ultimate reason an airplane's wings hold us aloft. Yet it is clear enough by now, beholding the awesome linkage between Einstein's theories and the mushroom cloud, that the explorations of science into both the macrocosm and the microcosm of existence can have implications reaching into our everyday lives. This is something that bids our attention to the Voyager's probe and makes us wonder what those puzzling rings of Saturn may reveal about planet Earth's mysteries.

The Salt Lake Tribune

Salt Lake City, UT, August 28, 1981

Voyager II's close-up look at Saturn is worth the attention it's getting. This will probably be the last space exploration spectacular earthlings can enjoy for several years.

If Voyager II's working parts, which malfunctioned earlier this week, can survive, it may beam first-ever, nearby pictures of the planet Uranus five years from now. That's the plan. But hard-working, remarkably successful Voyager II is showing signs of fatigue; cameras and other internal parts indicate a deteriorating condition. How well the 1986 visit to Uranus will match previous Voyager feats has become problematical. For now, however, there's the Saturn show.

Until the cameras failed Wednesday, it was every bit as fascinating and amazing as previous disclosures introduced by Voyager I. Discoveries as astonishing and perplexing to scientists about Saturn's famous rings are being repeated as Voyager I's twin space prober transmits its pictures taken from a different angle. Both the technology and the eyewitness data it is capturing are nothing short of awesome.

In less than 20 years, from the moment United States and Russian space craft started competing for achievement, man has pushed the limits of his knowledge about the solar system in which he spins farther than earth-bound speculation would have dared predict just decades earlier. Men on the moon, television views of Saturn rings may seem routine. In fact, they remain the most magnificent accomplishments of this scientific-technological age.

Yet the age has turned practical-minded, less grandly visionary. Space shuttle missions are scheduled and authorized. The possible value of living and laboring in full space environment will be worth exploring to the greater degree planned. But the truly tantalizing questions lie beyond orbiting earth stations, outside the solar system.

Above all else, the NASA Mariner, Pioneer, Viking and Voyager series are important for the observable proof they supplied that Earth occupies a unique position among the planets with which it shares the sun's life-giving and life-denying powers. Are there other planets, elsewhere in the universe, equally fortunate? Space probes modelled on, developed from improvements of Voyagers might find some answers.

In reiterating for a genuinely impressed public how competent the past NASA programs have been, new pictures from Saturn ought to remind those who made it possible how much more could be learned by renewing and continuing the same pure exploration effort.

The Charlotte Observer

Charlotte, NC, September 2, 1981

After a four-year journey spanning 1.2 billion miles, the Voyager 2 spacecraft reached Saturn exactly as planned last week. It was a dramatic achievement for America's space program. Unfortunately, it may be the last U.S. space probe of the decade.

Saturn, our solar system's second largest planet, has fascinated scientists for centuries. Its shimmering golden surface and mysterious rings — first spotted by Galileo in 1610 — remained elusive telescopic images until Pioneer 2 sent back photographs two years ago.

Last fall Voyager 1 transmitted more data about Saturn's rings, indicating that there may be as many as 1,000, instead of 6 or 7. Scientists were astounded by what they saw; one noted that in a mere week, we'd learned more about the planet than in all of previous history.

The new probe cruised last week within 63,000 miles of Saturn, some 14,000 miles closer than Voyager 1. Among the data scientists are now sifting through is more information on the rings, the regular bursts of radio static the planet emits, and Saturn's complex gaseous composition. All that will help them piece together more puzzles of the solar system.

Our ventures into space also have had important technological spinoffs, from better crop forecasting to heart pacemakers. But their significance goes well beyond that. Since America's space exploration began to flower in the early 1960s, we have gained a new perspective on our place in the universe and a new understanding of how little we really know.

That expansion of knowledge may soon be halted. The Reagan administration and Congress are pulling back support from the National Aeronautics and Space Administration, emphasizing the military aspects of the U.S.-Soviet space race over the scientific and human value of missions such as Voyager 2. Except for the much-delayed space shuttle (now scheduled to lift off next month) and a possible orbiting of Jupiter in 1987, the space program may be ending.

So let's hope Voyager 2 fulfills the rest of its mission in style. It is now heading for Uranus in 1986 and Neptune three years later. Perhaps it will provide an even more spectacular swan song for the space program, before hurtling away to spend another billion years or so cruising the far reaches of the universe.

The Houston Post

Houston, TX, September 14, 1981

The Voyager 2 spacecraft, now hurtling toward a 1986 rendezvous with the planet Uranus after a spectacular reconnaissance of Saturn, exemplifies the scientific tour de force that the U.S. program of unmanned space exploration has been. Carrying incredibly sensitive and sophisticated instruments, the two Voyagers, the Mariners and other space probes have roamed our solar system, explored alien worlds and beamed back enough pictures and other data to keep scientists busy for years.

The performance of these craft has been remarkably trouble-free considering that they had to travel for months and years, operate in unimaginable extremes of heat and cold, and maintain their communications links with Earth over hundreds of millions of miles. Oh, yes, there was the problem with Voyager 2's camera platform, which became stuck when the spacecraft flew behind Saturn. But the team of scientists and technicians who control our unmanned space flights reached out across 900 million miles and fixed it. Imagine!

Other missions are planned, but it will be five years before the next close encounter with another planet in our solar system. During this lag in way-out extraterrestrial activity maybe we should consider some more down-to-Earth applications for the expertise gleaned from two decades of unmanned space exploration. Do you suppose we could interest the people who designed and built all those superbly durable space vehicles in making some buses for our Metropolitan Transit Authority?

The Cleveland Press

Cleveland, OH, August 29, 1981

Many Americans may not have been aware of it, but the past two decades have been a veritable "golden age" in the exploration of the solar system. It will be a long time before there is another one.

Beginning with the first flyby of Venus in December 1962, the United States has sent 19 unmanned probes to five planets: five to Venus, six to Mars, one to Mercury, four to Jupiter and three to Saturn.

Actually, the golden age encompassed only 15 years. Mariner 2, the first of the missions, was launched in 1962; Voyager 2, the last, was launched in 1977.

Now receding past Saturn, Voyager 2 is scheduled to encounter Uranus in 1986 and Neptune in 1989. If, against all odds, its instruments are still working, it will send back the first close-up pictures of those unimaginably distant bodies.

Otherwise, however, the initial burst of planetary exploration is over, the victim of budget cuts and shift in NASA's priorities to Earth-orbiting space shuttles. The launch of a Galileo spacecraft from a shuttle in 1985, designed to go into orbit around Jupiter, is the only interplanetary mission firmly on NASA's schedule in this decade.

But what a productive era it has been. In 18 years we have learned more about the planets — and encountered more new mysteries — than in all previous human history. Analyzing the data from Voyager 2 alone will keep scientists busy for years.

Each planet, as well as numerous moons, has been found to be unique, from the deep rifts and ice caps of Mars to the inexplicable complexity of Saturn's rings, from the perpetual cloud cover of Venus to the extraordinary volcanic activity on Jupiter's moon, Io.

More questions than answers about the origin and history of the solar system have been provided by these 19 missions. But that is as it should be, for the human spirit thrives on challenge.

If nothing else, the Mariners, Vikings, Pioneers and Voyagers have given us a new appreciation of how special and fragile is the Earth — the only planet we know of harboring life, and the only life we know of impelled to search out the worlds beyond it.

San Francisco Chronicle

San Francisco, CA, August 27, 1981

VOYAGER 2'S DRAMATIC rendezvous with Saturn may have been somewhat marred by a balky camera platform, but the story of this little craft's voyage has brought back the excitement and sense of accomplishment that have been so much a part of America's space-exploration effort. Even if Voyager 2 did pull away from Saturn with a blind eye, the mission must still be considered a high-grade accomplishment. One well worth all the time, money and effort involved.

So, with the electric sense of space accomplishment rekindled, it was somewhat encouraging to hear President Reagan's new space agency chief, James Beggs, offer at least a qualified statement that the administration may yet budget enough money for the United States to fly an epochal mission to trap the world's first sample of a comet.

THE HUNT FOR Halley's Comet would seem a space challenge not to be missed. The Russians, the Japanese and a consortium of European nations are all readying unmanned comet flights, but our own comet mission has appeared to be in serious jeopardy due to the reluctance of budget planners.

The U.S. mission is more ambitious than the others; it hopes for a spacecraft that will actually trap samples of gases from the comet's head and bring them back inside the ship to be gathered by an orbiting Space Shuttle and then returned to laboratories on Earth for analysis. That vision took us back to some time-brittled clippings of this newspaper that told of the anxious watch by scientists during Halley's last visit in 1910.

OUR PLANET, some headline-writer of poetic soul noted, had been "switched by the tail of a celestial vagrant with the gentlest sort of caress." Well, that celestial vagrant will be swinging toward us again in 1985, and this time we will be able to reach out and discover what is inside. That's an opportunity that should not be lost to unrealistic budget cheese-paring. It will be another 76 years before the dazzling vagrant returns.

CHARLESTON EVENING POST

Charleston, SC, August 27, 1981

While Jet Propulsion Laboratory scientists are having trouble assimilating the flood of new information channeled back to earth by Voyager 2 as it passed its rendezvous with Saturn, we have difficulty comprehending how a one-ton space capsule, moving at speeds of more than 50,000 miles-per-hour, can travel almost one billion miles and hit its target right on the nose. The mission director's statement that it compares to "sinking a putt from 500 miles," only adds to our admiration of the feat.

The knowledge transmitted by Voyager will aid all mankind in understanding the universe in which the earth is such a small speck. Even if the spaceship does not survive to complete its mission, its contributions will be immense.

Back on earth, and back to reality, the Reagan administration is in the throes of making very hard fiscal decisions in an effort to produce a balanced budget by 1984. There is no threat to our national interests or security from Saturn. There is a very real threat posed by would-be conquerors right here on earth. The fascinating insight to the forces that formed this universe will not feed the hungry, clothe the poor, or aid the elderly. But we shall all be the poorer if space exploration is put on the back burner.

Ah, but a man's reach should exceed his grasp,
Or what's a heaven for.

Private Rocket Launched in Texas; White Issues Policy Statement

A rocket designed and built by a private company and intended for commercial applications was launched successfully on a 10 1/2 minute suborbital test flight Sept. 9, 1982 from a cattle ranch on Matagorda Island, Texas. The rocket Constega I, named after the wagon used by settlers in the trek west, was about 36 feet long. A booster purchased from NASA was used to launch the rocket. Space Services Inc. of America, the Houston-based company that had developed the rocket, said that it aimed to provide "low-cost, market-oriented" services for oil companies and other businesses with a need for satellites. The rocket lifted its payload to an altitude of about 200 miles. It fell to Earth in the Gulf of Mexico, over 300 miles distant from the launch site. Former astronaut Donald Slayton had been hired by Space Services and served as mission director. "Obviously, we are very happy with the result," Slayton said, adding, "All systems worked out exactly as they were designed to work." The success of the launch contrasted with a 1981 engine test in which a Space Services' rocket had exploded.

The White House issued a policy statement May 16, 1983 saying that business could purchase rockets developed by the government and then launch them, for a fee, from government launch facilities. Under the policy, companies would be able to purchase rocket parts and plans from the government, and they could also rent the government launch facilities at Cape Canaveral, Fla., and Vandenberg Air Force Base, Calif. The government "will not subsidize" the commercial utilization of the expendable rockets, the White House said, but it would offer its facilities at a price "consistent with the goal of encouraging viable commercial" launching.

The Orlando Sentinel

Orlando, FL, September 14, 1982

Somehow it seems appropriate that the recent successful launch of a "private enterprise" rocket took place in Texas, a state always associated with rugged individualism and frontier mentality.

Certainly the stubby 36-foot rocket is in keeping with the do-it-yourself line of thought most of us like to think represents the best of American entrepreneurial endeavor.

Historians may argue that too much credit is given to the pioneer inventors and industrialists of this nation. They may argue that many of them had help of one kind or another that is omitted in glowing accounts of individual enterprise.

And to a large degree, the Conestoga rocket sent 195 miles aloft from its launch pad on Matagorda Island owes an awful lot to the National Aeronautics and Space Administration. The rocket was made from spare and surplus parts obtained on the cheap from NASA. And several of Space Services Inc.'s key people came from NASA as well.

But it is in keeping with the American spirit that these guys should split off from the monolithic NASA and try to make a commercially successful venture on their own. They say they'll need about $15 million in private investments to make it profitable. And they also want to lease launch pads from NASA, an idea fraught with complexities in defining private-sector use of government property.

With the kind of skill and determination Space Services Inc. has already demonstrated, these are not insurmountable problems, particularly since NASA officials say they are more than willing to cooperate with their would-be rivals.

News reports of the Conestoga rocket launch said that it carried no payload on its first flight.

Wrong. It carried aloft one heck of a lot of faith in the fruits of individual skills and dedication, a most important cargo.

DESERET NEWS

Salt Lake City, UT, September 10-11, 1982

Launching of the Conestoga rocket this week sounds like it was far removed from what reporters are used to covering: the slick, expensive Kennedy Space Center launchings with banks of computer terminals, highly-trained astronauts, smooth public relations, and all the rest.

By contrast, the Conestoga was launched from a cow pasture on an island off the coast of Texas. A year ago, another Conestoga rocket blew up on the launch pad during an engine test.

Ah, well — they laughed at the Wright brothers, too.

The difference in the Conestoga launching and those at the Kennedy Space Center is that the former was paid by private enterprise, and the latter by the taxpayers. That alone should say something about why taxes are so high. In any event, the Conestoga made history as a private firm's bid to enter the satellite-launching business for the first time.

The sub-orbital flight of the Conestoga may well herald more ambitious projects in space by this and other private firms. Space Services, Inc., the rocket's developers, envisions ultimate orbital flights on a commercial basis. A company official says he hopes to attract business from oil companies interested in doing private earth resources surveys. He's also thinking of offering a cheap launch service for use by Third World countries.

Is there a role for private industry in the space business? Of course, provided proper controls are imposed to prevent cluttering in space. Potential industrial uses of space are so numerous that private enterprise would be derelict if it were not to explore these possibilities. Mapping of the earth's surface, satellite photography for potential new mineral sources, identification of bug-infested croplands, and satellite monitoring of oceans for oil spills are among the commercial uses of space. Space can also be used to make more perfectly rounded ball bearings and other products whose manufacture is improved by weightless conditions.

The Conestoga's most ambitious project in this week's test was to dump 300 pounds of water into space at the rocket's highest point, and to drop a 10-foot-long payload into the gulf at the end of its flight.

But look for bigger things. Remember the Wright brothers' critics who scoffed at an air machine that needed practically perfect weather conditions before it could fly.

Rocky Mountain News

Denver, CO, September 15, 1982

NEVER mind that the rocket was a vintage Minuteman purchased from NASA for $365,000. Never mind that the world's first privately engineered space launching was based on technology developed at a cost of billions by the government.

The successful suborbital test-firing of the Conestoga I from Texas' Matagorda Island was, indeed, the "victory for free enterprise" its backers hailed it as.

However, there's a catch. Conestoga I cost $6 million, and Space Services needs to raise $15 million to $20 million in new venture capital to launch its first satellite, planned for 1984, as well as find a permanent site from which to put heavy communications and other types of commercial satellites into orbit.

Even then, it will be in competition both with NASA's space shuttle and the European-built Ariane rocket for customers. But if a company with only seven permanent employees using off-the-shelf equipment could pull off a perfect rocket launch, who can say there isn't a role for private enterprise in space?

The Dallas Morning News

Dallas, TX, September 15, 1982

It wasn't an interplanetary space shot. It was just a suborbital flight and far less than a technical space spectacular. But the flight of Conestoga I from Matagorda Island opens yet another era of space development: the private-enterprise phase.

That may be the most important step into space yet. The U.S. space program was never intended to be a purely public enterprise. But since the beginning, it has been anticipated that private enterprise would join with the federal government in development of space, as was the case with other technologies.

Space Services Inc. of America, which launched the private space shot, is one of the pioneers in private development of the range of opportunities in space. The firm's goal is to place satellites into earth orbit for customers at a fraction of the cost charged by the federal government.

But this is only the first small step by private enterprise. Many Americans will live to see major advances in development of space potential by private enterprise and by partnership of private sector and government.

Success of the private enterprise flight, Conestoga I, should whet every American's appetite for accelerated development of space. The cosmos stands as the 20th century's next frontier.

The San Diego Union

San Diego, CA, September 14, 1982

It was a dark day for stockholders of Space Services, Inc., when the first attempt to launch a private enterprise rocket ended in an explosion on the pad last year. They should be feeling better after last Thursday's successful second try.

The 37-foot Conestoga I rocket arched over the Gulf of Mexico from a pasture in Texas. The enthusiastic launch crew, including former astronaut Deke Slayton, is now shooting for a date in 1984 to send a rocket into orbit.

It remains to be seen whether Conestoga I is the start of something big — whether there is a future for private entrepreneurs in the space-launching business.

Satellites have been providing commercial services for many years, especially in communications, but the National Aeronautics and Space Administration has had a U.S. monopoly on putting them into orbit. Its only competition has come from overseas.

There is no reason why NASA should be the sole proprietor of space-launching facilities in America. Indeed, the investors in Space Services, Inc., are convinced that private enterprise eventually will be able to provide launch services cheaper than NASA.

Who knows, the space agency may one day find itself in the same boat with the U.S. Postal Service, which has seen much of its parcel post business drained off by private concerns. Especially if private companies can promise next-day delivery into orbit.

The Kansas City Times

Kansas City, MO, September 13, 1982

Even as the NASA space shuttle Columbia was being fitted to its booster rocket and propellant tank at Cape Canaveral for its first operational mission Nov. 11, two competitors were challenging in the skies. The shuttle, despite the high hopes generated by four successful test flights, will not have the world satellite installation and servicing market to itself if plans work out for the European Space Agency, a 10-nation consortium, and a private Texas company called Space Services, Inc.

On consecutive days the Houston firm launched the first successful private-sector space rocket on a suborbital design test from Matagorda Island, Texas, and the European agency suffered its second failure in five shots with the Ariane rocket from a French Guyana base. Despite the setback when the Ariane failed to reach enough speed or altitude to orbit, the consortium has 24 contracts for a mid-1983 start of commercial flights and hopes to corner a third of the world business, using more powerful versions of Ariane.

SSI intends to put smaller rockets into low orbit at a fraction of the government agencies' costs, with most of its prospective customers oil companies interested in remote-sensing satellites that can pinpoint gas and oil reserves. The Texas outfit will need a new launch site and a better rocket than the 20-year-old Air Force Minuteman booster that was used.

Such previous NASA missions as the moon landings and distant planetary probes have been largely cost-plus projects, but the shuttle, as a utilitarian space bus and truck operation, is supposed to try to break even. Most of its cargo bay space on future flights has been reserved by the Pentagon, which gives NASA a healthy start, but for corporations with satellite aspirations the space agency may have to compete for business with both the Europeans and an American private space enterprise.

The Times-Picayune
The States-Item

New Orleans, LA, September 12, 1982

With its name recalling the westward trek of American frontier settlers but with its path the space age's southeastward and up, Conestoga 1 made a successful suborbital test flight rightly judged "very, very symbolic." The rocket is the brainchild of Space Services Inc., an American company determined to build a private-sector space program. SSI consultant Lee Scherer, former Kennedy Space Center director, said, "This is a test vehicle whose demonstration will show that a private concern can go do this."

Last year, SSI's first rocket, the Percheron, blew up during an engine test. Thursday the Conestoga lifted off from a small pad on Matagorda Island off Texas and in a 10-minute flight reached an altitude of 190 miles and a range distance of 320 miles, splashing down in the Gulf of Mexico 270 miles off the Mexican coast. This was the first of a planned series of Conestogas to be launched into low Earth orbit.

The purpose of the venture is to offer launches at a lower price than that charged by the public National Aeronautics and Space Administration and other national space programs to companies that use or want to use their own satellites. On the prospects list are oil companies that use low-orbit resource survey satellites and Third World countries looking for cheap launch facilities. In the planning stage is a larger rocket that could put communications satellites into orbit 22,000 miles up. The launch was watched by 57 SSI investors and several dozen guests and potential investors.

The SSI venture is a classic exercise in entrepreneurship and management. SSI buys its basic hardware readymade from government and the aerospace industry — Conestoga was powered by a Minuteman I missile second-stage engine, and for higher orbiting SSI is talking to NASA about buying an Atlas-Centaur — and contracts with regular space-equipment producers and consultants. Mr. Scherer and former astronaut Donald K. "Deke" Slayton, mission director, came from NASA, as did others on the operations staff.

Matagorda Island is not right for orbital launches, so the company is trying to lease a site on Hawaii next to an old missile range launch site, and part of the Atlas-Centaur deal involves leasing a little-used launch pad at the Kennedy Space Center in Florida.

SSI has come this far in only two years, and if its Conestoga rocket-wagon proves itself in orbit and its operations-scale costs prove competitive, it may indeed open a new frontier of space.

Pioneer 10 Exits Solar System

The unmanned Pioneer 10 space probe, which had been launched in 1972, crossed the orbit of Neptune June 13, 1983 and continued on its way into interstellar space. The event was hailed by U.S. space officials, who noted that Pioneer 10 was the first man-made spacecraft to voyage beyond the realm of the planets. Pluto was usually more distant from the sun than Neptune, but currently and for the next 17 years Pluto's eccentric orbit had taken it within that of Neptune.Pioneer 10 had been designed to perform a fly-by of Jupiter, and it had done that successfully at the end of 1973. The spacecraft was still radioing information back to Earth, its equipment far exceeding the expected level of performance.

The Charlotte Observer

Charlotte, NC, June 16, 1983

Men were fearfully beating drums in Indonesia Sunday to ward off the evil giant "Kalau Rau" that they thought had caused the first total eclipse of the sun in 350 years. On the same day, other men watched with pride as Pioneer 10, the spacecraft created to explore the universe, hurtled toward the edge of this solar system on its mission to the stars.

That juxtaposition of events is a reminder of both how far the human mind has taken us and how much remains a mystery.

It was nothing less than wondrous when, at 8 a.m. Monday, the little silver-and-gold craft slung itself out of the solar system, carrying a plaque identifying its source as Earth.

Thus, after hurtling through space for more than 10 years and surviving many dangers, the 500-pound craft began a new phase of its journey. Now, scientists say, Pioneer 10 will float through airless, frictionless space for infinity — long after the men who conceived of it are dust.

Until the signal fails or becomes too distant for Earth to hear, an 8-watt radio transmitter sending back coded beeps to scientists is humanity's ear to the universe — and our hope for greater understanding of our relationship to that awesome void of outer space.

The Saginaw News

Saginaw, MI, June 17, 1983

Nothing is forever? For earthly mortals, that may be so. But we mortals have created something that, if it is not forever, is about as close as you can come.

Pioneer 10 this week passed the orbit of Neptune, and thus left the Solar System on a journey to forever. Its lifespan now is measured in millions of years, but even that is not beyond the ken of our scientists. They have projected its path through the void for at least the next 862,000 years, beyond Proxima Centauri and Altair out to a star known only as D+25 1496. The map Pioneer 10 will follow is on a scale of 30 million million miles.

The national debt has gotten us used to the idea of unimaginable numbers. But we are speaking here of utter infinity. And as long as there is someone to listen, Pioneer 10 will keep talking to us, sending back messages — even images of stars — for eternity.

It is humbling to look into the heavens and contemplate our puny status compared to the vastness of it all. It is also a matter of pride that we could create something that can join the voyagers of the universe.

Whatever may become of its creators, Pioneer 10 will go on, a memorial to our ability to reach beyond our own tiny orb, the limited universe of our own human senses, and carry something of ourselves out to the stars. No, we are not alone, not any more.

The San Diego Union

San Diego, CA, June 13, 1983

An unheralded historical event occurred nearly 3 billion miles from planet earth at 5 a.m. today. At that moment, Pioneer 10 left the solar system and became the first human artifact to enter interstellar space.

In breaking away from the hold of the sun, the nine-foot spacecraft embarked on a journey into the unknown that may leave it as the only surviving memorial to our era.

The 11 years that have elapsed since the launching of Pioneer from Cape Canaveral may represent only a tiny fraction of a trail-blazing trip that could take billions, possibly trillions, of years. When the sun explodes into a "red giant" and vaporizes the earth in 5 billion years or so, Pioneer 10 could be continuing on its voyage through the expanding universe.

Where and when Pioneer's incredible journey will end are questions that cannot be answered. But if the hardy craft encounters other intelligent life in the vastness of space, its small aluminum plate carrying sketches of a man and a woman and indicating the location of our solar system may provide a footnote for a future history of the universe.

Some day, somewhere, scholarly ETs out there could learn of a civilization of earthlings that flourished on the third planet from a small star in a faraway galaxy long ago.

The Orlando Sentinel

Orlando, FL, June 15, 1983

'Beyond the solar system . . . ." You don't have to be much of a science fiction buff for that deceptively simple phrase to give you pause.

With Pioneer 10's passage Monday beyond the last recognized limits of our solar system, man has sent an extension of himself into what had been the exclusive realm of his imagination.

It is a stunning accomplishment that is all the more astounding because it was, well, unplanned. Launched March 3, 1972, Pioneer 10 was to visit some of our planetary neighbors and last about 21 months. Eleven years later it is still working beautifully. All but one of its 11 scientific instruments are faithfully transmitting data back home while the probe moves away from us at a steady 30,558 mph.

In addition to its scientific mission, Pioneer 10 carries a plaque denoting its origins and some recorded messages, including the sounds of a great whale and the rock music of Chuck Berry.

All of this amounts, as one observer noted so well, to the astronomical eqivalent of a note cast asea in a bottle in the very human hope that someone else is out there, that we are not alone after all.

Pioneer 10 has become another source of fascination for those of us given to gazing at the night sky. What will they think out there? What will they make of great whales and Chuck Berry?

The Hartford Courant

Hartford, CT, June 14, 1983

There was exhilaration in the moment Monday when the Pioneer 10 satellite, after 11 years in transit, poked beyond the most distant planets in our solar system. There was also loneliness.

It is still to be seen whether the tiny statellite is really a pioneer, or a refugee.

The satellite has performed beyond all expectations — it was supposed to last only 21 months. But it endured every storm in space, continued to send back information and astonishing photographs of the planets. Its signals — now received on Earth with the strength of one billion-trillionth of a watt — could continue for another decade.

There may be other life somewhere in space, and that life may encounter Pioneer 10, pick it up, turn it around and examine it, observe the gold-plated plaque bearing a picture of a man and woman, and draw some conclusions about the universe. But maybe not.

Maybe Pioneer 10, when its signals no longer reach Earth, will have lost its final contact with life in the universe for eternity. The planet Earth seems so rare and so special against the prospect of endless deadness in the rest of the universe.

The satellite is a kind of reproof to those who would endanger this unique place with orbiting military hardware, including laser weapons. These people are moved by the bizarre notion that weapons in space, which could plunge the human race towards obliteration, would make the planet safer.

Pioneer 10 might live up to its name and succeed in communicating the fact of human life to other intelligences. But at the worst, opposite extreme, it might be a refugee, bearing a moot message, from a planet that succeeded in destroying itself.

The Philadelphia Inquirer

Philadelphia, PA, June 18, 1983

Now that the idea of folks in galaxies far, far away and visits from extraterrestrials is becoming second nature, the glory of science has snapped reality back into its intimidating place: This week, the first particle of human creation left the solar system.

For Pioneer 10, a 570-pound satellite crammed with scientific instruments launched from Florida in 1972, it was the merest first step on a voyage that will extend trillions of miles and take billions of years. Having left the modest celestial intersection of the sun and its nine planets, Pioneer is now tooling across the immediate galaxy, the Milky Way, and can be expected to go on to the next in due time.

By then it will be an orphan. In the five billion or so years that this leg of the eternal trip will take, the sun will have burned out and Earth and its fellow planets will have died of natural causes. Think of it: Pioneer 10, mankind's first permanent artifact.

It was only about 175 years ago (that's a seventieth of the time till Pioneer approaches its first star, Barnard's) that travel and communication were limited to the capacities of feet (human or equine) and sails. Soon enough, flight was no longer for the birds, people began talking with friends across town or across the globe, and men were walking, yes, *walking*, on the moon.

But that progression has been nothing more than the twinkling of an eye for the likes of stars and planets and Pioneer 10. Any day now earthlings may receive news that a neighboring jillion-year-old star has exploded or died — news to Earth, that is, for even at the astonishing speed of light it will have taken decades or centuries for the flash to reach earthly clock-watchers.

Nowadays it takes 4½ hours for Pioneer 10 to get its news to the Earth — invaluable tidbits about solar winds and gravitational fields. The scientists who decode the messages expect to receive them for 10 or 12 more years, and nobody knows what they will contain. Probably they will provide an amazing array of data on esoteric astrophysical topics that, sorry to say, aren't sexy enough to capture the imagination of the folks back home.

But who knows? Columbus, also venturing where no one had been, expected to find China. Pioneer carries a metal plaque depicting a man and a woman and an attempt to convey to any unknown observers where it came from. Just maybe, in a thousand or a million or a billion years, some intelligent being will come across Pioneer and its plaque, understand it and try to return the civility.

And just maybe, a thousand or a million or a billion years *ago*, folks in a galaxy far, far away sent out their pioneer. Will it make that right turn by Betelgeuse and head for Earth? Will earthlings understand it? Will they respond? Or will the human race have self-destructed and missed the message by a thousand-year twinkling of an eye?

The Dispatch

Columbus, OH, June 14, 1983

AS *PIONEER 10* streaks to new distances beyond the solar system, America's space program prepares for new adventures closer to Earth: the blast off Saturday of the space shuttle *Challenger*.

On Monday, *Pioneer 10* became the first human vehicle ever to go beyond the orbit of the farthest planet from the Sun, in this case Neptune. Although Pluto is normally the outermost planet, its egg-shaped orbit has brought it closer to Earth. But even though the 500-pound craft is travelling at a speed of one million miles a day, it will not get beyond the Sun's gravitational effects "until our great-great-great grandchildren are grown," a space program official points out.

***Pioneer 10* has compiled a remarkable record of achievements during its flight, brought about, primarily, to its own endurance. It was expected to survive for only 21 months after its launch from Cape Canaveral on March 3, 1972. But it is still working after 11 years and, if all goes well, it will continue to send messages to Earth for another decade.**

It is impossible to predict what space travel will be like by the time *Pioneer 10's* last messages are received, but the astronauts of the U.S. space program are preparing for another step into the future with the launch Saturday (at 7:33 a.m.) of the *Challenger*. The *Challenger* will carry 21 experiments into space along with its five-member crew. Among the crew will be Sally Ride, the first female astronaut in the U.S. program. The crew will deploy two communications satellites in addition to conducting the experiments. The *Challenger* will return to Earth after a six-day mission and, for the first time, will land back at Cape Canaveral — just three miles from its launch pad.

It is, in all, a grand week for the space program, a program in which all Americans can be proud.

The Washington Post

Washington, DC, June 18, 1983

WHILE IT WON'T sharply affect the terms of life on this planet, the most extraordinary event of the past week was surely Pioneer Ten's passage beyond Neptune. It has now crossed the outer limit of the solar system and continues serenely on its course. Originally built to operate in space for 21 months, Pioneer is now in its 12th year of flawless operation. Its designers think that it will continue its transmissions for perhaps another decade.

Astronomers knew quite a lot about the solar system before the space probes, but they suffered the limitation of a single perspective. With them, science is suddenly able to see from other angles. Pioneer and its successors, the Voyager spacecraft, have now produced a wealth of information that will refine and expand understanding in ways that cannot be fully assessed for many years.

Because the physics of the very large and the very small is one unbroken web, the galaxy and the atom continually provide clues to each other's structures. Current theory of the origin of the universe comes chiefly from the study of subatomic particles and, conversely, particle physics often turns to astronomy for confirmation of its insights. You can never know what purpose it will serve to learn a little more about, say, the radiation belts around Jupiter. But experience suggests that whatever you find will tell you about much more than Jupiter alone.

Pioneer is continuing to provide news and commentary about the solar winds through which it is now sailing. In time it will finally cease to transmit. Then it will finally be lost to its inventors—lost, but also safe. It is now in a region where no accident can befall it. In some 10,500 years Pioneer will pass, at some distance, a star. In terms of the voyage on which it is now embarked that is fairly soon.

By the reckoning of the world that it has left, how long a time is that? Ten thousand years ago *homo faber*, as the academics sometimes call us—man who makes things—had acquired considerable skill with small tools of stone and bone. But we had no agriculture then, and were farther from Cheops' pyramid than Cheops is from us today.

Pioneer's journey seems altogether likely to run for hundreds of thousands of years and conceivably much longer. Even if *homo faber* manages by an uncharacteristic application of intelligence not to blow up this planet, astronomy says that the sun will eventually collapse in a natural death. Machines often outlive their inventors. But it is remarkable to think that at some unimaginable reach of space Pioneer may become the farthest memorial of a solar system that no longer exists.

The Times-Picayune
The States-Item

New Orleans, LA, June 15, 1983

It is a dizzying measure of technological advance that only 59 years after man first lifted himself off the Earth's surface in powered, guided flight, man could launch a spacecraft with reasonable confidence that it would fly out of the solar system. Now, 11 years after that launch, the spacecraft Pioneer 10 has left the solar system, accomplishing, according to space scientist Dr. James A. Van Allen, "one of the greatest of human achievements."

But Pioneer 10's future is even more dizzying than its distinguished past, which was full of pathfinding firsts. It still has some distance to go before it sees the last of the solar system. It may discover a suspected tenth planet, and it remains in the heliosphere, the sun's "magnetic bubble" that extends an unknown distance beyond the planets.

Its controllers expect its communications power will last another ten years, by which time it will be about 5 billion miles from the sun. Perhaps it will be able to send back notice of its final exit and information about true interstellar space.

After that, Pioneer's adventures will be beyond us in time as well as space. Interstellar space, according to another space scientist, is "one of the most benign environments one can imagine. There's absolutely nothing to stop it." The chances of Pioneer's being captured by another star are judged so remote that it is expected to outlive the star system it was sent from. In another few billion years our sun is expected to reach its cataclysmic old age — to expand and absorb its planets, and then explode. Pioneer should still be trekking through the Milky Way.

The vastness of interstellar distances measure the chances of close contact. Traveling at 30,588 miles an hour, Pioneer will glide for 10,507 years before making its closest pass by a star, and the closest it will come is 3.8 light-years — 244,000,000,000,000 miles. Interestingly, it will be Barnard's star, the closest star to our sun with a "dark companion." It was once generally thought that such companions might be planets, but now the view seems to be swinging to the suspicion that they are the burned-out halves of two-star systems.

In any event, Pioneer will not be able to tell us or our descendants. Perhaps our descendants, assuming we have any, will have found such things out already. The tantalizing question about Pioneer — another our generation, at least, will never get an answer to — is whether it will tell about us to someone out there. The explanatory plaque on its side, designed to do just that, has been compared to a message in a bottle set adrift in the cosmic sea. In the self-satisfied mood of the present celebration, we decline to consider the ramifications of that possibility.

The Courier-Journal

Louisville, KY, June 22, 1983

THE NEWS from space is routine. The awesome has become prosaic.

The Challenger space shuttle is doing a lazy backstroke through the wild blue yonder, its five-person crew gliding through major and minor scientific experiments as if they stood on terra firma testing the direction of the wind. Their amused and amusing nonchalance on the doorstep of the void reassures and somehow stuns those of us left behind.

What marks this, the seventh flight of the space shuttle, as something special is the presence of a woman aboard. Astronaut Sally Ride is America's first, though not the last. Her performance under pressure, indistinguishable from that of her fellow astronauts, should give the final shove to the silly, emotional barriers thrown up to keep women out of space. Those uncomfortable with women in non-traditional roles had better get used to it.

The shuttle lift-off was perfect. The placement of a couple of working satellites was on target. The ordinariness of every day in space is infinitely reassuring. This shuttle flight, like others before it and others to come, is a near-miracle become commonplace.

But even this anomaly pales before the simple fact of Pioneer 10. This little adventurer, at 570 pounds no larger than the conestoga wagons that explored the West, was shepherded into space 11 years ago, sent on its way to take a good look at some of Earth's planetary neighbors. It sent back to gaping U. S. scientists graphic, close-up pictures of Jupiter's violent "eye," and in recent months has been searching the outer rim of the solar system for evidence of a 10th, as-yet-unknown planet outside the orbit of Pluto. It is old, as spaceships go, and few scientists expected it to last even this long.

But Pioneer 10 kept going, and it is going still. It has punched its way through the sun's ephemeral envelope, into interstellar space. Every mile farther from the Earth it sails, so humanity pushes a mile into the unknown, where no one has been and no one will go for many years yet.

Pioneer's eight-watt transmitter still sends whistling beeps back to its creators, reporting on solar winds, cosmic rays, the emptiness of space. It survives and by surviving reminds us that, beyond the routine of orbit, lie the staggering possibilities of a universe we've only begun to investigate.

The Kansas City Times

Kansas City, MO, June 14, 1983

What a wondrously apt name Pioneer 10 has turned out to be for the tough little, 570-pound, nuclear-powered spacecraft which has passed out of our solar system in the 12th year of what seems likely to be a voyage into eternity. It now continues on beyond Pluto, normally the outermost known planet, having traveled more than 3 billion miles — farther than any other vehicle launched from Earth — and has sent back more information than any other space probe since it was launched from Cape Canaveral on March 3, 1972.

It is still speeding along at a million miles a day, its magnetism measuring device out of action but its 8-watt radio still sending faint signals that take 4 hours, 20 minutes at the speed of light to reach the NASA monitors at Mountain View, Calif. When it flew within 80,000 miles of Jupiter in December 1973, sending back striking pictures of the giant reddish-brown, orange and gray planet on which a pattern of cyclonic forces could be seen, its primary task was completed. But Pioneer 10 sailed onward and upward, past Saturn in 1976, Uranus in 1979 and Neptune.

All the data since Jupiter has been an unexpected bonus, and the spacecraft could keep transmitting for years to come. NASA officials see little chance that it will strike anything and be destroyed. Because of the likelihood that this tiny craft might fly endlessly, it was fitted with a gold-anodized plaque carrying illustrations of the unclothed figures of a man and a woman, just in case some distant alien intelligence, unknown to us, might retrieve this object. Imagine the wonder of such beings at making such a find, speculating on the nature and location of those who launched such a vehicle. And what a brilliant and suitable end it would be to Pioneer 10's trail-blazing journey.

Rockford Register Star

Rockford, IL, June 17, 1983

The sun king. The sun god. The sun worshippers.

All the centuries of blazing light dimmed the other day when a mere 11-year-old left the only world known to humankind, the solar system, and moved into the dim vastness and darkness beyond.

Pioneer 10, a spacecraft hurtled aloft 11 years ago, a product of earthbound scientists, is now so far from its Cape Canaveral, Fla., launching point that it takes four hours and 19 minutes to receive its radio signals. Those signals are traveling at the speed of light!

Having left the planets, the little spacecraft (only 570 pounds) is now outward bound to what seems a sparse but coldly glistening world of stars.

But Pioneer 10 is still moving. Speed: 30,558 miles an hour — each mile of it away from the sun. And Pioneer 10 still is transmitting. As it left the solar system, scientists awaited its message. And, sure enough! Pioneer 10 checked in to say it had just adjusted its data-gathering pattern for one of its instruments, as commanded.

The scientists broke out a bottle of champagne to celebrate.

"Today, Neptune. Yesterday, Pluto. And tomorrow, on to the stars," toasted Jack Meyers, spacecraft operations chief at the Ames Research Center in Mountain View, Calif.

Infinity lies ahead. Pioneer 10 just keeps going into the unknown.

Should it happen upon intelligent life, a shield attached to its exterior bears the forms of a human male and female plus a diagram of the solar system. Footprints in space. Clues to us, adventurers all, a race that has traded the Oregon Trail for a jaunt into stretching, limitless, uncharted space.

CHARLESTON EVENING POST

Charleston, SC, June 16, 1983

As the Pioneer 10 spacecraft spun out of the solar system Monday on its voyage into infinity the popular question was what was Pioneer's greatest accomplishment in a era of successful space efforts.

Pioneer 10, it has to be noted, was aptly named. It blazed a trail through asteroid belts for other spacecraft — another Pioneer and two sophisticated Voyagers whose cameras added substantially to man's knowledge of space. Pioneer 10's main mission, its primary project objective, was exploration of the planet Jupiter. It sent back startling pictures, plus a wealth of other information. Now, to top that feat, Pioneer 10 has traveled farther from the earth than any other man-made machine.

What is most gratifying, though, is the fact that Pioneer 10 has performed beyond all expectations. It was designed to last 21 months. Its odyssey already has lasted 11 years. With one exception — a device for measuring magnetic fields — all the research and communications gear turned on at the time of launch is still functioning. None of the redundant equipment has ever been activated. Not only that, but mission controllers expect to be able to maintain contact for the next eight or 10 years as Pioneer probes beyond the sun's domain. After that, Pioneer 10 will sail silently and endlessly through deep space. That is a tribute to the technological skills of those who built, launched and talked daily to Pioneer 10. It is a lofty target, too, for all who take quality control seriously.

The State

Columbia, SC, June 18, 1983

WE HAVE often been awed by achievements of our National Aeronautics and Space Administration — putting the first man on the moon, etc. — but the *Pioneer X* project is absolutely mind-boggling.

You can go to the movie *Return of the Jedi* and enjoy spectacular special effects and clean science fiction entertainment, but step outside the theater, look up and be astounded by the reality of *Pioneer X.*

It is no longer in sight, of course. But up there the 570-pound spaceship is streaking at 30,558 mph out of our solar system, far past our own planets, into interstellar space — the first man-made object to escape our universe.

That the little spaceship launched 11 years ago to check out other planets is still sending radio signals to its home station on Earth is truly amazing. It is expected to report for another 10 years, although it was designed to last only 21 months.

"The spacecraft will probably survive forever," Alan Fernquist, assistant flight director at NASA's Ames Research Center, said.

Forever. Think of that.

The Miami Herald

Miami, FL, June 17, 1983

A DELIGHTFUL scene in the movie *E.T.* shows Elliott and the neighborhood kids frantically pedaling on bikes to get their out-of-space visitor back to his ship. Frustrated with the slow pace, one of the boys quips: "Why doesn't he just beam up?"

American audiences understood instantly. It is a commentary about this nation's facile relationship with space that concepts such as "beaming up," space adventures, and extraterrestrials are commonplace.

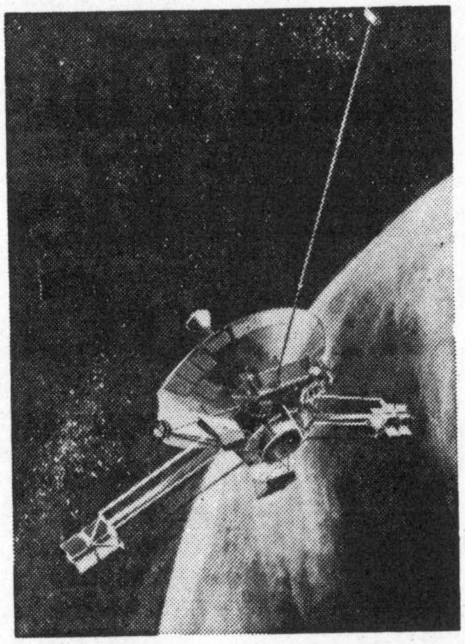

In the fantasy world of movies and television, anything is possible. Intergalactic star wars are routine. Laser duels are daily fare. Close encounters with beings from other planets are perfectly plausible.

In the real world, man's progress against the vastness of space is measured on a different scale. Landing men on the moon, the nearest interplanetary body to earth, was a major, dramatic accomplishment. So too is the extraordinary journey of Pioneer 10, the spacecraft that on Monday left earth's solar system and headed for the stars. Pioneer 10 is a milestone — or should it be "light-year-stone"? — in human achievement.

The 570-pound robot spacecraft embodies Earth man's brief history, his present accomplishments, and his dreams.

Pioneer 10 was designed to last less than two years. Instead it has functioned flawlessly for 11 years and may continue indefinitely. It was supposed to have been pulverized by an asteroid belt beyond Mars or destroyed by toxic radiation near Jupiter. Miraculously, it survived.

Now the craft has entered the pure environs of interstellar space. It could outlive man and Earth. Thus, for the moment, Pioneer 10 represents man's best hope to connect with other intelligent beings, if there are any within range.

When matched against the infinite imagination of man's mind and its fanciful quest for space adventure, Pioneer 10 pales as it does against the endless time and space of the universe. But as a measure of where man is and hopes to be, Pioneer 10 is a living monument.

Hail, Pioneer, and *bon voyage.*

The Houston Post

Houston, TX, June 17, 1983

It is gone, perhaps destined to drift endlessly across the universe, the first man-made object ever to venture beyond our solar system. Even before the Pioneer 10 spacecraft crossed the invisible boundary June 13 and began its trek through the Milky Way galaxy it had already established itself as a triumph of American science and technology. Its 11-year odyssey among the planets that orbit the sun has given us a wealth of scientific information and whetted our appetites for more.

The quarter-ton marvel of sophisticated instrumentation has demonstrated amazing durability and reliability since it was launched from Cape Canaveral, Fla., March 3, 1972. It has lasted far longer than the 21-month life for which it was designed and with any luck will continue to send data back to Earth for another 10 years before its signals fade away in the vast distances of space. It breached the asteroid belt that some scientists thought might be an impenetrable barrier. It gave us our first close-up pictures of Jupiter's Great Red Spot, revealing it to be a giant storm roiling the surface of that liquid planet. As it continues its voyage into the unknown, the craft will transmit further information, such as measurements of the heliosphere, a cloud of particles and gases emitted by the sun that extends far beyond the planets.

Three other spacecraft — Pioneer 11 and Voyagers 1 and 2 — will follow Pinoneer 10 on its journey beyond the solar system after they complete their missions by the end of the decade. Though the remarkable performances of these and other unmanned space probes have captured the imagination and support of the public, the U.S. space exploration program in the 1980s has felt the bite of budgetary austerity. But those in the National Aeronautics and Space Administration and the scientific community who have guided the nation's deep-space projects point out that their cost has been relatively modest. The scientists hope that this record of economy-mindedness will win funding for other missions scheduled or proposed for the rest of the century.

Our deep-space fleet has been a wise and prudent investment in the acquisition of knowledge about other worlds. But Pioneer 10's departure on the final leg of its journey is more than a historic first for humankind. A small plaque attached to the craft with drawings of a man and a woman, our solar system, a hydrogen atom symbol and a star map epitomizes our eternal optimism that we share the universe with other intelligent beings. If they exist, they might find Pioneer and divine something of the creatures who sent it. However that may be, scientists say that when the sun dies 5 billion years from now, Pioneer 10 and its sister ships will probably still be wandering among the stars, memorials to a little blue planet in the suburbs of the Milky Way.

U.S. Space Station Forecast; Changing Attitudes Cited

James Beggs, head of NASA, said July 18, 1983 that he expected President Ronald Reagan to give approval for a manned space station soon. "If the United States does not take this step, we will lose our pre-eminence in space," Beggs said. The NASA head characterized a space station as "the next logical step for long-duration work." Beggs cited an apparent shift in attitude by George Keyworth, the White House science adviser, as grounds for his optimism regarding a space station. Keyworth had previously voiced opposition to expensive new undertakings in space, but in an interview reported in the July 8, 1983 issue of Science, Keyworth said he thought that "the country would take a major thrust in space very soon."

NASA proceeded to lobby hard for the space station despite the reservations of other government agencies and science advisor panels through 1983 and early 1984, The Washington Post reported January 18, 1984. Among the chief reasons was that the program would allow NASA to preserve its large research and engineering staff now that the space shuttle program was fully established. According to the Post, the Defense Department and the Central Intelligence Agency had opposed the program out of fears that it would draw money from their own programs. The Post also quoted the chairman of the National Academy of Sciences' space science board, Thomas M. Donahue. Donahue said that when NASA had asked the board whether basic research in science would "require or be enhanced by a space station," the board's answer had been no.

The Orlando Sentinel

Orlando, FL, July 21, 1983

NASA Administrator James Beggs has been warning that the United States could lose its lead in space exploration if it doesn't move ahead with plans for a permanent manned space station. The problems, though, aren't so much technical as political.

A space station is the next logical step now that NASA has shown that the shuttle meets its primary purpose: shuttling. What's needed is a permanent place for research and further exploration of space.

It already is known that weightlessness on a space station would make possible new drug-making processes and the manufacture of computer chips more refined than anything possible on earth. With the space shuttle trucking goods to and from the station, it could serve partly as an orbiting factory. And it could serve as a base to launch craft deep into space. Such talk still sounds a bit like the material of a science fiction novel, but they are things that can be done now.

Some benefits won't be realized until the country is deep into the space-station project. But it makes sense to invest national resources in a field where we can be the leader rather than to prop up older industries doing jobs that other nations might be able to handle just as well.

A space station could be operating by 1992, but NASA needs the approval of Congress and the Reagan administration. And that won't come easy: It could cost as much as $10 billion. The first step is $200 million to get the project going.

Rep. Don Fuqua of Altha, chairman of the House Committee on Science and Technology, believes the cost can be controlled by involving other nations. He says Japan, West Germany and Italy are interested in helping build it.

His approach sounds reasonable. Space exploration is necessary but expensive. It will get more costly as the projects become more sophisticated. By sharing the cost, the United States can maintain its pre-eminence while delegating certain projects to other nations.

The test will come this fall when NASA will submit its fiscal 1985 budget. With $200 billion deficits, there will be political pressure to solve grass-roots problems by cutting back our efforts in space. But among the long-range benefits from exploring space will be new jobs created by new industries. Those benefits won't come unless we take the next step.

Houston Chronicle

Houston, TX, July 21, 1983

The public attention given to the space shuttle missions demonstrates that the U.S. space program has lost none of its ability to generate excitement. What has been missing since the end of the Apollo moon landings is a sense of the program's long-term goals and direction.

This week, however, President Reagan's science adviser asked the National Aeronautics and Space Administration to prepare a "grand vision" of the future that might include a U.S. space station, lunar bases and astronaut trips to Mars. This signal from the administration is a welcome recognition of the critical role America's space efforts will continue to play in the nation's scientific and technological development.

Such a vision, now being prepared by a NASA task force, will not mean that the agency will be given carte blanche to pursue whatever project it wishes — certainly not until the current budgetary problems are diminished. But the agency's desire to develop a permanent space station has met with strong support in Congress.

NASA administrator James M. Beggs says a space station will "open up commercial opportunities we have not dreamed of; it should improve our national security, provide more sophisticated science, and be a source of international cooperation." The space program's track record indicates that such benefits are not simply wishful thinking.

At this time, the Soviet Union already has a small station in orbit, and Europe and Japan have expressed interest in manned missions in space. NASA's "vision" of the future will help to give the agency the direction it needs and enable the administration to make the decisions necessary to ensure that the United States will continue to lead the way in space exploration and provide the nation and mankind with the benefits that will inevitably result.

THE KANSAS CITY STAR

Kansas City, MO, September 25, 1983

What should the next major space project be now that the shuttle has completed testing and is launched on an ambitious schedule of flights? For some time the National Aeronautics and Space Administration has believed that a permanent manned space station is the next logical extension of the shuttle program, and all at once the prospects for such a project have improved markedly. NASA may request $200 million for design work on a space station in its FY 1985 budget, to be submitted in September.

One thing which has cheered NASA planners is the turnaround of the White House science adviser, George Keyworth—formerly opposed to the concept—who recently suggested the space agency should come up with a "grand design" for the future. If the design funds are forthcoming, NASA estimates it could orbit a space station, with four to six persons aboard, by 1991 at a cost of $6 to $8 billion. The facility could be enlarged later by the addition of modules.

Such a space colony could serve as a sort of depot for the shuttle craft in their work of installing and servicing various kinds of satellites. The present Earth-launched flights, with a gliding return to a landing, must be of limited duration. But from a station in gravity-free space, a shuttle could set forth on a mission with minimal power from thruster engines.

Beyond these benefits, American industries are showing great interest in the types of experiments and manufacturing processes that can be carried out in a weightless environment, as graphically demonstrated by some of the shuttle flight experiments. Thus the space station could be a research laboratory and factory as well as a shuttle base. Commercial possibilities, not yet apparent, could be extensive.

A permanent space station would be an expensive undertaking, without question. But it would yield tangible benefits well beyond the mere prestige of sustaining a United States presence in space. And if this nation should forsake such an opportunity, there can also be no question that the Russians, Europeans and Japanese will move into that void before too many years.

The Dispatch

Columbus, OH, September 20, 1983

An interesting debate with long-range implications is going on in this nation's scientific community over the value of producing and deploying an orbiting space station. Proponents of the idea say it is the only way the United States can maintain its pre-eminence in space. Others are not so sure and believe that the space shuttle could do anything a station would be asked to do and at far lower cost.

Many NASA officials want the space station. NASA Administrator James Beggs argues that "if the United States does not take this step (of developing the station), we will lose our pre-eminence in space because the Soviets will not stop, the Europeans will not stop and the Japanese won't stop." He is pushing for congressional approval of $200 million in NASA's 1985 budget for the station. The total cost of the 4-to-6 person module would be between $6 billion and $8 billion.

Beggs is undoubtedly sincere in voicing concern about the consequences of not building the space station, but it must also be remembered that NASA in recent years has suffered from funding cutbacks and a general lack of enthusiasm for its efforts once Americans walked on the Moon. Many within the agency are looking for a project that will once again capture the nation's imagination and they think that the station might be that project.

Others contend that the space shuttle, which has established a remarkable record of achievement, has the capability to perform space station functions well into the next century. And they argue that the money that would be poured into the space station could better be used to fund other ventures, such as deep-space exploration and planetary probes, which would be as exciting as any station could be and probably far more valuable from a scientific fact-gathering point of view.

They received the support of a formidable ally recently when the National Academy of Sciences recommended that a space station not be developed and that the shuttle be adapted to meet national needs during the next two decades. "Our finding is that present systems are adequate," the academy wrote in a report on space capabilities.

This argument makes sense, and unless NASA can come up with a more compelling argument for the development of a space station, the nation would do well to make the most of what it already has. Exploiting the shuttle's potential would seem to be the best way for this nation to maintain its preeminence in space.

The Morning News

Wilmington, DE, July 4, 1983

IN A SPACE RACE as in any other, one tries to win. That's one thing to keep in mind when examining a statement by James M. Beggs.

The National Aeronautics and Space Administration chief predicts that President Reagan will approve plans for a U.S. manned space station. "If the United States does not take this step," he warned, "we will lose our pre-eminence in space."

Eminence is something most of us aim for. Hope for the prize, first place, spurred the United States to the prodigious efforts in the '60s that culminated in our historic moon flights. Our sense of competition was aroused by the Soviet triumph in putting Sputnik into space in October 1957. Part of our response was to bolster our schools through the National Defense Education Act.

Defense, security, freedom go hand in hand with jealousy, fear, envy. Sometimes these negative emotions can have positive results. They helped get us to the moon and regenerated scientific education in the nation. They helped buy other benefits, achieved through the space program, including some that will alleviate hunger and disease on the planet.

Now, 26 years after Sputnik, we must prop up our educational system again. Mr. Beggs' "grand design" for re-entry into the space race could be the brace to do it with.

The plan is to go step by step from the shuttles to manning, in 1991, a station in synchronized orbit. It might cost $8 billion. We must launch the program, said Mr. Beggs, "because the Soviets will not stop, the Europeans will not stop and the Japanese won't stop."

The NASA's chief told a Space Station Symposium Monday that the project would "open up commercial opportunities we have not dreamed of . . . improve our national security, provide more sophisticated science and be a source of international cooperation."

Ah, there's the word. A sense of competition can do a lot in the short-term race, the space sprint; it might win us eminence. Cooperation, where the United States and the other superpowers wholeheartedly join the planetary team, will do much more for the long haul.

The Houston Post

Houston, TX, September 23, 1983

Important firsts have been achieved, the machine has been proven, the once impossible made to seem routine — the space shuttle has progressed from dream to plan to test vehicle, to the tool it has become. And while upcoming missions might not be as spectacular as those filled with firsts and fantastic photos of astronauts floating in spacesuits and of an orbiter floating against the curved rim of the Earth below, the material returns and the nearly limitless possibilities the system opens are only now beginning to be seen.

Astronaut Richard Truly, who commanded the eighth shuttle flight on the six-day mission that began with a night launch at Kennedy Space Center and ended with a night landing on Labor Day at Edwards Air Force Base, Calif., summed it up in a few words. "Some of these flights are going to have more pizazz than others," he said.

Truly's flight had considerable pizazz of its own, but perhaps it didn't rank in that department with some of its predecessors. The element of wondering whether the shuttle was going to work had largely been replaced by a confidence in a system undergoing its eighth orbital flight. Fewer and fewer difficulties with the spacecraft have been reported on each mission, and this one was no exception. The activities on board were accomplished with expected competence. The communications and weather satellite was deployed for the government of India without a hitch, (although it developed problems later) and the Canadian mechanical arm proved itself, as did its operators, in its most demanding test yet. The first engineering tests of a satellite system that will vastly improve communications between NASA ground facilities and orbiting spacecraft were accomplished.

But the real pizzaz will come with the fruits of these flights, both past and future. Already the shuttle has been used as a platform for a process developed by McDonnell Douglas Corp. that could lead to a pharmaceuticals industry in space, one that could make rare and expensive medicines more abundantly available. That system was used on the eighth flight to sort live cells, a process that could contribute to using surgical implants to cure some diseases.

The list gets longer with each flight. New manufacturing processes are being investigated. Some of them no doubt eventually will enable us to do things that can be done only with great difficulty on Earth, or not at all. Knowledge of ourselves, our planet, our solar system and beyond will increase to an extent we can now only begin to imagine.

Spinoff from the space program has contributed in many areas. The shuttle takes us another step beyond spinoff — something developed for a specific task in the space program and found to have other applications. Now it is becoming a matter of direct development of things and processes with end uses a long way removed from space.

The shuttle is a step toward a space station — the next logical development. A space station should and indeed must be built if we are to retain a technological lead in an increasingly competitive world. It will be able to do many things the shuttle does better and more completely, because of its size and relative permanence in orbit. And it will complement and complete the shuttle, which until the space station is built is a little like a truck that ends each trip where it began, with no destination at the other end of the route to pick up or deliver cargo.

The pizazz we will see from future shuttle flights, and from a space station, will go far beyond spectacular launches and breathtaking pictures. It will make a good many changes, for the better, in the way we live.

DESERET NEWS

Salt Lake City, UT, July 24, 1983

Nearly 20 years ago, U.S. experts put together a long-range strategy for the exploration of space. Unfortunately, it ran into budget problems and other obstacles. But at least part of that strategy may be revived soon if President Reagan approves funding for a space station.

Space agency officials indicated this week that prospects are good for the first $200 million in funding beginning in fiscal 1986. The total cost is estimated at $6 billion to $10 billion to have a space station in earth orbit by 1992.

While the first reaction in these times of huge federal deficits is to cringe at the cost, the alternative is to let the program die on the vine. If the space station is not built, the U.S. will lose its supremacy in space within a decade, according to James Beggs, administrator of NASA.

Right now, the U.S. is enjoying great success with its shuttle program. A space station is the next logical step. Such a station would be built in sections in space, with the shuttle ferrying the necessary materials into orbit. The finished product would be a permanent station with four to six people on board. It would grow with the later addition of modules and people — just like adding on to a house.

The space station would offer the chance to do long-term scientific experiments in the weightless vacuum of space; would have some military benefits, and would allow the commercial production of new drugs and materials that cannot be duplicated on earth — a potential that has drawn serious support from private industry.

NASA's long-term strategy, formulated in the early Apollo days, involves the building of a shuttle system, then a space station, then a scientific base on the moon, and perhaps a manned flight to Mars and even the moons of Jupiter.

It still seems like a good plan. Those steps ought to be followed up because they provide development of the technical base that the U.S. needs to grow and compete with other nations.

But beyond that, the U.S. should respond to the deeply felt need to reach out to the next frontier. It has been part of the American character since before the earliest days of nationhood. We would all be poorer if this country lost its desire to keep finding out what's on the other side of the next hill — or the next star.

THE INDIANAPOLIS NEWS

Indianapolis, IN, July 20, 1983

News that President Reagan's space adviser, George Keyworth, has asked the National Aeronautics and Space Administration to prepare a "grand vision" for future space projects is a welcome reversal of policy.

Previously, Keyworth appeared in opposition to the development of a space station project. Other than the space shuttle program, which went on the drawing boards more than a decade ago, space exploration has been downgraded by several White House administrations since the conclusion of the Apollo moon landings.

In a recent speech in Seattle, Keyworth signaled a change of heart when he said: "Some people have jumped to the conclusion that I have a bias against a space station because I insist on a valid mission before we make any commitment to it.

"That's not true. But I think it's time for us to take a broader look — with more vision, much more vision — at where we expect the American manned space program to go over the next quarter century."

He went on to suggest the possibility of an orbital space station, a manned lunar station or even manned exploration of Mars. Then, in an interview with Science magazine, Keyworth said: "I think the country should take a major thrust in space very seriously. We've shown that the space shuttle works and is reliable. We have the technology to build a space station. It is only an intermediate step in a more ambitious long-range goal of exploring the solar system."

This week NASA is gathering together hundreds of industry, government, foreign and military planners to help prepare a plan of action to submit to the White House this fall. The proposal is expected to contain a funding request of between $60 million to $120 million for work on a space station.

Unquestionably, any "grand" space vision carrying a large price tag is going to come under close scrutiny with annual budget deficits projected at about $200 billion. But, continued exploration of the "high frontier" should not merely be regarded as an unnecessary luxury, even in lean times.

From the standpoint of international prestige, clearly the world looks up to those nations which have a presence in space as leaders in technology. Although the Soviet Union has failed to match this nation's spectacular feats of landing a man on the moon or having a space shuttle, intelligence sources indicate that Russia has made huge strides in the amount of throw weight its booster rockets are capable of and is ahead of the U.S. in the race for a major space station.

Space station technology has clear military implications.

The space program also has demonstrated technological spinoffs for development here on earth. Much of this nation's leadership in computer technology is attributable to the space program. A space station holds the promise of other commercial and scientific applications, from the development of new drugs, to manufacturing ball bearings, harnessing solar power, creating improved crystals for electronics and manufacturing satellites in space.

Prestige, military and commercial applications aside, exploration of space is perhaps the major legacy which this generation has to leave to history. It is an opportunity similar to that facing the Spanish and Portugese explorers who discovered and settled the New World. The "grand vision" of a space station may indeed prove to be the vision of a gateway to the stars.

The Evening Gazette

Worcester, MA, November 21, 1983

Now that space shuttles are cruising more or less regularly into earth orbit and back again, the National Aeronautics and Space Administration wants to give them a destination in the sky. It wants to take the next logical step of building a permanent, manned space station in orbit.

NASA is urging President Reagan to put start-up money into the fiscal 1985 budget to be submitted in January. With that beginning, the agency says its scientists, engineers, shuttles and astronauts can have a space station in operation by 1992. That would be in time for the 500th anniversary of Columbus' discovery of America.

It is not clear yet whether Reagan will go for the idea. He has supported the space shuttle program, which has been beneficial to the military as well as to commercial engineering developments. But space programs are expensive. The space station project bears an official seven-year price tag of $20 billion to $30 billion, which means it would probably end up costing well over $30 billion.

Can the president justify that sort of new commitment at a time when the government is running deficits of $200 billion a year and has no strategy for trimming them?

A factor bound to weigh heavily in his decision is that relatively few voters give the space program a high priority among government spending alternatives. People have generally liked the successful results of the space program, but most view it as an expensive luxury rather than an essential government activity like defense and various other service and benefit programs.

That public view of space projects may be changing, what with television coverage of shuttle missions and the popularity of movies like "The Right Stuff" and books like James Michener's "Space." The American Space Foundation, a Washington lobbying group, has grown from 2,000 to 20,000 members over the past two years.

Right now, however, the overwhelming argument against any costly new space program is those huge federal budget deficits that loom as far as can be seen into the future. The nation is in for economic trouble unless Congress and the White House agree on a way to scale back those deficits. After that has been accomplished, perhaps the government should think about building a space station.

The Times-Picayune
The States-Item

New Orleans, LA, November 26, 1983

The space shuttle Columbia is scheduled for launch Monday on a mission packed with firsts that point the way to full implementation of the shuttle fleet and to advancing toward a permanent manned space station.

Columbia this time carries the European-built Spacelab and the largest crew, six, ever sent into space in a single craft. Two of the crew are non-astronaut scientists who will work in the lab in shirtsleeves just as they would on Earth except for being weightless.

Spacelab is a self-contained cylinder built to fill the shuttle's cargo bay and connected to the shuttle by a tunnel and airlock. Like the shuttle, it can be reused, outfitted for different work. The crew on this flight will perform 72 investigations in life sciences, atmospheric physics, Earth observations, astronomy, solar physics and materials sciences. The work orders were drawn up by scientists in 11 European nations, the United States, Canada and Japan.

Columbia will stay in orbit for nine days, the longest shuttle flight yet, and can add two more days if necessary. "The mission is exceedingly important in demonstrating to the world that we can integrate and work with a large number of experiments at the same time," says James Beggs, head of the National Aeronautics and Space Administration.

"That has enormous significance for the future because in the future we hope to fly scientists, engineers and maybe even folks who have not been trained to the degree that we have trained this crew," Beggs said.

The further significance is that such missions are also dry runs for a permanent manned station that would be large enough to translate cargo-bay labs into spacious modules where not only experiments but manufacturing can be done — for example, of specialty products like crystals, metal alloys and medicines for whose production space is an ideal, even a necessary, environment.

Monday's flight will be the fourth and last of a shuttle this year. Ten flights are scheduled for next year, including at least three with spacewalks and one to retrieve and repair an ailing satellite. A third shuttle, Discovery, will be joining the fleet, and the program should move into full, productive swing.

The Christian Science Monitor

Boston, MA, September 2, 1983

Once again a hard-working astronaut team has shown that the US space shuttle can deliver.

With India's communications-weather satellite safely launched, another commercial contract has been fulfilled. Now the astronauts are continuing to explore the full capability of the shuttle Challenger – practicing using the manipulator arm with an 8,000-pound dummy payload, testing communications through the TDRS relay satellite, and preparing for the shuttle's first night landing.

Such pioneering space activity seems to be taken for granted now. As with the homesteading which really opened up the US frontier territories, such work-a-day activity seems less dramatic than the exploits of the first explorers.

But the social pioneering which this Challenger team represents is another matter. As the President noted in his call to the astronauts, this mission has given Guy Bluford an opportunity to be a role model – not only for blacks but for all Americans – in one of the most visible cooperative activities of our time. It is indeed a smoothly working team of Americans that is in orbit and not a social experiment. As Astronaut Bluford himself observed, it is a foretaste of what a truly integrated US society can be with each person contributing his or her talents without artificial distinctions of race. Thus we, too, salute America's first black astronaut in orbit.

Meanwhile, although the shuttle missions have largely gone well, the US manned space flight program still has an uncertain future. The shuttle effort continues to suffer from corner-cutting due to earlier funding restrictions. For example, there was not enough money to build an adequate reserve of spare parts. When critical parts have been needed for flight operations, they have been taken from the shuttle assembly line. Now, after taking Spacelab into orbit on the next mission in October, the shuttle Columbia is likely to be grounded for 18 months mainly to act as a source of parts to keep Challenger in service. This tends to confirm the concern of critics who say the shuttle system has so little reserve capacity that any serious mishap, such as loss of an orbiter, would severely cripple its ability to fulfill its commercial obligations.

Then too, the administration has yet to decide whether or not to proceed with a permanently manned space station. The Soviets are well along toward having such a facility. US industrial leaders, in fields which would use such a station, are urging its construction. The National Aeronautics and Space Administration has endorsed the project as the logical next step in US manned space flight. This is a long lead time item. If such a space station is to be available in the early 1990s, NASA will have to start developing it soon.

The outstanding performance of the shuttle system should not lull the US into complacency about its space capabilities. It could easily find itself being second best to the Soviets, or even the Europeans, in the 1990s unless a strong sense of direction and a challenging new goal, such as a space station, is given to the US effort.

Space Station Profiled; Design and Purpose Proposed

NASA for several years had been studying the feasibility of creating a permanently manned space station, according to press accounts January 26 and 27, 1984. The proposal received the blessing of President Ronald Reagan in his Jan. 26 State of the Union address.

According to NASA officials, the space station was to weigh about 100 tons, made up of four chambers large enough to accomodate a total crew of six to eight persons, with crews to be rotated every six months. The modules would be assembled in orbit between 240 and 300 miles above the Earth. As envisioned by NASA, one chamber would house the crew, a second would be a work and exercise chamber, a third would be an orbital laboratory, and a fourth would house the power supply. Vast solar panels would supply the station with 75 kilowatts of electric power. Two astronomical observatories would be attached to the power modules. The station would be equipped with a small "space car" for retrieving damaged satellites. At a news conference Jan. 26, NASA administrator James Beggs portrayed the station as a multi-purpose structure that could be used as a science laboratory, as a repair shop for damaged satellites, as a factory for producing medicines and metals under zero-gravity conditions, and as way station for passing spacecraft and jumping-off point for planetary exploration. NASA officials had targeted the project for completion in the early 1990s, with the official date for manning the station set at 1992, exactly 500 years after Christopher Columbus' discovery of America.

According to news reports, the space station program would ultimately cost at least $8 billion, with some estimates running as high as $20 billion. A request for a start-up authorization was expected to be included in President Reagan's federal budget for fiscal 1985, due to be presented to Congress February 1.

NASA's plans to build the space station were not justified on scientific, military or economic grounds, according to a congressional study released November 13, 1984. The 234-page report, by Congress' Office of Technology Assessment, said a "persuasive case" could be made for putting some structures in orbit, but not for the entirety of the program as envisioned by NASA. "Because the nation does not have clearly formulated long-range goals for its civilian space activities," the study said, the NASA concept "is not likely to result in the facility most appropriate for advancing U.S. interests into the second quarter-century of the Space Age."

THE PLAIN DEALER

Cleveland, OH, November 16, 1984

Before the nation goes any further, says the congressional Office of Technology Assessment, it should figure out what to do with a space station. The project is no longer one small step for man, but one giant stomp on the deficit. For $7.5 billion a year, the OTA rightly believes that the public deserves a larger role in NASA development of space station application.

The question is not exclusively one of democratic process. NASA's proposed station would be a multi-purpose project of more than 100 different uses, which should be democratic enough for just about everyone. OTA contends, however, that very few of those 100 uses justify permanent manned structures. It complains that NASA does not adequately distinguish between serious uses and possible ones and that the NASA approach stresses technological application over social and public ones. Further, OTA implies that project viziers tend toward specific applications rather than broader goals with specific objectives.

Enter now a lot of confusing talk about the maturity of the space program. OTA officials contend that if the space program is not fully mature, it is at least well into its adolescence, and that as such it should not be left to NASA alone. Maybe, but the risk of substantive public debate over the uses of a space station—even its size and design—is that no consensus be reached and that the project become enmeshed in politics and space-age neo-myth. There is no reason to think that a space station of any size is anything more than a tool, one best left to those who know how to use it.

Nevertheless, OTA has a point, which is that for the vast amounts of money spent, the public should not allow NASA an entirely free rein. Thus although it might be unwise to open space-station usage to debate, it is wise to suggest that NASA think in terms that extend beyond scientific and technological advocacy. (Shipping high school experiments up in the space shuttle is an effective popular symbol, but nothing more.) Further, there is a question about whether some space station functions wouldn't be redundant. Satellite servicing might make sense, but could an orbiting observatory be better maintained by the space shuttle? If so, then NASA should not spend the extra money to provide a station with that ability.

Boiled down, the OTA's desire is twofold: to ensure that the public has some influence on the uses of a space station and to encourage NASA to develop a sharper critical capacity with regard to space-station usage. OTA supports the idea of a space station, but worries that, left to the technocrats, it will not be a fully worthwhile investment because it will be weighted in favor of technology. Those are fair and timely cautions, ones that project managers and congressional financiers should carefully consider.

AKRON BEACON JOURNAL

Akron, OH, November 16, 1984

SOME INTERESTING ideas popped up recently at a NASA-sponsored conference in Washington about future uses for space technology.

There was discussion of a proposal for the 1990s to build a giant atom smasher — sort of a racetrack for subatomic particles — that would be 100 miles long, in Texas, where the space won't be missed. But physicist Edward Teller, who is often — to his dismay — called the father of the hydrogen bomb, said if you are going to plan a project that big, you should instead build the atom-smasher around the equator of the moon, to run some 6,700 miles. The moon, he says, has the advantage of being a natural vacuum even though, many will no doubt point out, it ain't Texas.

Such projects seem pretty ambitious. But the fact that the technology is available for such things means we are closer to the Buck Rogers era than we might think. More to the point, the conference heard more than 150 papers on lunar bases and other space projects that are probable early in the next century.

The world needs such vision, if only to take its mind off the petty squabbles that paralyze human relations around the globe. In that vein, the best of all possible scenarios would be for a joint venture in space by all nations, with special attention to early agreements between those that are on the verge of establishing permanent outposts in space.

A joint U.S.-Soviet station on the moon comes to mind as a way to combine resources in a common interest and for peaceful purposes. The more nations work together, the more they understand each other; and that makes them less likely to do something stupid. With the United States and Russia only decades away from a major presence in space, a primary goal should be to leave the tension and competitiveness between them at home, when they journey to the stars.

We've all seen the pictures: The Earth looks more peaceful when viewed from space. Perhaps space — and in the race to get there — is the place to start building a peace that can be brought down to earth.

Roanoke Times & World-News

Roanoke, VA, February 2, 1984

THE PRESIDENT's proposal for an $8 billion manned space station raised some eyebrows, given the size of budget deficits. Some were ungracious enough to suggest that he was trying to steal some "Right Stuff" thunder from John Glenn, a possible Democratic opponent.

There are good reasons, though, to give thoughtful consideration to a space station. Shorn of its romantic veneer, space remains an important frontier for scientific effort, perhaps industrial effort too. The U. S. program has been stripped down since the glory days of its moon landings, but there's still a large space establishment. If it is not to stagnate — not exactly the American way — it needs new directions.

A manned space station would be the logical next step. Yes, there already is the shuttle. But a station would give it a more definite mission; without it, the shuttle can only take a lot of Sunday drives. There are many kinds of scientific experiments that require a longer presence in space than the shuttle can give. That presence would have to be sustained, if commercial use — e.g., the manufacture of extremely pure drugs — is to be practical.

Other uses are envisioned for a station: as a repair shop for disabled satellites already in space, as a factory for building other space structures too big to be shot into orbit, and as a jumping-off place for journeys deeper into space. "It's a bridge to the future," says John Hodge, director of the National Aeronautics and Space Administration's space station task force.

A case also can be made against the project. The 1985 budget would contain only $150 million in start-up money, aimed at orbiting the station by 1992. But the estimated $8 billion cost might be only the beginning; the space program has had its share of overruns, and the shuttle ran well behind schedule. There's also concern that the station won't be devoted entirely to peaceful purposes.

Maybe a compromise could be worked out. A station wouldn't have to be manned year-around. John Pike, staff assistant for space policy at the Federation of American Scientists, says: "For more legitimate missions of a space station, people get in the way." And it's expensive to keep them there. He suggests an arrangement whereby the shuttle would carry people to the station only every few months.

Overall spending priorities have to be set. A manned station may be canceled or delayed; certainly there are more important things for Uncle Sam to do down here. But chances are that this project will make it into the budget in some form. The Soviets already have an extensive space-station program, and with Congress as well as the administration, that's usually argument enough.

THE SUN

Baltimore, MD, January 27, 1984

President Reagan's proposal for a permanently manned U.S. space station stirs visions of majestic spaceships wheeling through space in the classic film "2001." These are visions that stir men's souls, as they have stirred the president's. While there a few caveats, the visions may rest on a foundation of practical utility and necessity.

While NASA has yet to specify the exact shape or size — or even the purpose — of the space station, there are some obvious benefits. One is the capability for assembling larger, heavier payloads than a space shuttle can carry on a single flight. These might include larger satellites and sophisticated spaceships for long voyages. Immediate value might be found in commercial ventures — pharmaceutical and others — that are possible only in space; a more efficient base for launching satellites into high geosynchronous orbit, and special space science studies.

Just as important, a space station would keep the U.S. ahead of the Soviets in space. Work on the station would revitalize the now somewhat moribund agency. NASA once again would attract the nation's top engineering and scientific talent. Without this stimulus, the Soviets threaten to overtake and surpass the U.S. in space.

Space station opponents contend there are risks in committing the nation to an $8-billion venture with so few specific goals, while giving short shrift to unmanned space science projects that promise more for less. Critics also point out that NASA has been laying the foundation for the space enterprise for 25 years — at a cost of $23.5 billion for the Apollo moon-landing program alone — and that it is time for private industry to start paying its way.

Opponents of the space shuttle made the same arguments, but now that vehicle is on the brink of major successes. The $150 million the president is requesting would permit NASA to make preliminary studies of the space station. After these are complete, Congress will have better information upon which to judge the space station proposal.

DESERET NEWS

Salt Lake City, UT, January 15, 1984

Now that the U.S. space shuttle is becoming a dependable "truck" for carrying out projects in earth orbit, the door is open for the next step — construction of a permanent space station.

President Reagan appears ready to support that idea, although he had been lukewarm earlier. Starting with a modest $14 million down payment in fiscal year 1984-85, the $6 billion space station would become operational in 1991.

Seed money to start the program was included in a tentative spending plan described recently by Budget Director David A. Stockman at a briefing for Republican members of Congress.

It can be argued that in a time of staggering federal deficits, the nation cannot afford to undertake any new projects that are not absolutely necessary. But if the U.S. is to have a space program worth the name, it cannot simply stagnate, but must move ahead and exploit the scientific, commercial and defense possibilities of space.

The cost — $6 billion spread over seven years — is miniscule when compared with federal budgets of more than $900 billion each year.

The facility envisioned by NASA would contain an enclosed laboratory and laboratory platforms exposed to open space; power supplies, shuttle docking facilities, and living quarters for a crew of six. Prefabricated pieces of the station would be ferried into orbit by the shuttle and assembled in space.

In an effort to drum up support for the project, some top officials of NASA are singing an old song — the need for the U.S. to stay in front of the Russian bogyman.

But such arguments should be viewed in perspective. In the first place, there is no "race" as such. The space race declared by President Kennedy was to the moon. That was won many years ago by the U.S., leaving the Soviets so far behind that, as one observer put it, "They didn't even come in second."

Obviously, the U.S. cannot let orbital space become a Russian monopoly, because it does have military ramifications. But that doesn't mean getting panicky and spending huge sums in a kind of "star wars" mentality.

Any approval for a permanent space station must focus on practical and scientific justifications for the project. People will find that the space station easily can stand on its own in those respects.

Wisconsin State Journal

Madison, WI,
January 16, 1984

The National Aeronautics and Space Administration wants a space station. The Reagan administration, according to Budget Director David Stockman, intends to ask for planning money in the 1985 spending plan the president soon will send to Congress.

But does the United States really *need* a space station?

The National Academy of Sciences says no, not this century. It argues the duties a space station could perform can be handled by the U.S. space shuttle and existing rocket boosters.

The Pentagon says it has no use for the project, either.

According to Stockman, the president will ask for $14 million in 1985, the first installment of a $6-billion project.

Before any money is allocated, the burden is on NASA to show convincingly that this new project is necessary.

The Idaho STATESMAN

Boise, ID, January 31, 1984

We applaud President Reagan's decision to send an orbiting space station aloft as a national goal for the decade. This country was built upon the spirit of exploration. It will be good for us to revive it.

Mr. Reagan also was right to open the door to international participation in a space-station program. Space is a promise for all mankind, and all will benefit from this effort.

It has been sad to watch the U.S. commitment to space founder after the manned lunar explorations of the late 1960s and early '70s.

Those were exciting days, and the space shuttle effort that succeeded the Apollo project was at best a pale shadow of its predecessor.

To be blunt, there is little challenge these days in flying out of the atmosphere only to re-enter it. Columbus, after all, could only discover America once. At some point, Europeans had to take up homekeeping in the new land.

Mr. Reagan's decision is not without controversy, of course. Former astronaut Buzz Aldrin and others have urged a moon base instead of a space station. Military agencies will jockey for control of space technology. Still others have argued that the United States is wrong to spend resources in space when there remain poor and hungry people on Earth.

But the human condition requires not only clothing and feeding. Challenge, excitement and invention are sustenance for us, too.

Without them, life sinks to the level of survivalist doctrine. The alternative to looking up at the stars is to stare down at our own feet.

WORCESTER TELEGRAM.

Worcester, MA,
March 23, 1984

NASA is no shrinking violet when it comes to planning for future ventures.

The agency has orbited the earth, sent missions to the moon and other planets and now has a handle on shuttle trips into space.

The shuttle is key to the next two big steps, NASA says — a manned space station and a permanent moon colony.

Science fiction? Hardly.

James M. Beggs, NASA administrator, is telling people that the moon is a place humans can get to and camp on. He's giving speeches in which he lays out the plan to extend man's reach beyond the surface of the home planet.

Beggs even has a schedule that goes something like this: 1995 — NASA builds a space station 300 miles above earth. 2010 — working from that stepping stone, and from a second space station in orbit around the moon, the U.S. establishes a manned base on Luna. 2020-2030 — productive activity (manufacturing, mining and the like) begins on the moon. 2060 — a Mars colony has been established and is flourishing.

Far out? Predictions have always been a tricky business. Witness the predictions of automated homes, and a helicopter in every garage made at world fairs of the 1930s. But many once seemingly far-out devices are now part of ordinary life. And Beggs does have the vehicle necessary to make space accessible — the space shuttle.

The basic prediction — that space is a frontier that man will reach and conquer — has already proved true. 2010 may very well be Beggs' year. Stranger things have happened.

Arkansas Gazette.

Little Rock, AR, November 19, 1984

The nonpartisan congressional Office of Technology Assessment has examined the National Aeronautical and Space Administration's plan, backed by President Reagan, to build a space station in earth orbit at a cost of about $8 billion in 1984 dollars. It concludes that the station has not been justified on scientific, economic, social, political or military ground.

At the heart of these and other conclusions, which have been given to the Senate Committee on Commerce, Science and Transportation, is the reasonable view that if the United States is to embark upon a program of one or more stations in space the decision should be made not on narrow technical ground but in a broader context of public benefits. That is, the country needs to have a better idea of its goals and directions in space before it continues to pour enormous sums into the space program, of which a station would be a part. Annual spending on the space program now is about $7 billion.

That is a great deal of money and in the course of its report OTA tries to put it into some perspective. For illustration, it notes that the annual cost for space shuttle development is $2 billion; a similar sum would be needed for each of the four or five pressurized modules on a space station. A sum of $2 billion, OTA notes, would require a stack of $1 bills extending 140 miles above the earth, high enough to reach the orbiting space shuttle. Or, in down-to-earth terms, it would make a 10 per cent down payment on 300,000 houses costing $70,000 each.

OTA's point is clear enough: If the United States is to spend vast sums on a space program it should have more than gee whiz reasons.

Edwin (Buzz) Aldrin, a former astronaut, says he believes the Soviets are conducting experiments that probably will put them on Mars before the United States sends another spaceship to the moon. Bon voyage!

The Birmingham News

Birmingham, AL, January 29, 1984

It is probably fair to say that the least well-received portion of President Reagan's State of the Union address was that part in which he set the construction of a permanent, manned space station as a national goal.

This is unfortunate.

For, as the president noted, space remains man's great frontier. It offers both concrete potential for technological and commercial development and, perhaps more important, the opportunity of adding to the store of human knowledge and of further exploring the unknown.

The prospect of establishing a permanent space station is enormously exciting. The station would serve as both a manufactory for exotic items that can be produced only in space and as a jumping off place for further space study. It would bring us one step closer to fulfilling man's dream of true space exploration.

Yet the reaction to the president's proposal from many Americans was distinctly ho-hum. Even Republican Rep. Bill Dickinson of Montgomery, a strong Reagan supporter in many areas, responded that he "doesn't know if we need a space station right now" in light of the high federal deficits.

True, the president's program would cost some $8 billion over the next decade. But this would seem a relatively small price to pay for such a potential in progress, especially in comparison with the size of the overall deficit and with the very large sums devoted to the military buildup.

The economic advantages that the program would bring Alabama, through increased work at Huntsville, would be enough to court the favor of most Alabamians.

But there is more than economic gain involved with the idea of a space station. Perhaps most important of all, it gives the nation a great and worthwhile goal, much as the lunar missions of the '60s did, to accomplish what has not been done before, to use American spirit and ingenuity to push back the frontiers of knowledge and to reclaim the excitement of the space age.

We hope, as the idea sinks in, that more and more Americans will perceive the great good inherent in President Reagan's proposal.

The News American

Baltimore, MD, January 29, 1984

President Reagan's proposal to put a permanent space station in orbit has met with criticism, but not from a source we would have expected. Hardly a word, surprisingly, from people who wonder about the expenditure of $8 billion at a time when the federal deficit is the highest in history, and when the safety net is leaky. No, the criticism comes from a number of eminent scientists. They're not against a space station — they say that there's no need for a *manned* space station and thus no need to spend so much money.

For example, Dr. Von R. Eshleman, head of the Stanford University Center for Radar Astronomy, said NASA chose a manned station primarily because it could be "sold" to the people more easily: "The only reason to put men into space is that some people at NASA believe it's easier to sell the program to the public if they've got a Buck Rogers out there flying around the stars. Because of the enormous cost of putting men up there, that's all we'll get for the money. If we could use robotics, that same money could be used to develop the technology that would allow us to really begin exploring the universe."

And listen to James Van Allen, the pioneer space scientist who discovered what is now called the Van Allen Radiation Belt: "The conduct of work done by manned spacecraft is enormously inefficient," he said. "Using automated spacecraft we could do a great deal more in planetary exploration and a better job."

Indeed the station that NASA has in mind is going to be pretty much automated anyway. James Beggs, the NASA chief who once headed what is now the Westinghouse Defense and Electronic Center near BWI, says the automated equipment will handle routine space station operations; the crew will handle research and manufacturing.

No matter. We're for the space station. We're for it not because it will put us ahead of the Russians out there in the beyond (which it will, by 10 years, Mr. Beggs figures). We're for it because it would be derelict for this country not to push into the last frontier and not to seek who knows what treasures and who knows what new dimension for mankind.

The cost is immense. Most of the money will come from the American taxpayer. But Mr. Reagan's idea is to involve heavily the non-governmental sector. He wants business and industry, which will benefit from the work that will take place aboard, to invest heavily in its success. As we've said many times, this nation cannot afford to shrug its shoulder to the possibilities of space. Americans never have been afraid of pushing on beyond the frontier. And space, remember, is a frontier we have barely crossed.

The Register

Santa Ana, CA, November 18, 1984

The sight of astronauts hauling a satellite out of a bad orbit may not be as exhilarating as watching men walk on the moon, but it gave us more than a tickle of excitement just the same.

The moonwalk inspired a lot of poetry, and it certainly helped keep the doors of the federal treasury open to NASA. The ability of the shuttle to salvage an off-course communications satellite is exciting because it is, rather, precisely the sort of prosaic commercial application of the space program that helps move it along toward privatization — a closing of the door of the federal treasury.

Billions of people watched the moonwalk. The audience for the satellite retrieval was considerably smaller, but we expect it may have included a high proportion of executives — engineers and money men — in the communications and other industries that might profit from the industrialization of space and have been watching the progress of the "space truck."

The chances of the private sector substantially replacing government in the exploration and exploitation of space depend in some measure on the extent to which the government gets out of the way and lets it happen.

A good place to start might be something as simple as a review of antitrust laws that now make it difficult, if not illegal, for competing corporations in some industries to pool resources for research and development. Such cooperation among large corporations seems likely to be essential for private-sector space exploration, given the huge sums of money involved.

Interestingly, while the shuttle was retrieving the satellite, the congressional Office of Technology Assistance released a report that said it might be time to invite the public to help assess — and broaden — the goals of future space projects, specifically NASA's proposed (and Reagan - supported) permanent space station, because it is the public that must pay NASA's huge bills. Congressional committees, the report said, tend to listen only to scientists.

As long as average citizens — in their role as taxpayers — are footing the bill, perhaps they ought to have a hand in designing the space program.

The best hope for the future of space exploration, however, is to turn it over as much as possible to the private sector. It is the best way to assure that it will meet the most needs of the most people at the least cost.

Under that condition, average citizens — in their role as consumers — would really have a hand in designing the program.

The Morning News

Wilmington, DE, January 25, 1984

PRESIDENT Reagan is expected to announce tonight that he approves an initial commitment of $200 million toward establishing a permanent manned U.S. space station. The final cost could be $9 billion.

This, Washington observers say, may be one of the very few initiatives Mr. Reagan will propose in his State of the Union address. Some folks, still unpersuaded about the virtues of space activity, will be skeptical about its cost-benefit ratio. They also will be dubious about such projects as the space telescope, now scheduled for launching in 1986 at a cost of $1.2 billion — $400 million above original estimates.

The advances in practical knowledge made possible by our ventures in space seem to justify even those astronomical sums, but a timely report this week reinforces that argument: an account by a German news agency out of Moscow that fir seeds taken into space in the Soyuz-Appollo program and subjected to weightlessness have produced trees that grow substantially faster than other seeds.

If we can induce faster growth for trees — and possibly other plantings — it could hold promise of progress in forestation, prevention of erosion and other pursuits. A thought of Garden Earth from Project Space is not beyond the realm of imagination.

The Seattle Times

Seattle, WA, January 10, 1984

A NEW congressional study reports that the Soviet Union is nearing construction of a permanent space base that will serve as an eventual springboard for Soviet settlements on Mars and the moon.

That finding could be the decisive factor in a question facing President Reagan as he prepares his State of the Union speech to Congress this month.

This state's senior senator, Slade Gorton, R, wrote the president recently, urging him to begin immediately a "national effort to launch a manned, civilian space station" as the next phase of the U.S. space program. Gorton is chairman of the Senate subcommittee on science, technology and space, which conducted hearings last year on the space outlook.

In his letter to Reagan, Gorton noted that the space shuttle has a maximum orbiting capacity of roughly 21 days, and said a space station clearly is necessary as a required prelude to extended space travel.

That's what the Russians have in mind, according to the congressional Office of Technology Assessment, which finds that "the Soviet space-station program is the cornerstone of an official policy that looks not only toward a permanent Soviet human presence in low-Earth orbit but also toward a permanent Soviet settlement of their people on the moon and Mars."

Gorton is unquestionably right in pointing out that a space station and laboratory is the "next logical step" after the shuttle and the lunar landings. It would be folly not to proceed further along the path that began with a panicky crash program in the aftermath of the Soviet Sputnik in 1957.

No crash program is indicated at this juncture. There are too many other claims on the federal budget. But the U.S. space program should be allowed to go forward to the "next logical step" at reasonable funding levels.

The Courier-Journal

Louisville, KY, November 17, 1984

THIS WEEK has been a time of both jubilation and concern for NASA, the National Aeronautics and Space Administration. The jubilation, of course, is over the successful mission of the space shuttle Discovery, whose crew retrieved two communications satellites that earlier had been launched into the wrong orbits. The concern is over a new report by a congressional agency that questions NASA's ambitious plans for an $8 billion permanent manned space station.

The report by the Office of Technology Assessment has been momentarily eclipsed by the remarkable feats of the Discovery crew. But the issues raised in that report must be addressed before Congress gives NASA the green light to proceed with the space station. Specifically, the Office of Technology Assessment charges that the space station has been over-designed by NASA to perform more than 100 kinds of activities, with little concern for priorities and budget limitations.

The report acknowledges that "a persuasive case can be made" for putting some sort of manned structure into orbit. But NASA's plan, it said, is a "Christmas-tree proposal" that seems designed more to keep the space agency's engineers busy than to achieve clearly defined scientific objectives.

Similar complaints have been directed over the years at other ambitious NASA projects — from the Apollo moon-landing project to unmanned probes to distant planets. Scientists, like other mortals, frequently disagree about budget priorities. What one scientist may see as a fully justified or even necessary undertaking may be dismissed by another as a monumental waste of money that could be used for far worthier investigations.

Non-scientists are split on these issues, too. How many times have you heard angry mutterings about how the billions spent to put a man on the moon could be used to build roads or feed the poor here on earth? Yet the vicarious adventure of space exploration has also thrilled millions of Americans — and people of other lands. Close-up pictures of the rings of Saturn and rocks retrieved from the lunar surface aren't just the property of scientists; they also belong, in a sense, to all of mankind.

The question, then, isn't whether to continue space exploration but to establish priorities, recognizing that a nation staggering under huge budget deficits cannot afford everything.

Admittedly, a permanent manned space station would offer opportunities both to conduct long-term experiments and to manufacture substances that can only be properly produced in a zero-gravity environment.

Soviets plan station, too

The Pentagon doubtless has an interest in such a project, too. The Russians have more experience than the United States in lengthy manned space flights and reportedly have their own plans for a permanent manned station. Talk of peril to the national security — a "space-station gap" — is sure to be dragged into the argument before long. And of course a manned space station would be a handy platform for testing some of those "Star Wars" missile defenses that President Reagan is so keen to build.

But the scientist-bureaucrats at the Office of Technology Assessment shouldn't be dismissed as so many kill-joys. Their agency was set up by Congress to ask tough questions that might not occur to non-scientist lawmakers. It's now up to NASA to provide persuasive answers.

The Orlando Sentinel

Orlando, FL, February 1, 1984

President Reagan is about to ask Congress to finance the country's next spectacular in space, a manned space station by the early 1990s. Over the long term, the project could cost at least $8 billion. To get started, the tab is $150 million for next year. It deserves congressional support.

There's something else that Congress should do, though. It should be deciding what the country wants in the long run from the space program. As things stand now, the U.S. space program has no definite long-range goals. That shortcoming is made clear by the confusion over what the next step into space ought to be and by a sharply critical report from the Office of Technology Assessment.

That said, a space station makes sense even as an end unto itself. For one thing, it could be used to make a number of highly technical products, some of which simply can't be made in significant quantity in Earth's gravity. A space station could serve as a service station and as an observatory with a view of the cosmos much clearer than possible from Earth. And it would provide information on how to live in space for long periods.

That's where the long-range plan comes in. Learning how to live in space would mesh logically with a vision of manned exploration of the solar system. Soon there needs to be a decision on whether the United States intends to move in that direction or whether it wants to try some other means of space exploration. Some proposals would keep man on Earth and let robots do the work in space.

The United States must keep pushing ahead with space exploration. The situation today is not unlike that in the 15th century when Europeans wondered what lay beyond the blue horizon. Using good instincts and the technology of their day, they discovered and prospered. Man can prosper today by navigating the stars and by experimenting in space.

THE MILWAUKEE JOURNAL

Milwaukee, WI, February 3, 1984

The Congress and the electorate should be open minded but skeptical about President Reagan's ambitious proposal, made in his State of the Union speech, to build a permanently manned space station.

The most obvious question about the project is its sky-high cost, estimated at $8 billion over eight years and perhaps $30 billion over the next 16 years. The actual cost, given the history of overruns, could be higher.

Reagan is making this extraordinary proposal while simultaneously seeking an additional $8.4 billion in economic aid over five and one-half years for Central America, plus $1.6 trillion over five years for various military programs. Meanwhile, the administration is underfunding needed social programs.

The military programs already announced will contribute to budget deficits that threaten to weaken this country's economic strength and thus a key source of its national security. Spending another $8 billion for a space station could weaken it further. Not surprisingly, the Office of Management and Budget opposes the plan.

The program would be worthwhile if its advantages were both clear and overwhelming. But are they? Reagan speaks of the project as a way to "build on America's pioneer spirit and develop our next frontier," saying that a space station will permit "quantum leaps in our research in science, communications, and in metals and in life-saving medicines which can be manufactured only in space." The National Aeronautics and Space Administration says the station could be used, among other things, as a base for manufacturing techniques that require a gravity-free environment.

However, paired with those exciting prospects is a danger that a space station will be transformed into a base for weapons, thus sparking an arms race in space. That possibility is not frivolous, given Reagan's interest in star-wars technology.

NASA's enthusiasm for the project is not fully shared by the Space Science Board of the National Academy of Sciences, which has said it saw no need for a manned space station for another 20 years. Even the intelligence and military establishment are said to be cool to the concept.

Simply stopping the exploration and exploitation of space is not only repugnant but possibly suicidal. In the modern era, the US must keep pace in science and technology. But whether that requires a manned space station at a cost of $8 billion-plus is another question. Lofting an unmanned station, or a series of satellites, might make more sense. All those possibilities should be closely examined and set against the nation's other urgent needs.

Giotto to Rendezvous with Comet; Other Probes Planned

The European Space Agency launched its Giotto spacecraft July 2, 1985 set for rendezvous with Halley's Comet in March, 1986. The launching, Western Europe's first interplanetary probe, took place at the Kourou Space Center in French Guiana. Giotto was placed by the agency's unmanned Ariane 1 rocket into a "parking orbit" above the equator, where the 2,112-pound spacecraft would "kick" itself from Earth orbit to begin a 360 million-mile journey toward destruction, some 310 miles from Halley's Comet, on or about March 13, 1986. The spacecraft was expected to be sandblasted into destruction at that point by billions of high-speed dust particles careening off the comet's nucleus—but not before, scientists hoped, the craft had completed experiments on the image, chemistry, magnetism and other aspects of the comet and its tail. The spacecraft was named for the 14th-century Florentine painter Giotto di Bondone, who saw Halley's Comet in 1301 and painted a representation of it in a fresco still exisitng in the Scrovegni chapel in Padua, Italy. Visible from earth every 76 years, the comet was last seen in 1910.

The Soviet Union in 1984 had dispatched two Vega spacecraft for rendezvous with Halley's Comet from a distance of 6,200 miles, and Japan was preparing to launch Planet A for a photographic look, some 160,000 miles away, at the comet's tail and the cloud of gas enveloping the comet as it circled the sun. The United States also was planning observations about 50 million miles from the comet. They would be made from the space shuttle, or from a recoverable satellite deployed 100 miles from the shuttle, by ultraviolet telescopes that would operate continuously for a week at the time the comet was expected to present its brightest and biggest scenes to Earth.

Chicago Tribune

Chicago, IL, September 18, 1985

Why do dirty little cosmic snowballs made of dust, ice and frozen gases come sweeping around the solar system, leaving a trail of glowing particles across the night sky and touching minds on Earth with primordial wonder and mystery? How can a comet's head no bigger than a mile across generate a turbulent tail thousands of miles wide and half a million miles long? Where do comets go after they have swept back past the planets? How can they return with such predictable preciseness and how can their icy substance survive the sun's glare?

In search of answers and in hopes of stealing a bit of thunder from next spring's encounters with Halley's Comet, the United States has flung a little satellite across the tail of the Comet Giacobini-Zinner. Already, the data the International Cometary Explorer (ICE) has sent back are resolving some age-old questions and hinting at even greater mysteries still unsolved.

ICE itself is a small wonder, a cheap shot intended as a consolation for scientists deeply disappointed that the Carter and Reagan administrations could not spare money to go chasing Halley's Comet when it makes a once-in-76-years appearance in 1986. The Soviet Union, Japan and the European Space Agency have satellites to rendevous with Halley's. The U.S. will not.

Instead, for $3 million, NASA was allowed to take a little working satellite already in orbit to monitor solar gases and redirect it toward the less-famous, incoming comet. In 1983, NASA scientists fired some tiny rockets on board the five-foot tall ICE, swung it around the moon five times to generate enough energy to break out of Earth's gravity, and sent it 44 million miles away to the precise encounter with Giacobini-Zinner. ICE has already been working for seven years in space and traveled more than one billion miles.

Now ICE's messages back to Earth have described new phemonena about electrically charged particles of gas in the comet's luminous tail, about the shock waves generated as the comet's gases buck the mysterious solar wind and what happens when sunlight heats the surface of the comet's head.

But there are still many mysteries left about the comets that come from deep in space and may be remnants of matter left over from the formation of the sun and its planets 4.5 billion years ago. Thanks to scientific progress, comets are no longer seen as evil omens. But thanks to science, we have caught a glimpse of even greater wonders that await more exploration.

The Tennessean

Nashville, TN, April 6, 1985

THE Soviet Union is planning an elaborate program of unmanned planetary exploration which is the envy of American scientists involved in lunar and planetary studies.

The Soviets plan a 1988 mission to Phobos, a moon orbiting Mars, a 1989 flight to orbit the moon over its poles and a 1986 two-stage spaceship that would explore both the planet Venus and Halley's Comet.

The U.S. space agency has done quite a bit of unmanned exploration of some of the planets, but this work seems to be overshadowed by the manned missions, the space shuttles and other projects. The pursuit of academic knowledge of the universe, which is the most important part to some people, seems to be secondary in the U.S. space effort.

"The Soviet program seems to take more of their national attention than it does in this country," said Mr. Mike Duke, a lunar and planetary scientist at the Johnson Space Center in Houston. Mr. Duke talked with a group of Soviet scientists at a conference on lunar and planetary studies at the space center last month and learned quite a bit about the Soviet plans.

He said the Phobos mission would put a spacecraft within a few hundred feet of the surface of the Martian moon, which has a low gravity field. The craft is to fire laser beams at the moon's surface and an instrument will analyze the reflected light to determine the moon's composition.

Mr. Duke said this process works in the laboratory, but that nobody has tried it before in space exploration. Thus, the Soviets may be preparing to score another first in space.

Mr. Duke also said the Soviets are studying the possibility of probing an asteroid named Vespa in a joint project with the French in the 1990s. Vespa is one of the largest pieces of rock in the asteroid belt orbiting the sun between Mars and Jupiter.

It seems the Soviets may be getting ahead of the U.S. in this kind of exploration. Unmanned probes may not be of much interest to Washington because they do not lead directly to military advantage. But the country that neglects the pursuit of knowledge for its own sake loses much. The U.S. space agency's unmanned probes have made some spectacular discoveries in the past. But they seem to have been put on the back burner in recent years. They should be permitted to do more.

The Times-Picayune
The States-Item

New Orleans, LA, July 7, 1985

The European Space Agency's Giotto space probe blasted off Tuesday on a course that will make it a possibly suicidal shot across the bow of Halley's Comet next March. It joins two Soviet probes already headed toward Halley's future vicinity and a Japanese craft to be launched in August. It is a stately space ballet the budget-conscious Reagan administration chose to sit out, but American instruments are on the Soviet craft and U.S. antennas will help track Giotto.

The comet, which comes by every 75-some-odd years, is now only a planet away, 400 million miles out between Jupiter and Mars, zipping toward the sun at 62,000 miles an hour. It won't be visible to the naked eye until next January, and, unfortunately, will be seen best in the Southern Hemisphere.

Our best eyes will be those in space. The two Soviet craft will pass the comet within 6,000 and 2,000 miles respectively, and the Japanese will get no closer than 120,000 miles. Giotto will cross Halley's path at about 300 miles. This will fly it through the comet's sizzling atmosphere, and the encounter may destroy its instruments.

They will have had time, however, to do much work. Aboard are television cameras and sensors that should give us our first direct information on what a comet is. Our current model dates only from the 1950s: a "dirty snowball" that gets its dramatic tail when the sun's heat vaporizes its surface and the solar wind blows off particles and gasses.

The current supposition is that comets are deep-frozen debris from the birth of the solar system, and since star formation and the presumably rarer formation of planetary systems are still imperfectly understood, the information will be of importance.

But not all the mystery, romance and theorizing about comets will be solved. The spacecraft will not, for example, tell us where comets come from or why. The current theory about that is that billions of them circle the solar system and that gravitational perturbations produced by passing bodies nudge them into long ellipical orbits of the sun. There is a new theory that this nudging is done by a yet undiscovered planet, imaginatively dubbed Nemesis, that has rained showers of comets onto Earth and caused the extinctions recorded in paleological evidence.

There are troubles with that and similar theories, though it is generally agreed that a comet or a fragment of one hit Earth with extraordinary force, thankfully in a vast, uninhabited region, in Siberia in 1908. But Giotto might be able to throw light on a theory that comets carried to Earth, and perhaps other planets in our system, the chemical building blocks of life. There are some problems with that theory, too, but the Earth is known to be steadily dusted with tons of comet dust.

In any event, the scientists will revel in their data and we, who may not get the show our imaginations are preparing us for, will revel in the views our age's extraordinary machines can send back from another age's object of helpless wonder.

New Orleans, LA, September 17, 1985

Beyond the world of astronomy and space exploration, mankind's first interception of a comet may not be long remembered. But within the arcane world of astronomy, it has understandably set off a flurry of excitement. The comet is not the storied Halley's, which visits the inner solar system once every 76 years or so and is scheduled to do so again next March, but its little known relative, the Comet Giacobini-Zinner.

Last Wednesday officials of the Goddard Space Flight Center at Greenbelt, Md., resourcefully redirected a wandering half-ton satellite designed to study solar physics through the tail of Giacobini-Zinner. The low-budget comet probe turned out to be a phenomenal success.

The close encounter occurred some 44 million miles from Earth as Giacobini-Zinner streaked away from the sun. The relatively small comet was named for Michel Giacobini, who first discovered it in Nice, France, in 1900 and Ernst Zinner, who rediscovered it in Bamberg, Germany, in 1913.

The visit by the rugged little spacecraft, the International Cometary Exlorer, to the gaseous tail of Giacobini-Zinner was as much a triumph of American determination and ingenuity as anything else.

In 1981, after it became clear that the United States would not be sending its own spacecraft to observe Halley's comet, Dr. Robert W. Farquhar, Goddard Space Center flight director, came up with the idea of redeploying a satellite named International Sun Earth Explorer toward Halley's path. The satellite had completed most of its objectives and was some 900,000 miles from Earth and in a position for a flight by Halley's comet.

But astronomers determined that the tiny craft's 5-watt radio would not be audible on Earth. At that point, Dr. Farquhar decided that the satellite, renamed International Cometary Explorer, should be aimed instead at Giacobini-Zinner, and the rest is record-book history.

A series of ground-controlled intricate maneuvers sent the durable little satellite looping around the moon, where it gathered gravitational force, and through the center of Giacobini-Zinner's million-mile-long gaseous tail.

The craft survived its ride through Giacobini-Zinner's tail, which it found to be much wider (about 14,000 miles) than scientists had theroized and full of magnetic forces and high-energy particles. Scientists at Goddard are gloating over the little satellite's windfall of information.

Meanwhile, in-space observations of Halley's comet will be left to the Soviet Union, Japan and the European Space Agency, whose scientists are encouraged by the glowing success of another U.S. space first.

The Oregonian

Portland, OR, January 5, 1985

The successful launching of the first artificial comet was an exercise in international cooperation, a step that will be continued when Halley's Comet bursts through the orbit of the planets in 1986. Cooperation in space research must be encouraged, lest military competition become a dangerous alternative.

The United States, West Germany and Great Britain cooperated in lofting a cloud of barium vapor that glowed when hit by solar winds, providing clues as to the way the solar system was formed.

Much more such information is expected to come from the international effort to observe Halley's Comet when it returns. The Soviets have launched Vega 1, an 8,000-pound satellite, which is the first of two such craft scheduled to visit the comet in March 1986, along with two other satellites, one from Japan and the other representing the 11-nation European Space Agency.

The United States will not send an observing satellite, Congress having pinched off the project and the National Aeronautics and Space Administration electing not to press the issue. There will be a U.S. observation from a telescope aboard the space shuttle, while the Soviet Union has promised to share fully its data and help the Europeans track their spacecraft, called Giotto, named for a Florentine painter who depicted the comet in a fresco done in the 14th century. Japan's craft is named Planet A.

The comet, which last visited the inner solar system in 1910 and was named for British astronomer Edmund Halley, who first predicted its periodic return, will also be under observation from thousands of ground-based telescopes. Like the skyships, they will try to observe the comet's growth and decay and the shock wave where the solar particles pile up much like standing waves seen in wind tunnels.

There are countless ways nations can cooperate in exploring the planets and the galaxy. Such cooperation will not only reduce the costs, but will provide new scientific viewpoints, resulting in discoveries that would be missed if only those nations that can afford the expensive efforts participate.

THE EMPORIA GAZETTE

Emporia, KS, July 15, 1985

HALLEY'S Comet has run into no celestial detours and is expected to show up in our neighborhood later this year.

Last week's mail brought a University of Kansas news release that gave some useful information about the comet.

David Beard, distinguished professor of astronomy and physics, warned that this visit by the comet might not be its most spectacular. "The thing is," he said, "it's going to be buried in the southern sky most of the time this trip."

Mr. Beard said that December would be the best time for Kansans to look at the comet. It should be visible to the naked eye and be about 60 degrees above the horizon, near the constellation Aries.

"It won't be very colorful or bright," Mr. Beard said, "because it's still pretty far from the sun, and we don't get very close to it at that time." A comet seems more spectacular the closer it is to the sun and the closer it is to the viewer, he added.

The comet will be behind the sun in February and will not be visible. When it reappears in March, it will be closer to the sun and to earth than it was in December, but that won't do us much good here in Kansas. The comet will be lower in the southern sky — so low it will be hard to see this far north.

As a sidelight, Mr. Beard said that many people who said they remembered seeing Halley's Comet on its last visit in 1910 were not remembering that comet at all. That was also the year of the Great Comet, which Mr. Beard called "the most spectacular comet of the century."

If you remember seeing a comet in 1910, which comet you saw depends on the time of year you saw it. Halley's was visible in the spring, and the Great Comet in the winter.

ST. LOUIS POST-DISPATCH

St. Louis, MO, September 17, 1985

NASA was handed a lemon and it made lemonade: That is the essence of the space agency's brilliant success in the flight of its International Cometary Explorer (ICE) satellite through the tail of a comet. No other satellite has ever come as close to a comet.

Back in 1981, budgetary cutbacks made it impossible for NASA to join Soviet, West European and Japanese efforts in sending unmanned satellites to closely study Halley's comet in 1986 when it makes its once-every-76-year visit. NASA scientists then hit upon a novel idea: Why not employ a "used" research satellite, one already in orbit, for the mission?

There was one available, a 1,000 pound joint American-European satellite called International Sun Earth Explorer 3. It had been used to measure solar winds and it had performed most of its experiments. However, intercepting Halley's comet would prove difficult and the satellite's 5 watt transmitter was too weak to be heard given the range. So as an alternative, the satellite was aimed at a smaller comet, Giacobini-Zinner (named after the turn-of-the-century French and German astronomers who discovered it). Using a series of complex loops around the moon to build up speed and to change direction, ICE was launched toward Giacobini-Zinner in December 1983.

The passage through the comet's tail proved much more interesting and provided more data than NASA had even hoped for. The new information included the sheer size of the comet's tail and the amount of particles and magnetic forces trailing the comet. As one NASA scientist noted, "A good experiment generates more questions than answers." By that standard, this scientific mission was a great success.

THE ARIZONA REPUBLIC

Phoenix, AZ, September 18, 1985

MAN'S infatuation with comets — expected to zoom to new heights in March when Halley's Comet makes its next glowing apparition — has gotten a premature boost, thanks to Yankee ingenuity.

Scientists at the Goddard Space Flight Center in Greenbelt, Md., directed a small, half-ton satellite to mankind's first rendezvous with a comet, at 44 million miles from Earth, visible only through powerful telescopes.

For 20 minutes the International Cometary Explorer zipped through the Comet Giacobini-Zinner's 14,000-mile-wide tail picking up valuable new scientific data about those enigmatic wanderers from deep space with icy nuclei, fast-moving dust particles and tails thought to consist of gases.

It likely wouldn't have happened at all had the United States not canceled a probe of the more famous Halley's Comet, which will be examined in minute detail by five unmanned spacecraft — two each from the Soviet Union and Japan and one launched by a consortium of 11 Western European nations. The cost of an American mission could've run as high as $1 billion, money the Reagan administration thought was better spent elsewhere.

As a result, National Aeronautics and Space Administration scientists chose a satellite launched seven years ago to study solar physics and decided to retarget it in 1982 on a trajectory that zipped it around the moon five times for momentum, flinging it on a successful course for its close encounter with the comet.

For its $3 million price tag, America once again can claim an all important first, and get a head start on the other expeditions designed to unlock the biggest question of all: whether comets have preserved the primordial elements that created Earth and the life on it.

THE SACRAMENTO BEE

Sacramento, CA, September 14, 1985

In the best scientific tradition, the unprecedented passage of an American-built spacecraft across the tail of a comet some 44 million miles from Earth came about almost as an afterthought born of earlier disappointment. Because federal budgetary constraints precluded a U.S. mission to intercept Halley's comet when it makes its every-76-years pilgrimage across our solar system beginning late this year, scientists in the National Aeronautics and Space Administration decided to try for a consolation prize.

A half-ton craft launched in 1978 to study solar winds had completed most of its work, so in 1982 scientists at the Goddard Space Flight Center in Maryland began steering it toward a rendezvous with the comet Giacobini-Zinner, which passes through the solar system every 6½ years. After a series of improvised maneuvers that took it five times around the moon, to pick up added energy from that body's gravitational forces, the International Cometary Explorer (or ICE, as it's called by most scientists) was then launched on a path that, earlier this week, sent it hurtling through the 14,000-mile tail of a comet whose core is believed to be little more than a mile across.

That feat was astounding in itself. But what was so gratifying to scientists was that ICE apparently suffered no damage in the intense magnetic field of the comet's tail. The craft has sent back data described as "rich in detail" that will help expand knowledge of space physics, and particularly of the structure and dynamics of comets. That's especially important to American scientists, who are still brooding that unlike Japan, the Soviet Union and the European Space Agency, this country is not sending a spacecraft for an encounter with Halley's comet.

What all this will lead to can scarcely be imagined, especially by the vast majority of people who can barely grasp what's involved. But that mystery is what makes nature so wondrous, and it's what motivates scientists to keep probing and tinkering in the eternal quest to unravel ever more of that mystery. May the forces continue to be with them.

The U.S. decides, after all, to join the scientific effort to intercept Halley's Comet

AKRON BEACON JOURNAL
Akron, OH, December 1, 1985

MAYBE COMETS aren't what they used to be.

Oldtimers speak with reverence of Halley's Comet: Brighter than a full moon it was, back in 1910, when it lit up the night skies. By comparison, this year's edition is a pale imitation.

The comet passed by Earth last week almost unnoticed — too cloudy, too distant, and too much in competition with man-made light. It will be back in the spring, after its swing around the sun, but even then more an object of curiosity than of awe.

Still . . . something quite remarkable accompanies Halley's Comet on its round-trip of the solar system. It is what makes us human — the ability to wonder where the comet goes when it leaves us; whether it passes only barren terrain or comes within view of other beings who wonder also.

Halley's is a special case because it ties us to the universe, to the unknown, and to our own past. It is the only really bright comet that is also predictable. And its cycle — 76 years — is roughly that of a human lifespan, which deepens its meaning.

It has been observed on every return at least since 240 B.C., named for British astronomer Sir Edmund Halley who predicted its 1758 return although he did not live to see it.

The comet's return is also linked to historical events: the destruction of Jerusalem in A.D. 70, the conquest of England by William of Normandy in 1066, the fall of Constantinople in 1453. Other comets appeared to herald the death of famous rulers and the crumbling of empires.

It is said that in 1456 Pope Callixtus III excommunicated Halley's Comet as an agent of the devil. Roman emperor Nero, on the advice of his astrologer, murdered anyone who might succeed him when two comets appeared six years apart, the latter being Halley's in A.D. 66.

Modern man has generally taken a more enlightened view of comets (although there was that one incident in Oklahoma when a sheriff arrived barely in time to prevent a cult from sacrificing a virgin during the 1910 appearance). What concerns us now is the origin of these "dirty snowballs" and what they can tell us about the creation of the universe.

But maybe we should speculate about what Halley's Comet might find when it streaks back by Earth in 2061. Will mankind be able to meet it in deep space for a closer look, or will Earth be — as some cynics believe — just another barren planet by that time, the victim of nuclear winter?

That fate is up to us. We can only be certain that Halley's Comet will be back, whether we are here to see it or not. The legacy of Halley's Comet is humbling: We can marvel at the heavens and, given time, even come to understand them. But we can't, with any confidence, order the stars about. Comets obey their own immutable laws. Even in this relatively poor showing, Halley's Comet is back for another run. Pope Callixtus, who condemned it, and Nero, who tried to circumvent its power, are long gone.

The Des Moines Register
Des Moines, IA, September 19, 1985

The year of the comet got off to a thrilling start when a hand-me-down American spacecraft whisked with surprising ease through the tail of a little comet called Giacobini-Zinner, 44 million miles away. The first comet intercept by a man-made object was a real achievement, on a shoestring budget.

The spacecraft was launched in 1978 on a mission to monitor solar wind. It wasn't until 1982 that U.S. scientists decided to turn it into a cometary explorer by whipping it around the moon in an elongated orbit, then slinging it on a 20-month journey to intercept a pinpoint hurtling through space.

The space agency decided on this when it was denied money for a mission to Halley's Comet. By intercepting Giacobini-Zinner, the United States upstaged the Soviets, who will intercept Halley's in about six months.

Comets are perhaps the most mysterious objects in the solar system. Scientists say it will take months to sift the data from Giacobini-Zinner's tail, and the information most likely will raise as many new questions about comets as it answers.

A little knowledge always whets the appetite for more, and the success of this mission only increases the disappointment that the United States doesn't have a mission to Halley's, too.

Voyager 2 Visits Uranus; Discovers New Moons, Rings

The U.S. spacecraft Voyager 2 made its closest approach to the planet Uranus January 24, 1986, passing within 50,679 miles of the planet's cloud tops. Uranus is the third largest planet in the solar system and the seventh planet from the Sun. The spacecraft discovered new moons and rings around the gas-shrouded planet and provided the first evidence of the planet's magnetic field. It brought the total of known Uranian satellites to 15 and of known rings to 10. As had been the case in Voyager's earlier encounters with Jupiter and Saturn, some of the most intriguing images were those of the planet's satellites, which offered odd and unique examples of planetary physics.

Voyager 2 had been launched in 1977 toward Jupiter, in what was the last major U.S. unmanned planetary project. In 1979 the craft made a dramatic encounter with Jupiter, transmitting 17,00 photos back to Earth. The craft had then used the pull of Jupiter's enormous gravity to propel it toward Saturn, where it transmitted another 15,000 images in 1981. It was then flung on toward Uranus and will ultimately fly by Neptune in 1989 before leaving the solar system forever. With each planetary encounter the spacecraft had transmitted back to Earth a stream of data and dramatic photographs that had vastly enhanced knowledge of the solar system.

The new flood of discoveries began Jan. 16, with the announcement that Voyager 2 had detected six more moons around the planet. The objects, ranging from 20 to 30 miles in diameter, appeared to be roughly in the same orbit, suggesting that they might have been part of one large moon at one time. The moons were outside the orbit of Uranus' nine known rings. New discoveries announced Jan. 22 revealed the movement of windblown clouds on the planet, a brownish haze at its sunlit pole, and the presence of two more small moons. An orange-brown haze visible at the pole strengthened conjectures that the planet's outer, blue-green gaseous mantle was primarily composed of methane, which would tend to form such a smog under the action of sunlight. The first clear evidence of the planet's magnetic field was reported Jan. 23, in the form of radio emissions caused by charged particles spiraling along the magnetic lines of force.

Chief project scientist Edward Stone told the press, "The images already tell us the Uranian system is unlike anything we have seen in the solar system."

The Boston Globe

Boston, MA, January 24, 1986

Today mankind's deepest reach into the solar system, spaceship Voyager, passes within 50,000 miles of the planet Uranus – a stirring tribute to the coordinated efforts of thousands of scientists and technicians over more than 10 years, 8½ of them since Voyager was launched from Earth on Aug. 20, 1977.

Most attention will properly be devoted to pictures and other data Voyager sends back from this close encounter with a distant relative. Predicted but still unseen moons may be recorded. A longstanding doubt about the length of the planet's rotation – 12, 16 or 24 days – will probably be resolved. There will be a series of close looks at the known moons and perhaps some clues to their makeup. The Uranian atmosphere and weather patterns will be traced in far better detail than can be gleaned from earth-bound telescopes.

Well and good. A better understanding of the solar system deserves to be highly valued for its own sake.

Voyager also says much about the indispensable role of cooperation in such projects. Society has come a long way since the time when Uranus was discovered by a single astronomer, William Herschel, 205 years ago. Today's efforts cannot be conceived, much less achieved, without great combinations of highly trained persons with wide ranges of skills, working together in neatly timed sequences to put Voyager at precisely the right place at the right time.

Voyager has arrived at Uranus in a most spectacular way, having passed close by Jupiter in 1979 and Saturn in 1981, sending back highly detailed pictures from those neighbors. Using the planets' gravity as slingshots, Voyager was boosted to today's encounter. Even with the accomplishments firmly in the record books, it is difficult for ordinary citizens to fathom the complexity of the calculations built into charting that course – a decade ago.

The trip also reaffirms the great value of unmanned space flights with highly instrumented craft that make enormous use of limited funds for expanding America's and mankind's knowledge of the outer planets. Voyager's arrival stands as an unintended commentary on the poor allocation of resources associated with NASA's other great effort, the shuttle program, which has led to the abandonment of so many other scientific projects.

Fittingly, Voyager has yet another great stride to take beyond those that have so far carried it a looping three billion miles. If all goes well, the spacecraft has three more years, during which it will reach Neptune for a first glimpse at that planet – and then on forever into the universe. A grand tour.

Sunday Journal-Star

Lincoln, NE, January 26, 1986

Only those whose capacities for human wonder have been technowashed are not awed by the latest treasures showered on the world by the United States' space probe, Voyager 2. Rarely has the word "spectacular" been more appropriate.

Today Uranus, the seventh planet from the Sun and 1.84 billion miles from Earth, is less of a dark mystery.

Television and sensor data from a 1,200-pound spacecraft launched in 1977 and never intended for the achievements it is ringing up will not necessarily tell us why Uranus circles the Sun sideways or why one of its poles constantly is in sunlight for 42 years before switching positions for 42 years with the other pole. But coming within 50,679 miles of Uranus Friday, Voyager 2 already has shown Earthlings that some of the planet's 14 — at least — icy moons have craters, faults and valleys.

Even more thrilling information should be known when the rush of radio and imaging data is interpreted. And then it will be on to Neptune, 3½ years away, for Voyager 2, and still more discoveries, if its cameras and single functioning radio hold out.

Part of the awe about this adventure honors the human mind. Astonishing mental powers were required to design a machine so cunning that it could be made to send back volumes of pictures and information about Jupiter (in 1979) and Saturn (in 1981), and then brilliantly be slung toward the surface of Uranus, exploiting the gravity of the those closer planets.

Had the United States depended entirely on a rocket motor to fly a satellite directly to Uranus, the trip would have taken 30 years.

Into these reflections, though, comes a sobering further thought: The successes of Voyager 2 and the technology behind it may further strengthen a treacherous, even pernicious, popular conviction. That is the belief whatever the problem we face, science will provide the answer, that science will always and ultimately save us.

We know of no serious scientist who would agree with such an encompassing premise. Yet it is on just that seductive philosophy that President Reagan and his allies are building their fabulous, and fabulously costly, Strategic Defense Initiative. The life or death of the nation should rest on more tenable foundations.

The Dispatch

Columbus, OH, January 30, 1986

If all our national ventures worked as well as Voyager 2, our problems would be far fewer.

The performance of the swift, efficient spacecraft last weekend as it made its closest fly-by of Uranus, the third largest planet, provided one of the most exciting and productive intervals in our space probing. It is unfortunate that Voyager's great success was closely followed by the disastrous *Challenger* flight Tuesday.

With its highly maneuverable camera eyes and near-perfect instrumentation, Voyager gave Uranus a face and a personality for earthbound scientists and negated some of their pet theories.

It found, for instance, that Uranus has not five moons but 15; has 10 full rings encircling it and 10, maybe more, arc-shaped pieces of rings; has a strong magnetic field and a cloud of hot hydrogen thousands of miles long; and that it is inexplicably warmer on its dark side than on the side toward the sun.

Its many moons — far more interesting bodies than had been thought — have gorges and ridges and craters and mile-high mountains.

While the initial returns from Voyager electrified the scientists, much of the data collected are yet to be sent back home from onboard recorders. Eleven experiments in all collected information on everything from dust particles around the rings to the planet's electromagnetic field.

Securing enough material to keep scientists busy for years is an important haul since these outer-space calls can't be made capriciously. The current juxtaposition of Saturn and Uranus, which gave Voyager the gravitational push to reach Uranus in a relatively short time, only happens every 175 years.

If Voyager 2 hadn't been superbly designed and engineered, the real facts about Uranus could have been denied Earth scientists for another generation.

Voyager is off now for a rendezvous with Neptune in 1989. We hope it has a successful trip.

FORT WORTH STAR-TELEGRAM

Fort Worth, TX, January 31, 1986

One doesn't have to be an astonomer to share the excitement generated by the latest astounding images transmitted by the Voyager 2 spacecraft from 2 billion miles out into the imagination.

The new discoveries made by Voyager 2 during its sojourn in the vicinity of the planet Uranus — the additional moons, the surprising geological formations on those moons, another Uranian ring, the planet's tilted magnetic field and others — will contribute enormously to our understanding of the cosmos.

Granted, it will take many years before the scientific community can put all the Uranian findings into perspective. But it is clear already that some working assumptions about the planet are due for some signficant adjustments. It may even be that some fundamental hypotheses of the physical sciences will be challenged.

The new Voyager discoveries are a tribute to the genius of those who designed the spacecraft and to the civilization that produced such manifestations of the genuis of humankind.

But at the same time the findings are humbling, for they tell us how small our world is and we as individuals are in the totality of being. And they underscore the minuteness of all our knowledge in relation to all there is to know.

Houston Chronicle

Houston, TX,
January 23, 1986

Voyager 2, which went up back in 1977, is outdating the encyclopedias and astronomy textbooks as it sends back graphic pictures on its journey through the solar system, first revising our view of Jupiter in 1979, then Saturn in 1981. Another big chapter is being covered now on Uranus.

Voyager has already found nine previously undetected moons circling Uranus, adding to Oberon and Titania (spotted by telescope in 1787), Ariel and Umbriel (first seen in 1851) and Miranda (found from a Texas observatory in 1948). The photos of the satellites and of the planet are revealing more than telescopes have since the planet was first seen in 1781. The big day comes Friday as the robot astronomer makes a six-hour fly-by within 50,600 miles of the planet out there 1.7 billion miles away from the ground crew on Earth that designed the spacecraft and is controlling the multimillion-dollar project.

Then on to Neptune, which is on the agenda for 1989, when the textbooks no doubt will have to be revised again.

Syracuse Herald-American

Syracuse, NY, January 26, 1986

It is not the wont of editorial writers to stand in awe of anyone or anything. It is their habit, it seems, to debunk other people's awe — to cut things down to size. The bigger the target, the heavier they load up.

But the journey of Voyager 2 — which dashed past the planet Uranus this past week — is truly an awesome event, the kind of thing that brings out the "Gee whiz" in even the most hardened cynic. Consider all that has happened on this tiny planet since Voyager left it behind forever in 1977.

All the time Americans were held hostage in Iran, Voyager continued on its way. Ronald Reagan was elected and re-elected president; the Soviet Union had four leaders; Olympic boycotts were exchanged. If you had been a passenger on Voyager, you never would have heard of AIDS, KAL 007, artificial hearts, Three Mile Island, Mount St. Helens or the phone company breakup.

▽ ▽

While we were preoccupied with all those earthly concerns, our mechanical surrogate kept speeding away from us, faithfully sending back many times more information about Jupiter, Saturn and now Uranus than we had learned since those planets were discovered. If the spacecraft's spectacular success continues, it will rendezvous with Neptune in another three years before leaving our solar system and wandering the infinite near-emptiness of space for the rest of eternity.

Voyager is now so far away from us that it takes nearly three hours for a radio signal, traveling at the speed of light, to reach earthbound receivers. Yet pictures — taken at light levels 360 times dimmer than Earth's by a camera moving at more than 40,000 miles an hour (you photographers can imagine the difficulty of that task) — make the 1.8-billion-mile passage across the void of space and arrive with remarkable clarity.

▽ ▽

And as space endeavors go, Voyager is a bargain. It all has been done for a total project cost so far of $600 million — less than the federal government goes into debt in a single day. By comparison, a single launch of the space shuttle costs $200 million.

Contemplating the enormity of the universe is humbling. We think of the sun as huge — more than 100 times the size of Earth — yet it is medium-small as stars go, a rather nondescript point of light among billions in the vastness of space. All of Voyager's achievements — exciting as they are — serve to emphasize how far we are from realizing the dream of interstellar travel. If Voyager had left Earth at the beginning of humankind's recorded history and made a beeline for the *nearest* star at its present speed, it would still be less than a tenth of the way there.

It is possible Voyager may indeed encounter another solar system sometime in the future, another planet like Earth, perhaps even other intelligent life forms. But don't wait for a postcard.

Wisconsin State Journal

Madison, WI, January 24, 1986

It was more than eight years ago that Voyager 2 began its long odyssey through space.

The scientists and engineers who designed the 1,200-pound spacecraft gave it a lifespan of four years, give or take, when it was launched in 1977 to take advantage of a planetary configuration that occurs only once every 176 years.

Happily, those scientists seriously underestimated the durability of Voyager 2.

Shortly after noon today, Voyager 2 will speed past the planet Uranus, more than *2 billion* miles from Earth. It will pass 64,000 miles above the planet's cloudtops and 18,000 miles from Miranda, one of Uranus' five largest moons. That's a near-miss by space standards.

Of the 12 known moons circling Uranus, seven have been discovered by Voyager 2, which has also shed new light on the planet's nine mysterious rings.

Scientists stand to learn much more about the solar system's third-largest planet today when Voyager 2 spends six hours in the Uranian system before using that planet's gravity to "slingshot" its way toward a rendezvous with Neptune in 1989.

Voyager 2 has only one radio in operation and the "scan platform" that moves its cameras and instruments has been partly jammed since it flew past Saturn in 1981, meaning it may be a dead spacecraft by the time it reaches Neptune.

But then again, no one expected Voyager 2 to live to see Uranus.

THE PLAIN DEALER

Cleveland, OH, January 24, 1986

Two billion miles from the world of its creation and hurtling onward into eternity, a ton of U.S.-designed technology this afternoon will flash within 50,000 miles of Uranus. With each faint whisper of Voyager 2's distance-weakened instrumentation, mankind's knowledge of the oddly tilted seventh planet grows. The data that this speck of human genius has gathered will be meat to generations of scientists. At least seven newly discovered Uranian moons will need names. The nature of the planet and its charcoal-black rings will be debated and, perhaps, determined. The wonders of the system's third-largest orb, the very existence of which was unknown to earthlings until two centuries ago, will become the stuff of children's picture books.

Perhaps the best news is, there may well be more to come. There seems to be no stopping this little scientific sojourner. Launched in 1977 by a team from Cleveland's NASA Lewis Research Center, Voyager was designed to get as far as Saturn. That accomplished handily, NASA reprogrammed its computers and, using Saturn's gravity as a slingshot, whipped Voyager onward to the outer planets. So, on the first day of September three years from now, if it can avoid the detritus of interplanetary space, Voyager is to encounter Neptune, some 2.8 billion miles from the Sun. That's not bad for a 12-year-old workhorse.

DESERET NEWS

Salt Lake City, UT, January 25, 1986

Much of the publicity and most of the money in the U.S. space program goes for shuttle flights. But the long travels of a small, unmanned Voyager 2 spacecraft are giving science a huge bonanza at comparatively cut-rate prices.

Launched at modest cost in 1977, Voyager 2 was supposed to take pictures and scientific measurements of Jupiter and Saturn — which it did in magnificent, breathtaking fashion. Yet it has gone on to accomplish far, far more than was originally planned.

As it hurtles past Uranus, the most remote object ever examined by a spacecraft, Voyager's instruments are telling scientists more about the strange planet than was learned in all the 205 years since it was discovered by telescope in 1781.

Already Voyager has discovered nine small new moons of Uranus, to go with the five previously known. The thousands of pictures and the computer data will take years to analyze. Scientists are still working on the material gained from Jupiter in 1979 and Saturn in 1981. That's just for starters; there are bound to be a lot of surprises, as there were in the flights past Jupiter and Saturn.

When it leaves Uranus, the complex but sturdy Voyager 2, already 1.8 billion miles from home, will head for distant Neptune, flying past there in 1989 for more of the same kind of study.

Voyager's accomplishments so far have been likened to sinking a 1,560-mile golf putt. One can only stand in awe at skill of the engineers, and even more at the incredible beauty of the solar system being exposed by Voyager's cameras.

The Hartford Courant

Hartford, CT, January 29, 1986

Even as we mourn the fall of Challenger, we think of Voyager 2, the unmanned U.S. spaceship that passed by Uranus last week and introduced Earth's creatures to this planet some 2 billion miles away. Yes, 2 billion miles.

For the past eight years, Voyager 2 has been traveling at a speed of 33,000 miles an hour and relaying a treasure-trove of information on the solar system, in particular Jupiter and Saturn.

Within a few days of the rendezvous with Uranus, scientists learned that the rings around that planet are composed of chunks of dark material at least three feet wide. Voyager measured winds blowing with the force of 220 miles an hour and temperatures of 350 degree below zero Fahrenheit. The planet's day was estimated to run in cycles of about 16.8 Earth hours.

Voyager's next major destination is Neptune, a 2.5-million-mile space-trot from Uranus. As we explore more the solar system, we learn more about ourselves. Yesterday, we witnessed a tragedy. Last week, a spaceship approached Uranus 69 seconds ahead of schedule.

THE TENNESSEAN

Nashville, TN, January 22, 1986

THE Voyager 2 spacecraft, which was launched at Cape Canavaral on Aug. 20, 1977, to explore the outer reaches of the solar system, has executed one of the most remarkable missions the space agency has carried out.

Voyager 2 is a companion craft to Voyager 1, which was launched a few weeks late on Sept. 5, 1977. Voyager 1 made a name for itself with probes of Jupiter and Saturn, sending back 18,000 pictures from Jupiter which greatly changed science's concepts of that planet and its five moons. But it seems that Voyager 2 is the one destined for lasting fame.

Voyager 2 also flew past Jupiter, and then sped on to Saturn. It arrived there in August, 1981, four years after leaving Earth, and — with its TV cameras turning and instruments functioning under the command of its on-board computer — sent back pictures which revealed much new information about Saturn's rings and other phenomena.

At that time Voyager 2 was almost a billion miles from Earth, just 2.5 seconds ahead of a perfect schedule. An error of 20 seconds would have thrown the craft off its timetable and possibly ruined its mission.

At the present time — nearly eight and a half years after it left Earth — Voyager 2 is still on time, 1.8 billion miles from Earth and exploring the planet Uranus this week. From there, it is scheduled to go on to the next planet, Neptune.

Voyager 2's cameras are still whirling and its instruments are still working. Shortly after it sailed past Saturn Voyager 2's camera platform jammed. But it soon became unstuck, and engineers at the mission headquarters in Pasadena, Calif., said they didn't know what the problem had been and if it was really solved. But apparently it was, because that has been more than four years ago and the spacecraft's cameras are still clicking away.

The precision of the space agency's unmanned probes is astonishing. Without human beings on board to repair its cameras, Voyager 2 has been aloft going on nine years, sending back enough data to fill several books.

The unmanned spacecraft is probing the outer limits of the solar system, nearly two billionmiles away from Earth. And that is only a tiny step on the way to the nearest star, which is light years away. It is fortunate that people can extend their powers of observation into the depths of the universe by instrumentation without having to go there in person. ■

Rockford Register Star

Rockford, IL, January 28, 1986

Excitement has swept through the Jet Propulsion Laboratory at Pasadena. It's the one-of-kind excitement one scientist called "The crescendo of discovery!"

It was not over anything so instantly transcient as the Super Bowl or as earthily mundane as U.S. carrier forces risking war by flaunting military power off the coast of Lybia.

The excitement orginated in the radio signals from Voyager 2, our determined and faithful scout to the frontier of space. Two minutes before noon last Friday, Voyager had arrived on station 50,000 miles off Uranus, third largest planet in the solar system and billions of miles from Earth.

The Voyager "imaging team" was delightedly gathering pictures of Uranus and its 14 moons, some of which show craters filled with what appears to be frozen water.

Equally exciting to these privileged receivers of Voyager's signals was data that will give them the ability to measure the the climate and terrain of the planet.

Through the team's interpretation of Voyager's pictures and the detection of electrical impulses, we may now all know how long a day is on Uranus and generally what kind of a day it is.

Voyager 2 has been creating excitement for its mother Earth since its launch in 1977. During it's odyssey into the inconceivable reaches of space, it is mostly forgotten by its Earth-bound launchers until it reaches its next assignment. Then comes the "crescendo of discovery" when Voyager tells us new things about our distant, but sun-related neighbors.

These discoveries too often are upstaged by glitzy sports events and threats of war, but they merit celebration. They merit celebration of man's success at reaching beyond this small, crowded and seething planet and into the frontier wherein his destiny lies.

WORCESTER TELEGRAM.

Worcester, MA, January 24, 1986

It takes a lot to excite ordinary Americans about space travel. First there were astronauts orbiting Earth, then men walking on the moon. Now with the space shuttles blasting off and landing almost routinely, people hardly cock an ear when the countdown reaches 10 seconds.

A historic moment in space flight occurs today. Voyager 2, the space probe that teamed up with its sister ship Voyager 1 to give us first-ever close-ups of Jupiter and Saturn flies past Uranus at 1 this afternoon. Two hours, 44 minutes and 50 seconds later, pictures will reach NASA's Jet Propulsion Laboratory in Pasadena, Calif.

No doubt tonight's network news will show pictures of this until-now mysterious neighbor in the solar system. There will be technical explanations that will go over the heads of most Americans. But the numbers should hit their mark: Uranus is 1.8427 billion miles away; it took Voyager two light years to get there, traveling at speeds like its present velocity of 45,000 miles an hour.

Astronomers and astrophysicists will need years to digest the data that will reach Earth today. And this craft, which was supposed to be all washed up years ago, heads for Neptune next, with a planned arrival in August 1989.

Its travel, so far, has been interplanetary. Our awe can be nothing short of intergalactic.

The Toronto Star

Toronto, Ont., January 24, 1986

Little more than 150 years ago the Beagle was sailing Earth's oceans. Naturalist Charles Darwin rode his floating science lab thousands of miles around the world for five years. He observed places, things and peoples never known, and he revolutionized human knowledge with new, and still controversial, theories: nothing less than the origin of mankind.

Today it's the Voyager sailing through space. No human is aboard, but none is needed. This space lab the size of a mini-van is travelling places it was never intended to go and conducting dozens of experiments with computerized instructions fed it from scientists on Earth.

A tidal wave of new and valuable information has already been acquired. Thousands of photographs and calculations are being transmitted home. The technology that has been developed on Earth to guide the spacecraft since it left U.S. soil in 1977 is staggering in itself: Consider that it was never intended to sail past Jupiter, let alone take photos of Uranus and beyond. This week it discovered Uranus has two more moons than anyone thought. New theories about the Northern Lights may emerge. And, as with all worthwhile explorations, each new answer raises a galaxy of new questions.

The cynics may discount the Voyager as a waste of money in a world where millions of people are starving. But it will be a sorry day when mankind loses the drive to explore new horizons. And it will be a sorry day when mankind's greatest explorations are concentrated on ways to destroy himself.

Space Station Plan Cut Back; 21st Century Program Unveiled

NASA unveiled May 14, 1986 a new draft of its $8 billion manned space station that it would assemble in the 1990s in conjunction with other nations. Scaled down somewhat from an earlier version, the station was now projected to begin with five modules, instead of seven, and to accomodate a crew of six to eight, instead of 10. Also, it was revamped to permit tending through periodic crew visits, "rather than having a permanent crew on board at the beginning," NASA administrator James C. Fletcher explained in introducing the new design. This reflected "what I understand to be the desire of some in Congress," he said. Assembly was scheduled to begin in 1993 and would require 14 shuttle flights to complete. Maintenance, resupply and crew rotation were expected to require about eight to ten flights a year thereafter. "We have set our sights on the future," Fletcher said. "But make no mistake, that future could be in jeopardy if we do not respond effectively to our immediate challenge—to restore this nation's launch capability. Fletcher was referring to the grounding of the shuttle fleet following the January, 1986 Challenger tragedy.

A space program for the early years of the 21st Century was unveiled May 22 by the National Commission on Space, a 15-member presidential panel headed by Thomas O. Paine, who served as NASA administrator during the Apollo program. The commission's report envisioned a new generation of space vehicles in the immediate future, a return to the moon by the year 2005 and colonization there by small groups of about 20 people, eventually expanding the effort to mining, manufacturing and scientific bases. This would be followed by colonization of Mars by 2035. Spaceships would ferry passengers to and from Earth in regular six-month voyages, "like ocean liners," the commisssion's report said. Fletcher hailed the report May 24 as a "bold course for our nation."

Roanoke Times & World-News

Roanoke, VA, November 18, 1985

'WHERE THERE is no vision," says Proverbs 29:18, "the people perish." At what point, though, does vision become visionary? When does a grand-sounding scheme go out the window — or into space — and become unrealistic, impractical?

That kind of question is constantly being asked of the U.S. space program. There were those — including President Kennedy's science adviser — who opposed the Apollo venture as unnecessary. It turned into a national triumph when Americans were first to step onto the moon. Many advances in scientific knowledge resulted from that program, not to mention the technological spinoffs that found application in people's everyday lives.

Still, it cost U.S. taxpayers $24 billion to put men on the moon. Was it worth an expensive crash effort to beat the Russians there? As Americans neared that finish line, the U.S.S.R. said it wasn't even in the race. A Soviet astronaut has yet to set foot on another body in the solar system. But to all appearances, this has not hampered that country's space program.

The Soviets' way isn't necessarily better. They have had their own space failures; they also take less care for the safety of their personnel. Yet they share the American fascination for putting people into space. They recently demonstrated their ability to keep individuals in orbit for months at a time, an assignment that seems not much better than an equal stay in a forced-labor camp.

To what purpose? The cost of shooting people aloft and keeping them functioning in an alien environment (then bringing them down safely) can easily outpace the benefits of the investment. Machines and electronic devices can be lifted into space much cheaper than can humans. It is also argued that such devices can perform a comparable range of tasks just as efficiently as people can.

Now the debate is on as to whether the United States should put up a permanent space station where people might carry on scientific experiments, manufacture exotic products, and — shhh! — get the enemy in our gunsights. President Reagan, who early in his first term slashed the space budget, proposed last January that the National Aeronautics and Space Administration start work on a space-station project that would cost $8 billion over the next eight years, perhaps $20-30 billion by the century's end.

Congress' Office of Technology Assessment has issued a 234-page study saying that the sort of station now being planned can't be justified for scientific, economic or military reasons. The agency suggests keeping the long-term concept. But it adds that the short-term investment is excessive in "an extended period of unusual national financial stringency." It also implies that if the public took a close look at the way its money is being spent on space, it would quickly bring things back down to Earth.

Space rhetoricians commonly deride talk of cost-effectiveness as timid and shortsighted. They are fond of invoking the example of Columbus (whose name the space station project bears): Where would we be, they ask, if he and Queen Isabella had not been so bold?

Well, we'd be here. Eventually, somebody else in Columbus' time would have made that voyage. But the comparison limps badly. We're already operating on that new frontier of space; more discoveries remain to be made, but we pretty well know they're not going to change our world. We have other priorities besides space, and we are entitled as a nation to set limits on public funds spent on a program like the space station, when unmanned projects are a feasible alternative.

We are also entitled to ask the private sector — which, since the tax cuts, has a lot more venture capital — to show with its own money whether it believes the space station is worthwhile. So far, as with development of the space shuttle, the answer is no. It will continue to be no as long as the venture capital comes from the U.S. Treasury.

ST. LOUIS POST-DISPATCH

St. Louis, MO, April 27, 1985

NASA has reached a milestone on its road to the development of a permanently manned space station in Earth orbit. The space agency awarded $138 million in contracts to McDonnell Douglas and Rockwell International for the "Phase B definition" studies of the NASA space station.

Calling the project "the next logical step" in orbital flight, NASA stressed that the final design is not completely set, but the design that is currently favored involves a so-called power tower, a 300-foot-long spine that supports wings of photovoltaic cells and a mirror-like device that uses solar heat to drive vapor-powered electric generators. Together they should be able to generate 75 kilowatts to a maximum of 300 kilowatts of electrical power. Also attached to the spine will be several modules (cells 60 feet long and 15 feet wide) that will house the space station crew and the work areas. The cells are designed for delivery to the station by means of the space shuttle, and at least two of them will be built by Japanese and Western European firms. The station would orbit 250 to 300 miles above the Earth.

NASA has stated that it would like to start "cutting metal" on the system's components by 1988 and to have the entire station operational by 1992 — the 500th anniversary of Columbus' discovery voyage to the New World.

While one could make the argument that basic research alone could justify the NASA space station, the planned facility should have practical missions as well. NASA is already proposing that it could be used as a site for satellite repair. Still, NASA should be very careful to make sure that the cost and scope of this project do not get out of hand.

The Toronto Star

Toronto, Ont., December 9, 1985

Next month Canada's new science minister, Frank Oberle, will be meeting with U.S. officials to pin down Canada's role in the world's first orbiting manned space station. So far we've expressed interest in participating, but we haven't said what we'd like to do or made any firm commitments. We should join, provided we're given a key role and assurances that the station will be for peaceful purposes only.

Why? Because the giant, floating factory and laboratory to be launched in the mid-1990s promises to be the single largest source of new knowledge about emerging technologies — such as automation and robotics — for the rest of this century and beyond. These technologies can be applied to industries such as mining, energy, forestry and marine engineering to make them more competitive. And many new spin-off industries would be created in the process.

Moreover, the station would be used for scientific experiments to develop new drugs, advanced micro-chips for faster and more powerful computers, and new materials that can't easily be made on earth. And it would be a base for the manufacture of special materials and for remote sensing of the earth's geology, environment and resources. The potential for space manufacturing is staggering — the U.S. Centre for Space Policy estimates commercial space revenues will be close to $70 billion (Canadian dollars) by the year 2000.

Japan and the 11 nations of the European Space Agency are eager to get on board. We can't afford a station of our own, but our participation could bring $2 billion into our economy and create 9,000 jobs over 10 years, according to federal estimates. The big question is: Will Canada have a significant role that will plumb the benefits that are to be had — or will we just be signed on to install the carpeting?

That will depend on whether the federal cabinet is prepared now to put some big bucks on the line. A group of scientists and businessmen affiliated with the Toronto-based Canadian Institute for Advanced Research says the government should be prepared to spend $60 million a year — a third more than current space spending — over the next 10 years. With that we could build a sort of service garage for the space station — a platform for servicing and repairing the satellites.

They argue that if we can't participate on this scale, we shouldn't participate at all. That's the only way we'll have a significant degree of control over any of the space station's operations and spur our own high-tech industries.

Canada should be a full partner in an international project like this, which is intended for the peaceful use of space. Canada and the U.S. National Aeronautics and Space Administration (NASA) have signed a memorandum of understanding that specifies co-operation for "peaceful purposes." But that's from NASA and not the White House. Oberle should ensure this condition is built into any agreement we sign, along with guarantees about access to the station once it's built.

This isn't Canada's only opportunity in space but it's one we shouldn't pass up. We should carve out a role that will build on our research base in areas like robotics, drive our industries to become more innovative and keep our best scientists in Canada. At present there are more than 40 Canadians working on artificial intelligence projects in the United States.

☆☆☆

If the government is serious about promoting space technology, it should also establish a national space agency. Right now responsibility for space-related activities — from communications to space surveillance of crops and forests — is spread throughout a whole raft of departments and agencies. If any of our partners in the space station were interested in some of our technology, there's no one-stop shopping. And companies can't gear up their research and development to meet national needs because planning is so fragmented. Such an agency needn't cost more than is already being spent on space interests across the government.

Canada was third in space and has established a reputation with products like the Canadarm. Building on this tradition, the cabinet should lead us into a new frontier.

Roanoke Times & World-News

Roanoke, VA, June 30, 1985

WHEN THE the Supreme Court pronounced lust to be normal and healthy (see our editorial of June 23), some folks remarked that from now on the sky's the limit.

NASA seems ready to meet the challenge. The space agency has commissioned a group to tell it how to make a space station comfortable and efficient for its occupants. That, says psychologist Yvonne Clearwater, includes finding a way to make sex convenient and private for those astronauts who want to indulge.

"If we lock people up for 90-day periods, we must plan for the possibility of intimate behavior," the environmental psychologist concludes.

Among the problems posed for space-station designers is the task of making sleeping compartments that could accommodate two people in privacy sufficient for such intimate behavior.

One answer would be compartments with adjoining, lockable doors that could be opened to form a playhouse for two. Another would be a compartment for "a married or significantly relating couple." A "significantly relating couple" is what lots of folks in Southwest Virginia would refer to as people who are living in sin.

We must soon deal with the question of whether Americans want to tarnish the squeaky-clean image their astronauts have borne ever since John Glenn made his solo orbit of the earth. Should NASA require that the space-station crew be made up exclusively of married couples, or should it allow for some who are just significantly relating?

And what about the effects of weightlessness on such intimate behavior?

That question might be answered by sending up a group of volunteers from a Masters and Johnson clinic. But it would probably be just as instructive to send a group of normal, healthy Americans into orbit with their court-sanctioned lust. Our guess is that they'd find a way. Love always does.

Rockford Register Star

Rockford, IL, July 19, 1985

The popular fantasy of sexual intimacy that is out of this world is being taken literally by an advisory group to the National Aeronautics and Space Administration.

A spokesman for the Habitability Research Group, charged with telling NASA how to make a permanent space station comfortable and efficient for its occupants, says eventually accommodations will have to be made for "normal, healthy sexual appetites."

The space station is expected to be established in 1992, with crews of six to eight men and women occupying the facility for 90-day periods. Between now and then, there is bound to be controversy over the idea of planning for intimacy, especially since the advisory group envisions such goings-on not just by married folks, but perhaps also by a "significantly relating couple."

With respect to unmarried couples carrying on, so to speak, in space under government auspices, "there are people who are going to be upset," concedes Yvonne Clearwater of the research group in an article she penned for Psychology Today magazine. However, she added, "it's not NASA's job to serve as a moral judge."

Perhaps not. But it is, in fact, NASA's job to deal with the technical aspects of space ventures. On this score, we suspect that research remains to be done on this intimacy idea.

What problems will the lack of gravity pose? Will the headache excuse for abstinence ring true?

Yes, space travel poses scientific challenges.

The Pittsburgh PRESS

Pittsburgh, PA, December 5, 1985

Success has become routine on space shuttle flights, but the accomplishments on these orbital missions continue to amaze and fascinate most earthbound mortals.

The just-ended flight of the shuttle Atlantis added a new wrinkle in man's quest to conquer space. It laid important groundwork for a permanent space station, expected to be established early in the next decade.

Two astronauts, working outside the shelter of their capsule, proceeded flawlessly through construction techniques expected to be used in building a space station.

They erected a 45-foot tower and a 400-pound pyramid-shaped module by snapping together pre-cut aluminum pieces. The method of construction, requiring no tools, was similar to that used in assembling children's Tinkertoy sets.

The two astronauts were exhilarated with their accomplishments on what one of them described as their face-to-face encounter with the universe.

Nor was the construction project the only achievement on this 23rd flight of the shuttle program. During their seven days in space, crew members also carried out an experiment involving development of a space-based factory for purification in a gravity-free environment of a hormone that may prove useful in treating certain forms of anemia.

In addition, they successfully launched three communications satellites, one for Mexico, another for Australia, and a third for RCA American Communications, Inc. The latter is the most powerful domestic communications satellite ever built.

The six men and one woman on this second flight of Atlantis, including Mexico's first astronaut, have earned congratulations.

The Toronto Star

Toronto, Ont., March 21, 1986

The best news to come out of the Washington summit meeting between Prime Minister Brian Mulroney and President Ronald Reagan is that Canada will join the U.S. in building and operating a manned space station. According to details provided in Ottawa by Science Minister Frank Oberle, Canada's role in the non-military project will be to spend $800 million over the next 15 years to build a service centre — something like an earthling's auto body shop — on the space station.

The service centre will be used for such duties as servicing laboratories, factories and scientific instruments, lifting equipment on and off space shuttles, and setting up telescopes and other instruments. Building and operating the centre offers Canadian science and industry an unparalleled opportunity to take part in the planet's most advanced research and development. And the spinoffs from that R & D, especially in automation, robotics and artificial intelligence, will surely have an impact here on Earth.

Before those dreams come true, however, some tough and important decisions remain to be made. Between now and next March, Canada-U.S. negotiations will set out Canada's participation more specifically. Guarantees are needed to ensure that Canada will have a role in management of the space station and that Canadian scientists and businessmen will have access to the station's factories and laboratories.

While those bilateral negotiations are proceeding, Mulroney's government has some important unilateral decisions to make as well. One is to create a national space agency so that responsibility for space-related activities can be concentrated and co-ordinated in one place with one responsible cabinet minister, instead of scattering the activities among a half-dozen or more departments. While administering the government's part in the station, a national agency would also be responsible for securing the partnership of the universities and private industry in our space projects. Indeed, the government and the new space agency should make sure that the research and development opportunities presented by the space station deal are done not by government but by our burgeoning young, talented and capable high-tech firms.

The Globe and Mail

Toronto, Ont., December 6, 1985

Last year, U.S. President Ronald Reagan and the National Aeronautics and Space Administration made an intriguing proposal to Canada: tell us by January, 1986, whether you want to help us build an orbiting station in space and, if so, what jobs you would expect to perform.

Canada has agreed in principle; now comes the tough part. Within a month, it must decide whether to commit for years what is now 40 per cent of its space budget to join in the construction, maintenance and operation of a space station occupied by rotating crews of astronauts and scientists. Will the rewards be worth the investment? Put another way, can this country afford not to be involved in what may be the leading edge of technological research and development for the next quarter-century?

The mind leaps from one extreme to another, from a wide-eyed Tom Swiftian excitement to a cautious calculation of U.S. motives in building this station: there is nothing, after all, to stop the military from making use of any breakthroughs made in creating this peaceful enterprise. But this is not Star Wars — it is 2001: A Space Odyssey. When the Canadian Institute for Advanced Research asked a panel of experts chaired by former University of Toronto president James Ham to study the space project, they produced a report so enthusiastic it sang.

The evolution of the station, to which the United States will commit an initial $8-billion, will take 25 years. During that period, the report says, "intelligent" systems will be developed to control much of the station's operation, and robots will be designed to inspect and repair the station and retrieve materials from co-orbiting space platforms.

These advances "will be adapted for use on earth in order to improve the productivity of U.S. industries." To remain economically competitive, Canada will need access to such advances — and to gain access, it will have to be a significant participant in the development.

The report urges Canada to express interest in developing a space-based service station, anchored to the main unit, which would refuel ships, repair satellites and see to the needs of those on board. The facility would cost an estimated $600-million over 10 years, but the spur it would give to our mastery of such fields as robotics and artificial intelligence would be invaluable. As well, there might be opportunities aboard the station to develop new materials (such as long-wearing, self-lubricating ceramics) with applications on earth as well as in space.

This isn't just a matter of saying yes or no. Canada would have to draw together experts currently scattered across the country and, with luck, entice back some of the estimated 40 Canadian researchers working in the field of artificial intelligence in leading U.S. laboratories. It would need a central science agency, along the lines of NASA, to control programs now divided among the departments of defence, communications and energy and the National Research Council.

If successful, the project would create not just material spinoffs, but "an exploitable knowledge base" of people who "have acquired knowledge in a certain field or sub-field and the ability to apply it in a variety of useful ways." Canada, an exporting country, can preserve its current standard of living only by being on the cutting edge of technological development — by increasing productivity as its response to cheap imports, using sensors and robots to take over dangerous jobs in the resource industries and developing and patenting products for the international market.

If we miss this one, the report says, we get left behind. The argument is as simple — and as complex — as that.

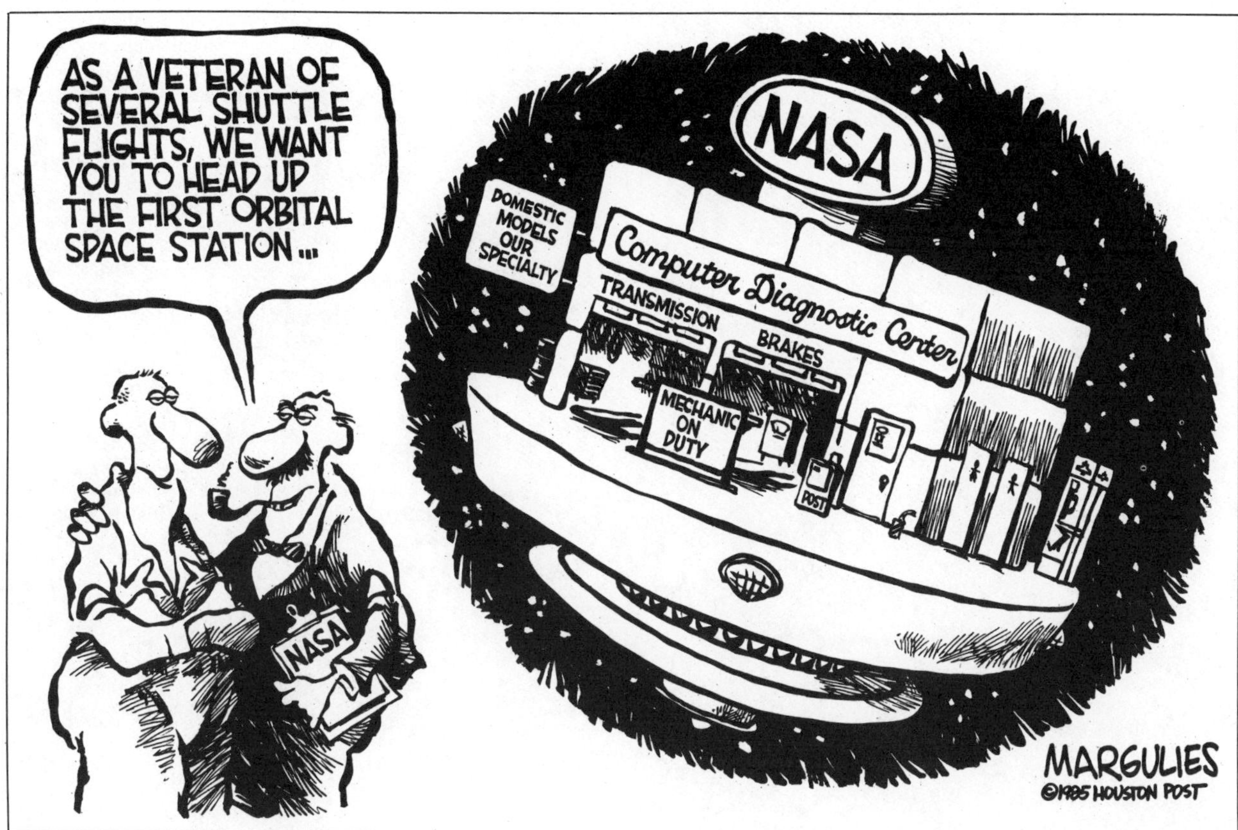

Houston Chronicle

Houston, TX, May 18, 1986

In his 1984 State of the Union Address, President Reagan called for the deployment of a permanent U.S. manned space station within a decade. White House budget planners need to be reminded of that commitment.

Officials at the Office of Management and the Budget are considering a cut of hundreds of millions of dollars from NASA's space station budget for fiscal 1987. Such a cut would delay construction of the station until 1997, reduce the efficiency of NASA's design team and harm the United States' efforts to retain the lead in space exploration and commercialization.

In a magazine article detailing his support for a U.S. space station, President Reagan said it is high time for the nation to take "the next bold step" in space. He said a permanent U.S. space station is an essential part of U.S. strategy for "conquering the next frontiers in space," promoting international cooperation and enabling U.S. industry to expand its operations beyond Earth.

He's right. In endorsing the space station, President Reagan said the economic and scientific potential far outweighs the estimated $8 billion cost. The phenomenal human and scientific benefits of the U.S. space program to date back him up.

Houston's Johnson Space Center has been given the central role in developing the U.S. station. The center has already attracted to the Houston area a variety of private businesses that hope to capitalize on space research and development.

Delaying the space station program would be unnecessary and inexcusable. It would dampen the economic hopes Houston has pinned on the development of private industry in space, weaken the U.S. leadership in science and space exploration, and damage the security of the nation and its allies.

The Houston Post

Houston, TX, January 13, 1986

The space station concept recently unveiled by NASA makes a lot of sense. It is a scaled-down proposal, both in terms of money and activities. And, as the space agency has said again and again, it's the next logical step in space and should be supported.

The modular plan provides some flexibility in the ultimate configuration of the space station, for which Johnson Space Center is the project manager. It also offers a considerable degree of protection against cost overruns and delays.

The facility is to be built in cooperation with the European Space Agency, Canada and Japan. NASA says its construction will require 14 shuttle missions, beginning in 1993 — seven years from now. It should be finished by 1996.

New NASA administrator James C. Fletcher used the space station announcement to warn that unless we solve our immediate problems the space station and the entire space program's future could be in danger. He rightly said it is vital to restore the nation's launch capacity after the Challenger disaster and unmanned rocket failures.

Without a safe and reliable space shuttle, we cannot build a space station, and never dream of pursuing the avenues such a facility will open to us. A space station can serve as a platform for space-based science and manufacturing, an assembly and launching point for further exploration and development, and will be a key to development of technology as yet unenvisioned.

It will require effort — and money. The money to replace Challenger should be made available without further effect on the space station. And that station should be built as soon as possible. Without these steps, we run the risk of becoming what one observer called the Portuguese of the space age — of being unable to profi as we should from the spectacular discoveries we have made.

Index

T

U

V

W

Y